Fundamentals of Neurophysiology

SPRINGER
STUDY
EDITION

Fundamentals of Neurophysiology

Third, Revised Edition

Edited by
Robert F. Schmidt

With Contributions by
J. Dudel W. Jänig R.F. Schmidt
M. Zimmermann

Translated M.A. Biederman-Thorson

With 139 Illustrations

Springer-Verlag
New York Berlin Heidelberg Tokyo

Robert F. Schmidt
Physiologisches Institut der Universität Kiel, Olshausenstrasse 40/60, 2300 Kiel, Federal Republik of Germany

Josef Dudel
Physiologisches Institut der Technischen Universität München, Biedersteiner Str. 29, 8000 München 40, Federal Republik of Germany

Wilfrid Jänig
Physiologisches Institut der Universität Kiel, Olshausenstrasse 40/60, 2300 Kiel, Federal Republik of Germany

Manfred Zimmermann
II. Physiologisches Institut der Universität Heidelberg, Im Neuenheimer Feld 326, 6900 Heidelberg, Federal Republik of Germany

Library of Congress Cataloging in Publication Data
Grundriss der Neurophysiologie. English.
Fundamentals of neurophysiology.
Translation of: Grundriss der Neurophysiologie.
Companion v. to: Fundamentals of sensory physiology edited by Robert F. Schmidt. 2nd corr. ed. 1981.
"Springer study edition."
Bibliography: p.
Includes index.
1. Neurophysiology. I. Schmidt, Robert F. II. Dudel, Josef. III. Fundamentals of sensory physiology. 2nd corr. ed. IV. Title. [DNLM: 1. Nervous System—physiology. WL 102 G889-
QP361.G7413 1985 612'.8 85-9965

Typeset by Bi-Comp, Inc., York, Pennsylvania.
Printed and bound by Halliday Lithograph, West Hanover, Massachusetts.
Printed in the United States of America.

9 8 7 6 5 4 3 2 1

ISBN 0-387-96147-X Springer-Verlag New York Berlin Heidelberg Tokyo
ISBN 3-540-96147-X Springer-Verlag Berlin Heidelberg New York Tokyo

PREFACE TO THE THIRD EDITION

Again rapid advances in the brain sciences have made it necessary, after only a few years, to issue a revised edition of this text. All the chapters have been reviewed and brought up to date, and some have been largely rewritten. The major revision has occurred in the chapters on the autonomic nervous system and the integrative functions of the central nervous system. But in the discussion of the motor systems and other subjects as well, recent insights have necessitated certain conceptual modifications.

In the description of the autonomic nervous system, the role of the intestinal innervation has been brought out more clearly than before. In addition, there is a new presentation of the physiology of smooth muscle fibers, and more attention has been paid to the postsynaptic adrenergic receptors, because of the increasing therapeutic significance of the α/β receptor concept. A substantial section on the genital reflexes in man and woman, including the extragenital changes during copulation, has also been added.

The text on the integrative functions of the central nervous system has been expanded to include, for the first time, material on brain metabolism and blood flow and their dependence on the activity of the brain. Reference is also made to recent results of research on split-brain and aphasic patients and on memory, as well as on the physiology of sleeping and dreaming.

The list of sources for further reading in this new edition has been made as current as possible; apart from a few "classics" indispensible for an understanding of the development of neurophysiology, most of the references cited were published in the last 5 to 10 years. With this list, the reader has immediate access to the original literature.

Many of the illustrations have been improved or replaced, and some new ones added. We are very grateful to Mrs. Renate Lindenbaur, Stuttgart, for her assistance in this task. We also thank the publishers of my book *Mediainische Biologie des Menschen,* Piper-Verlag in Munich, for permission to reproduce some of its illustrations. The test questions at the end of each section have been retained, with revisions as necessary, to provide the reader with a simple means of checking his progress.

As previously, this book is designed to present both the established background and the most important new results of brain research. The text is concise enough to be absorbed within a reasonable time by nonspecialist students of physiology, whether their primary fields are medicine, psychology, zoology, biology, pharmacology, or other natural sciences. No prior knowledge of anatomy or physiology is assumed; as each term is introduced it is defined and explained in context. The book should therefore be well within the comprehension of anyone studying at the university level. Together with its companion volume, *Fundamentals of Sensory Physiology,* it provides an extensive introduction to neural and sense-organ function. In particular, we hope the approach, while based on solid fact, will also do justice to the stimulating questions in brain research and to problems still unsolved.

On behalf of all the authors, I again take pleasure in thanking all who have helped to bring this edition into being. We are particularly grateful to Dr. Marguerite A. Biederman-Thorson, Oxford, England, for her excellent translation, to our technical and secretarial colleagues for their tireless efforts, to my wife for her help in compiling the index, and to Springer-Verlag New York, for their close collaboration and painstaking expertise in the production of the book.

Würzburg, Summer 1985 Robert F. Schmidt

CONTENTS

1
THE STRUCTURE OF THE NERVOUS SYSTEM

R.F. Schmidt

1.1 The Nerve Cells

Neurons. The building blocks of the nervous system are the *nerve cells*, also called *neurons*. It is estimated that the human brain possesses 25 billion cells. Like all animal cells, each neuron is bounded by a cell membrane that encloses the contents of the cell—that is, the cytoplasm (cell fluid) and the nucleus. The size and shape of these neurons vary widely, but the structural plan always includes certain elements (Fig. 1-1): a cell body, or *soma*, and the processes from this cell body, namely an *axon* (neurite), and usually several *dendrites*. The neuron diagrammed in Fig. 1-1 has one axon and four dendrites.

The classification of the neuronal processes in terms of an axon and several dendrites is made on the basis of *function:* the axon links the nerve cell with other cells. The axons of other neurons terminate on the dendrites and also on the soma. The axon and the dendrites normally divide into a varying number of branches after emerging from the soma.

The branches of the axons are called *collaterals*. The axons and their collaterals vary greatly in length. Often they are only a few micrometers long, but occasionally—for example, in the case of certain neurons in humans and other large mammals—they are well over a meter long (see Sec. 1.3).

Figure 1-2 illustrates various types of neurons. Notice in particular the great variation in the dendritic formations. Some neurons (for example, neuron C) have profusely branched dendrites, but in others (A and B) the ratio of soma surface to dendrite surface is somewhat larger.

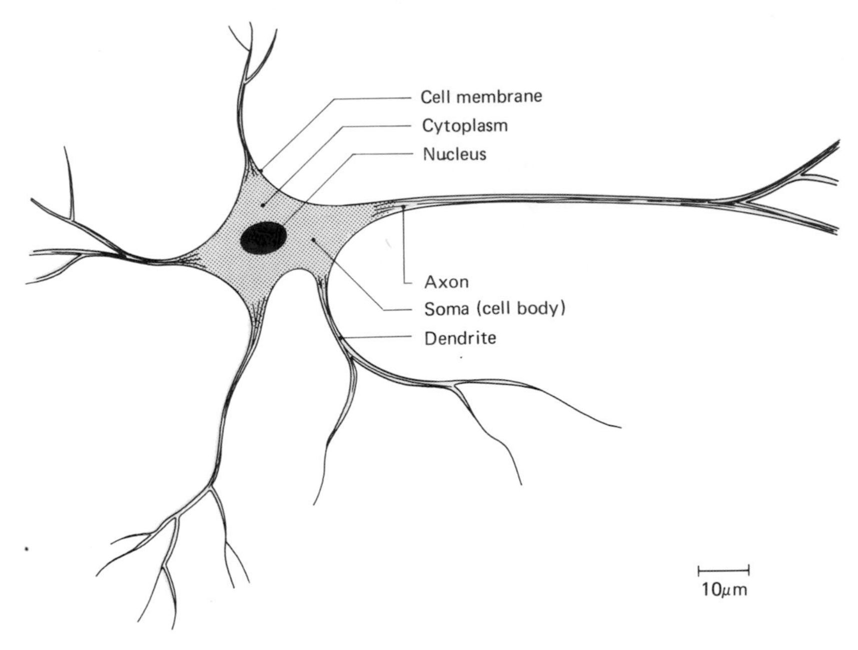

Fig. 1-1. Schematic diagram of a neuron showing the soma, the axonal and dendritic processes, and the cell contents; the approximate dimensions are indicated by the scale.

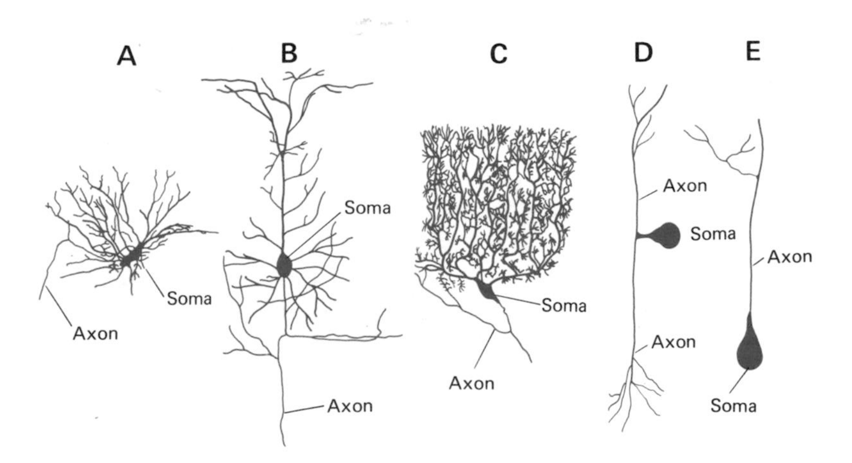

Fig. 1-2. Examples of the variety of shapes of neurons. See text for discussion. After Ramon y Cajal.

Finally, there are also neurons (D,E) that have no dendrites. The diameter of the neuronal cell bodies is in the order of magnitude of 5 to 100 μm (1 mm = 1,000 μm). The dendrites can be several hundred micrometers long.

Synapses. As already mentioned, the axon and all its collaterals join the nerve cell with other cells, which can be other nerve cells, muscle cells, or glandular cells. The *junction of an axonal ending with another cell is called a synapse.* Figure 1-3 shows examples of neuronal junctions. If an axon or an axon collateral ends on the soma of another neuron, then we speak of an *axo-somatic synapse*. Correspondingly, a synapse between an axon and a dendrite is called an *axo-dendritic* synapse, and a synapse between two axons is called an *axo-axonal* synapse. If an axon ends on a skeletal muscle fiber, then this particular synapse is called a *neuromuscular end plate* or neuromuscular junction. Synapses on muscle fibers of the viscera (smooth muscles) and on glandular cells have no special names.

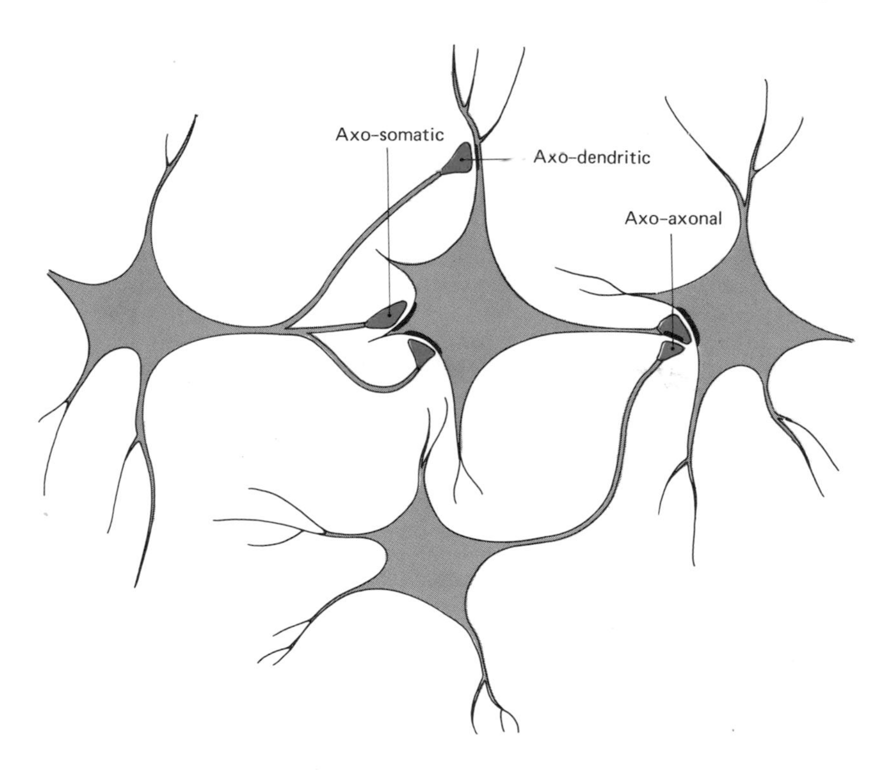

Fig. 1-3. Schematic diagram of synapses, named according to their locations. See text for discussion.

Effectors. Most neurons are connected by synapses to other neurons to form neuronal circuits. However, as already mentioned, the axons of a small number of neurons do not contact other neurons but, instead, connect with muscle cells or glandular cells. The striated skeletal muscles, the smooth muscles of the viscera and blood vessels, and the glands are thus the executive organs, or *effectors,* of the nervous system. We will deal with the structure of the effectors in the relevant chapters.

Receptors. To react properly to its environment and to supervise the activity of the effectors, the nervous system also needs sensing elements that respond to changes in the environment and the organism and transmit these responses to the nervous system. For this task, the body possesses specialized nerve cells. *Specialized nerve cells that respond to certain changes in the organism or in its environment and that transmit these responses to the nervous system are called receptors.*

Each receptor cell is specialized to respond to a particular form of stimulus. Receptors in the retina of the eye, for example, ordinarily respond only to light—electromagnetic radiation with wavelengths between 400 and 800 nm (violet to red). Receptors of the ear, on the other hand, are specialized to transduce sound waves (propagated fluctuations of air pressure) with frequencies from 16 to 16,000 Hz (Hz: cycles per second). Thus arises the important concept of sensory *modality.* Vision represents one modality, audition another, and taste yet another. The entire neuronal apparatus within each sensory modality is specialized for reception and processing of the corresponding type of stimulus. The term *adequate stimulus* is sometimes used to denote the type of stimulus to which a receptor is specialized to react. Other stimuli, corresponding to other modalities, if strong enough can sometimes also excite a receptor; for example, one may see "stars" if hit in the eye.

It is through the receptors that the nervous system senses the events occurring in our environment and in our bodies. In functional terms, the receptors provide information on

1. our distant environment (eye, ear: teleceptors),
2. our immediate environment (skin receptors: exteroceptors)
3. the attitude and the position of the body in space (the receptors in the labyrinth and those of muscles, tendons, and joints: proprioceptors), and
4. events in the viscera (interoceptors or visceroceptors).

(For a more detailed discussion of receptor physiology see *Fundamentals of Sensory Physiology,* R.F. Schmidt ed., Springer, 1981.)

The following questions, and those that follow each of the sections in this book, enable you to check your newly acquired knowledge. When answering them you should avoid as far as possible checking back in the text. The answers can be found beginning on p. 323.

Q 1.1 Which of the following statements are correct (one or more may be correct)? Note your answers on a sheet of paper and compare them with the answer key on p. 323.

a. Receptors respond to all environmental stimuli.
b. Each receptor has an adequate stimulus.
c. Receptors are specialized nerve cells.
d. The receptor is much more responsive to stimuli of other modalities than to adequate stimuli.
e. Muscles and glands are the effectors of the nervous system.

Q 1.2 *Neuromuscular end plate* means the junction of an axon with a

a. smooth muscle fiber.
b. glandular cell.
c. skeletal muscle fiber.
d. nerve cell.
e. Statements a to d are all incorrect

Q 1.3 Draw a diagram of a neuron and label its various parts.

Q 1.4 Draw diagrams of and label the three typical junctions that are possible between two nerve cells.

Q 1.5 The cell bodies (somata) of the nerve cells range in diameter from

a. 400 to 800 nm.
b. 5 to 100 μm.
c. 0.1 to 1.0 mm.
d. 16 to 16,000 Hz.
e. more than 1 m.

1.2 Supporting and Nutritive Tissue

While functionally the neurons are the most important building blocks of the nervous system, they are not the only cells of the brain. The neurons are encased in a special supporting tissue of *glial cells*, known collectively as the neuroglia. Moreover, there is a dense *network of blood vessels* penetrating the entire nervous system. The glial cells are more numerous than the nerve cells, but on the average they are smaller. As a result, neurons and glia each account for just under half of the volume of the brain and spinal cord. The remaining 10 to 20% of the brain volume is occupied by the *extracellular space* (see below) and the blood vessels.

Functions of the Glial Cells. There are various types of glial cells. One of their functions in the nervous system is equivalent to that of the connective tissue in the other organs of the body. However, the glial cells are not homologous with connective tissue, but rather are related embryologically to neurons. In addition to this *general supporting function,* glial cells form the *myelin sheaths of nerve fibers* (see Sec. 1.3), and they may also participate in the *nutrition of the nerve cells.* It has been suggested that they play a role in certain processes of nervous excitation, but this view is still a controversial one. Since glial cells, in contrast to neurons, retain the ability to divide throughout their lives, they also serve to *fill up gaps caused by loss of nervous tissue.* Such proliferations of glial cells (glial scars) are often the site of origin of spasmodic discharge within the brain; such spasms may take the form of epileptic seizures.

Extracellular Space. When examined under the light microscope, the neurons and the glial cells look as if they abut each other in the nervous system with no interspace, like bricks laid without mortar. Under the electron microscope, however, one can see that a very narrow gap (average width: 200 Å = 20 nm = 2×10^{-5} mm) separates the cells. All these interspaces are linked with one another to form the fluid-filled *extracellular space* of the neurons and the glial cells (Fig. 1-4A). At several points in the brain, known as the ventricles, the extracellular space widens to form large cavities, containing the cerebrospinal fluid. The composition of this fluid is very nearly the same as that of the "interstitial" fluid between the cells; the small differences between the two fluids, and the reasons for them, will not be discussed here.

One fact of great functional importance must be emphasized here. *The neurons exchange ions and molecules only with the fluid in the extracellular space;* there is no direct exchange from one neuron to another or between neurons and glial cells. The gaps between the cells (interstitial gaps) are wide enough to permit essentially unimpeded diffusion of ions and molecules within the extracellular space.

The extracellular space also surrounds the extremely thin branches of the blood vessels in the brain, the capillaries, and an exchange of material takes place between these and the extracellular space. Figure 1-4B shows diagrammatically the path of the oxygen (O_2) and the nutrients from the blood into the neuron and of the carbon dioxide (CO_2) and other metabolites from the neuron into the blood. A drug injected intravenously must, therefore, pass through the wall of the blood vessel (capillary membrane) and then through the cell membrane before it can take effect in a neuron (though there are chemicals that can act directly at the outside of the cell membrane). The capillary membrane of the cerebral blood vessels seems to be impermeable to many substances, which is why pharmacologists speak of a "blood–brain barrier."

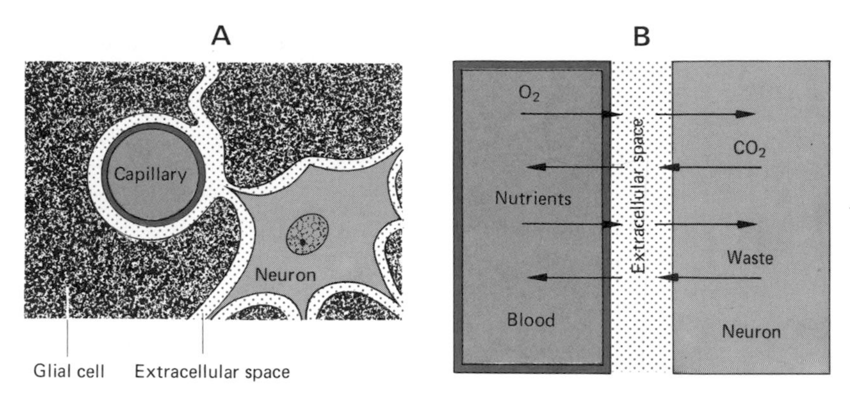

Fig. 1-4. The route by which neurons exchange materials with the blood. **A.** Rough diagram of the spatial relationships between capillary, neuron, glial cells, and the extracellular space surrounding them. **B.** *Arrows* indicate the diffusion of nutrients and waste products, including oxygen (O_2) and carbon dioxide (CO_2), into and out of a neuron via the extracellular space.

The neurons of the central nervous system (CNS), particularly those of the higher centers of the human brain (cerebral cortex), depend on a maintained supply of oxygen. An interruption in the blood flow (for example, cardiac arrest or strangulation) for 8 to 12 s is enough to cause unconsciousness. After 8 to 12 min irreversible brain damage usually has been sustained. When the body's oxygen supply is cut off only by an interruption of breathing (for example, during diving), of course, consciousness can be maintained for a minute or more, because the oxygen in the circulating blood can be utilized.

Q 1.6 Which of the following statements is/are correct?
- a. Glial cells form the connective tissue of the nervous system.
- b. The fluid in the extracellular space and in the ventricle of the brain is called plasma.
- c. Complete lack of oxygen leads to irreversible brain damage only after several hours have elapsed.
- d. The extracellular space encloses all neurons but not the glial cells.
- e. The glial cells constitute the blood-brain barrier.

Q 1.7 When nervous tissue is destroyed by illness or injury,
- a. the resulting space is filled with cerebrospinal fluid.
- b. blood vessels fill in the cavity.
- c. the neurons are replaced by division of the adjacent nerve cells.
- d. the missing tissue is replaced by glial cells.
- e. an air-filled cavity is formed at the site of the defect.

Q 1.8 Which of the following paths by which oxygen might enter the nervous system is the most important?
- a. From the blood capillary directly into the neuron.
- b. From the capillary via a glial cell into the neuron.

c. From the capillary via a glial cell into the extracellular space and thence into the neuron.
d. From the capillary via the extracellular space into a glial cell and thence into the neuron.
e. From the capillary via the extracellular space into the neuron.

1.3 The Nerves

The *central nervous system* consists of the *brain* and the *spinal cord.* All the remaining nervous tissue is referred to as the *peripheral nervous system.* The nerves in the periphery of the body are bundles of axons that are enclosed in sheaths of connective tissue. Their structure, origin, and classification according to morphologic and functional considerations will now be described.

The Nerve Fibers. Every axon in a peripheral nerve lies within a sort of tube formed by special glial cells, the Schwann cells (Fig. 1-5). The axon and the Schwann sheath surrounding it are together called the *nerve fiber.* The number of nerve fibers bundled together to form a nerve varies considerably. In a nerve thick enough to be seen easily with the naked eye, they are numbered in dozens or hundreds. In still thicker nerves, there may be tens of thousands of fibers.

About one-third of all nerve fibers are more elaborately encased; during growth, the Schwann cell winds around the axon several times, so that between the axon and the Schwann-cell body a sheath is formed consisting of a lipid–protein mixture called *myelin* (Figs.

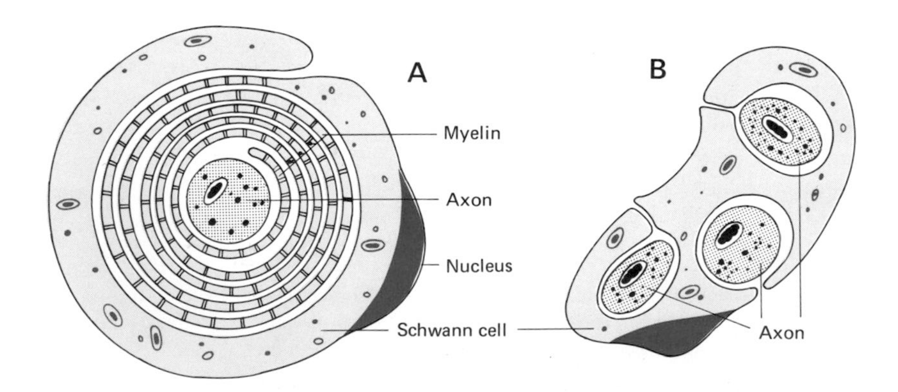

Fig. 1-5. Cross sections through one myelinated (**A**) and three unmyelinated (**B**) nerve fibers, showing the two types of sheath (myelin and Schwann-cell, respectively).

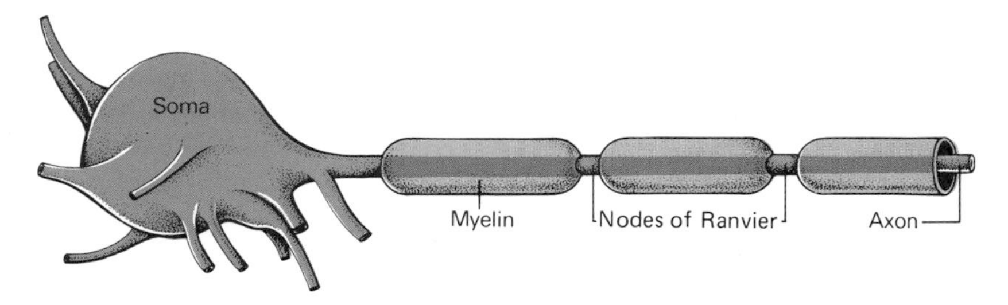

Fig. 1-6. Schematic three-dimensional diagram of a neuron with a myelinated fiber. The dendrites have been cut off. The myelin (*red*) of the sheath is interrupted at regular intervals by nodes of Ranvier. The Schwann cells (see Fig. 1-5) are not indicated separately.

1-5A and 1-6). In cross section such a nerve fiber resembles a wire covered with a thick layer of insulation. Nerve fibers "insulated" in this way are called *myelinated* (occasionally "medullated") nerve fibers.

Unlike the insulation of a wire, the *myelin sheath* is discontinuous; as illustrated in Fig. 1-6, the sheath is interrupted at regular intervals. Under the light microscope these unmyelinated parts look like constrictions or nodes, and are therefore called the *nodes of Ranvier* after their discoverer. In myelinated nerve fibers a node of Ranvier occurs approximately every 1 to 2 mm.

Nerve fibers without a myelin sheath are called *unmyelinated* fibers. These too, however, are ensheathed by Schwann cells; often a single Schwann cell can enclose several unmyelinated axons (see Fig. 1-5B). The arrangement of the Schwann cells surrounding the myelinated fibers is more uniform, with each Schwann cell occupying approximately the space between two constrictions.

Physiologically, the myelinated nerve fibers differ from the unmyelinated fibers chiefly because of the different velocities at which they are capable of conducting action potentials. For reasons explained in detail later, the conduction velocity is very high in myelinated nerve fibers and low in unmyelinated fibers. Within each group the conduction velocity also depends on the diameter of the axons: as the diameter increases the conduction velocity increases. Consequently, the various proposed anatomic and physiologic classifications of nerve fibers coincide reasonably well. Myelinated fibers are often referred to as *A fibers* and unmyelinated fibers as *C fibers*. Table 1-1 shows the commonest classification according to diameter, in which the myelinated fibers belong to Groups I, II, and III and the unmyelinated fibers make up Group IV.

Table 1-1. Classification of Nerve Fibers.

Fiber Group			Average Diameter (μm)
Myelinated fibers	I	A fibers	13
(diam. = axon	II	A fibers	9
+ myelin sheath)	III	A fibers	3
Unmyelinated fibers (diam. of axon)	IV	C fibers	≦1

Functional Classification of Nerve Fibers. Apart from the conduction velocity and the diameter, a number of other functional characteristics are used to categorize nerve fibers. The most important of these are illustrated in Fig. 1-7. The nerve fibers of the receptors are called *afferent nerve fibers,* or, more succinctly, afferents (left side in Fig. 1-7). They lead to the CNS and transmit information from the receptors about changes in the environment or the body. Afferent nerve fibers from the viscera are termed *visceral* afferents, while all other afferents in the body—from muscles, joints, skin, and sensory organs of the head (eyes, ears, etc.)—are called *somatic* afferents.

Transmission of information from the CNS to the periphery takes place by *efferent nerve fibers,* or, more succinctly, efferents (right side in Fig. 1-7). Efferents to the skeletal muscle fibers are called *motor* efferents. All the rest belong to the vegetative, or autonomic, nervous system and are therefore called *autonomic* efferents. The latter supply information to the smooth muscles of the viscera and in the walls of

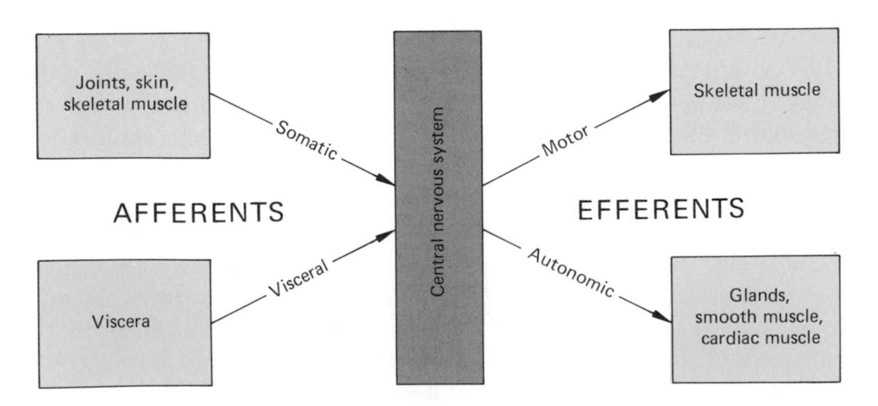

Fig. 1-7. Diagram of the classification of nerve fibers by origin and function. See text for detailed discussion.

the blood vessels, as well as to the cardiac muscle and to all the glands in the body.

Classification of Nerves. In the preceding paragraphs we considered only the functional classification of individual nerve fibers. But, as we have already seen, a nerve contains large numbers, often many tens of thousands, of nerve fibers. In practically all nerves (as, for example, in the ischiadic nerve, which innervates the greater part of the leg), afferent and efferent nerve fibers are bundled together. The types of nerve fiber contained in the nerve depend on the area (skin, muscles, intestines) it serves. It is now important to learn the names and the composition of these various nerves.

The nerves to the skin, the skeletal muscles, and the joints form the group known as *somatic nerves*. The nerves leading to the viscera are called *splanchnic nerves* (synonyms: autonomic nerves, visceral nerves, vegetative nerves; these terms are sometimes used with slightly different meanings, but we will not go into that here). A *cutaneous nerve* is thus a somatic nerve. It contains somatic afferents (afferent nerve fibers) from the receptors of the skin, but it also contains autonomic efferents to the blood vessels, the sweat glands, and the skin hair. A skeletal muscle nerve, usually called a *muscle nerve* for short, is also a somatic nerve. It contains motor efferents as well as somatic afferents from the receptors of the muscles and autonomic efferents to the blood vessels. A *joint nerve* or *articular nerve* is also a somatic nerve with somatic afferents from the receptors of the joints and autonomic efferents to the blood vessels of the joints and the joint capsule. The thick nerves (for example, the ischiadic) are usually *mixed nerves* that later branch into their component cutaneous, muscle, and joint nerves. Finally, we must mention that the *splanchnic nerves* contain visceral afferents and autonomic efferents.

Axonal Transport. The chief function of the nerve fibers is to transmit information from one nerve cell to the next or to effector cells (muscle and gland cells). The information is usually carried in the form of brief electric pulses, the action potentials. These are the subject of the next chapter. In addition, the axons are conduits for the transport of substances from the cell body (soma) to the synapses and in the reverse direction, from synapses to cell body. The various processes involved are together denoted by the key term *axonal transport*. The substances (e.g., amino acids, proteins, various nutrients) are of vital importance to the axon; if an axon is separated from its cell body (for example, when a nerve is accidentally cut) the axon dies, whereas the cell body ordinarily survives.

Axonal transport is sometimes extremely rapid. Protein molecules and synaptic transmitter substances (see Chapter 3) are transported from the soma into the synapses at a speed of approximately 40 cm per day. This transport is an active, energy-consuming process. It is apparently accomplished by a system of "microtubules," which probably act as a type of conveyor belt along which the substances to be transported are "pushed" toward the periphery. Retrograde transport, from the periphery to the soma, is about half as rapid, with a velocity of approximately 20 cm per day. Some viruses and toxins (e.g., poliomyelitis viruses, responsible for infantile paralysis, and tetanus toxin, which causes "lockjaw" when released by bacteria in wounds) "misuse" retrograde axonal transport to move from peripheral sites such as skin wounds into the nerve cell body. Once within the soma, they can produce their pathogenic effects. Other poisons interfere with axonal transport and thereby damage the nerve as severely as if it had been transected. The consequences can include muscular paralysis, disturbances of sensation, and pain.

Q 1.9 Which of the following statements is/are false?
- a. Cutaneous, muscle, and splanchnic nerves form a group known as somatic nerves.
- b. Unmyelinated fibers are always larger in diameter than the myelinated type.
- c. "Somatic afferents" and "somatic nerves" are synonymous terms.
- d. A cutaneous nerve has no motor efferents.
- e. A muscle nerve also contains autonomic efferents.

Q 1.10 By "nodes of Ranvier" we mean
- a. the points where an axon branches into its collaterals.
- b. the indentations in the Schwann cells caused by the unmyelinated nerve fibers embedded in them.
- c. the regular interruptions in the sheath in the case of myelinated nerve fibers.
- d. the gaps, filled with cerebrospinal fluid, between the cells of the CNS.
- e. the point of transition of the receptor into the afferent nerve fiber.

Q 1.11 The diameter of myelinated nerve fibers is in the order of magnitude of
- a. 0.1 to 1.0 μm.
- b. 1 to 20 μm.
- c. 20 to 100 μm.
- d. 0.1 to 1.0 mm.
- e. 1 to 10 mm.

1.4 The Structure of the Spinal Cord

Of the two parts of the CNS, the *brain* and the *spinal cord,* the latter is phylogenetically by far the older and is relatively simple and stereotyped in structure. We will now study the structure of the spinal cord and, at the same time, gain an initial impression of the arrangement of the neurons in the CNS.

The Structure of a Spinal Segment. The brain and the spinal cord are enclosed in bony protective casings (Fig. 1-8), the *skull* and the *vertebral canal,* respectively. As a result of this design, the soft tissue of the CNS is optimally protected from mechanical damage. The spinal cord is subdivided into segments, one segment for each vertebra. This one-to-one arrangement, developed during evolution, is obscured as an individual grows; the growth of the spinal segments falls behind that of the vertebrae so that, as the longitudinal (sagittal) section in Fig. 1-8 shows, in adult humans the spinal cord ends at approximately the level of the top lumbar vertebrae, although the segmented structure is fully retained.

The uniform structure of the spinal cord in the longitudinal direction—that is, the way it is built up of spinal segments—is matched by a uniform structure within each segment. Figure 1-9 illustrates a typical cross section. The cell bodies of the neurons are located in the inner region of the spinal cord, and the ascending and the descending nerve fibers are located in the outer regions. In a fresh section, the cell bodies (unstained and viewed with the naked eye) appear gray in color. Therefore, this region of the spinal cord, which is butterfly-shaped in cross section, is called the *gray matter.* The anterior (ventral) part of each butterfly wing is called the *ventral horn,* the side (lateral) part is called the *lateral horn,* and the posterior (dorsal) part the *dorsal horn.* The part of the gray matter located medially (toward the center) from the lateral horn is called the *pars intermedia.*

The butterfly-shaped gray matter within a spinal segment is surrounded by the ascending and the descending nerve fibers, which form the outer regions of the segment. The myelin makes the nerve fibers appear white in cross section; therefore, these regions are referred to as the *white matter* (see Fig. 1-9). The ratio of white to gray matter is not the same in all sections of the spinal cord. In the cervical and the thoracic segments, which are located closer to the brain, the proportion of white matter in the overall cross section is particularly large because all the ascending and the descending pathways pass through these segments, while only the pathways from the lower regions of the body run in the lumbar and the sacral segments.

Spinal Roots. In each spinal segment, nerve fibers enter the spinal cord dorsally and emerge from the spinal cord ventrally. Figure 1-10 is a cross section through such a zone with *dorsal* and *ventral roots.* All the afferent fibers, the somatic as well as the visceral afferents, thus enter the spinal cord in the dorsal roots. All efferent nerve fibers,

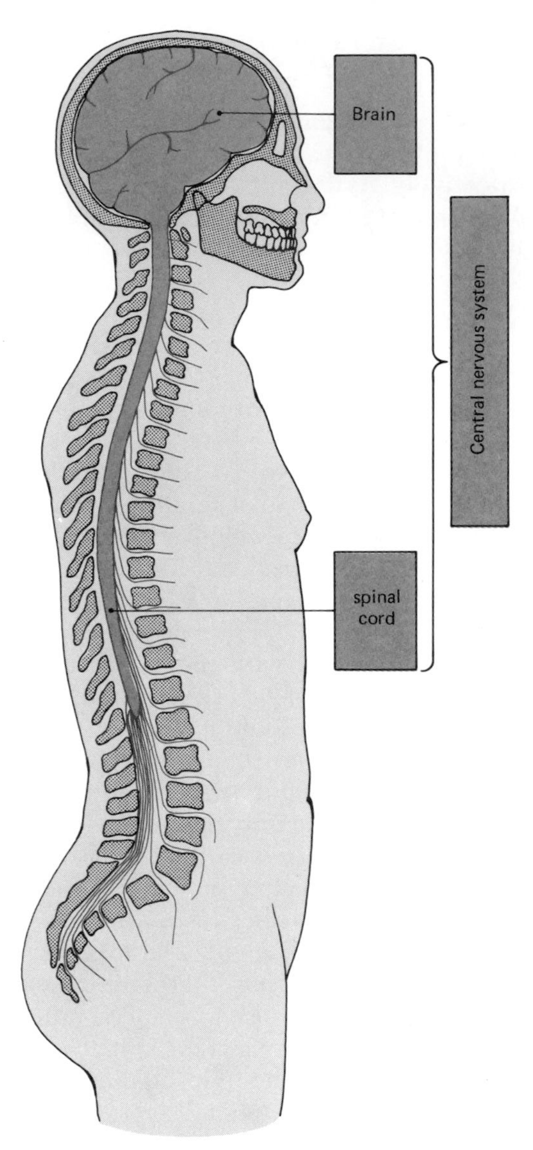

Fig. 1-8. Diagrammatic longitudinal section in the midline (sagittal section) through skull and spinal column. The roots shown leaving the spinal cord (see Figs. 1-10 and 1-11) are those of the somatic and visceral nerves. Each vertebra is associated with a spinal-cord segment bearing one pair each of ventral and dorsal roots.

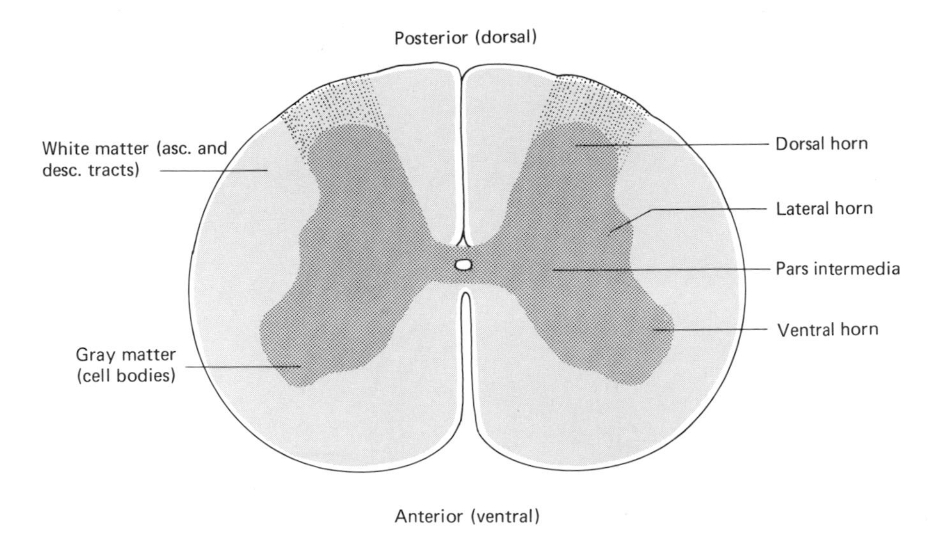

Fig. 1-9. Cross section through the spinal cord at the level of a lumbar segment. In other sections of the spinal cord, the shape of the gray matter and the proportion of gray to white matter differ somewhat from those shown here (see text).

that is, the motor and the autonomic efferents, emerge from the spinal cord in the ventral roots only.

The cell bodies of the efferent fibers are located in the gray matter of the spinal cord. The cell bodies of the motor efferent fibers, which lead to the skeletal muscle fibers, are situated in the ventral horn. Because of their position, these cells are called *ventral horn cells*, and because of their function they are called *motor* ventral horn cells, or *motoneurons*. Their axons, that is, the motor nerve fibers, are also often called *motor axons*. (A description of the position of the somata of the autonomic efferents is given in Chapter 8.)

In contrast to the efferent fibers, whose cell bodies are located in the gray matter of the spinal cord, the cell bodies of all the afferent fibers are situated outside the spinal cord, close to the places where the roots enter the vertebral canal. Such a local accumulation of nerve cells outside the CNS is called a *ganglion*. The accumulation of the cell bodies of the dorsal root afferents is called a *dorsal root ganglion*. The neurons in the dorsal root ganglion have three special characteristics: their axons divide shortly after emerging from the ganglion into the centrally projecting (dorsal root fibers) and the peripherally projecting (afferent fibers) branches (see Fig. 1-10); the soma has no dendrites; there are no synapses on the soma. Figure 1-2D shows a dorsal root ganglion cell.

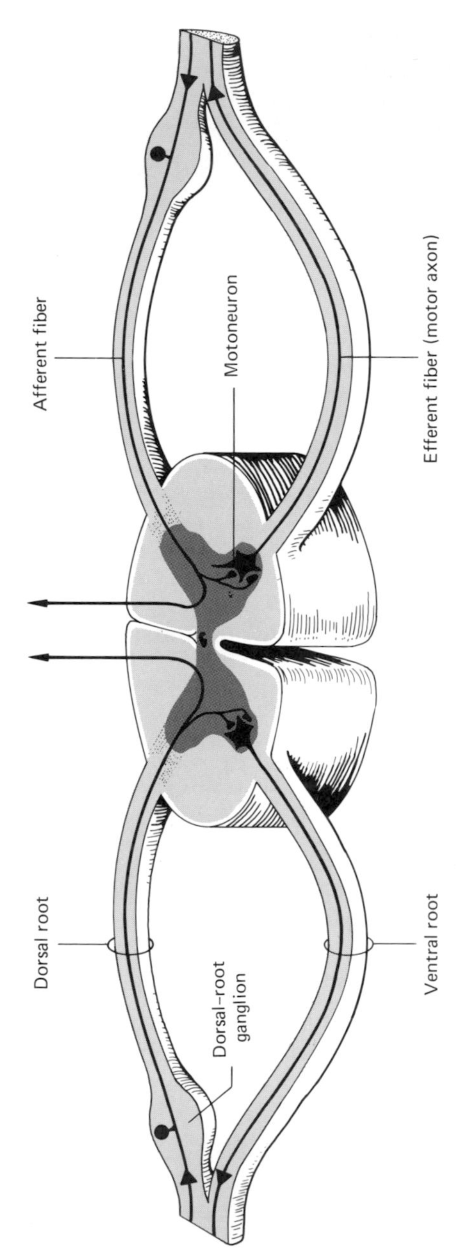

Fig. 1-10. Diagrammatic cross section through the spinal cord at the level of a root entry zone.

These structures are summarized and further clarified by the three-dimensional drawing of two spinal segments and their roots in Fig. 1-11. While still in the vertebral canal, the individual ventral and dorsal root filaments of each spinal half-segment come together to form common ventral and dorsal roots, respectively. In the dorsal root, the spinal ganglion appears as a distinct thickening. On both sides of the cord, the ventral and the dorsal root join to form the spinal nerve, which then emerges from the vertebral canal at a gap between the two associated vertebrae. Only after the spinal nerves emerge from the vertebral canal do their various somatic and autonomic nerve components separate from each other through a complex process of interlac-

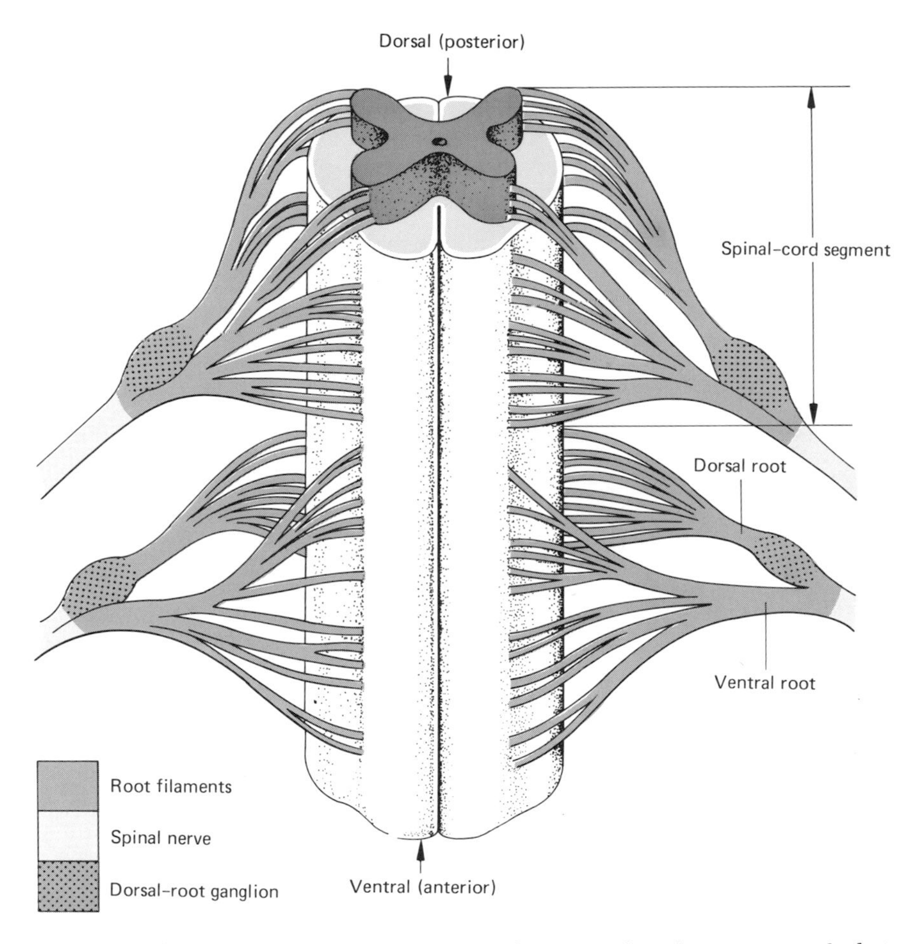

Fig. 1-11. Three-dimensional representation of two spinal-cord segments with their roots. See text for details.

ing and branching. The nerves emerging from the spinal cord serve the entire body, with the exception of the head, which is served by 12 paired *cranial nerves* (the term "paired" refers to the fact that there is one nerve on each side of the body). Relevant details and special information will be given later.

The anatomy presented in this chapter is sufficient for Chapters 2 to 4. In the later chapters, anatomic information will be introduced when needed.

Q 1.12 Draw a diagram of a nerve cell and label the various parts of this cell. Draw a diagram of, and label the possible connections between, two nerve cells.

Q 1.13 The cells of the CNS (brain and spinal cord) are separated from one another and from the blood capillaries by a narrow gap. This gap is called
a. myelin sheath.
b. neuroglia.
c. node of Ranvier.
d. extracellular space.
e. cerebrospinal fluid.

Q 1.14 What types of afferent and efferent nerve fibers are contained in the ischiadic nerves of human beings?

Q 1.15 Which of the following statements is/are correct?
a. Each spinal segment has two ventral roots.
b. One half of a spinal segment corresponds to each vertebra.
c. The motoneurons are located in the dorsal root ganglia.
d. The gray matter of the spinal cord owes its coloration to the myelin sheaths.
e. The cell bodies of the dorsal root ganglia cells have no synapses.
f. Each spinal segment has one dorsal root.

Q 1.16 Which of the following descriptions of nerve fibers contains mutually exclusive or contradictory concepts?
a. Myelinated, afferent, soma in ventral horn.
b. Unmyelinated, afferent, diameter 10 μm.
c. In mixed nerve, efferent, motoneuron.
d. Visceral, efferent, soma in dorsal root ganglion.
e. Afferent, visceral, unmyelinated.

2
EXCITATION OF NERVE AND MUSCLE

J. Dudel

A difference in potential, the *membrane potential*, usually exists between the inside of a cell and the extracellular fluid surrounding it. In many types of cells (for example, muscle cells or glandular cells), the function of the cell can be controlled by the magnitude of this potential. It is, in fact, the specialized role of the nervous system to propagate changes in membrane potential within its cells and to transmit them to other cells. These changes in potential can be regarded as units of information that help the body to coordinate the activity of various groups of cells. In particular, the body can feed the information impinging on it from the environment to a center where it is processed, enabling the body to adapt itself in a suitable manner to its surroundings. At the basis of all these functions is the membrane potential and the changes in this potential that spread out through the cells. How the membrane potential is generated and what conditions govern its changes will be discussed in detail in this chapter.

2.1 Resting Potential

Measuring the Membrane Potential. The potential difference between the interior of a cell and the fluid surrounding the cell—the *membrane potential*—can be measured by connecting one pole of a voltmeter to the inside of the cell and the other pole to the extracellular space. The appropriate apparatus is diagrammed in Fig. 2-1A. The voltmeter is connected through the electrodes to the test preparation, a cell which is maintained in a bath solution. Usually glass capillaries

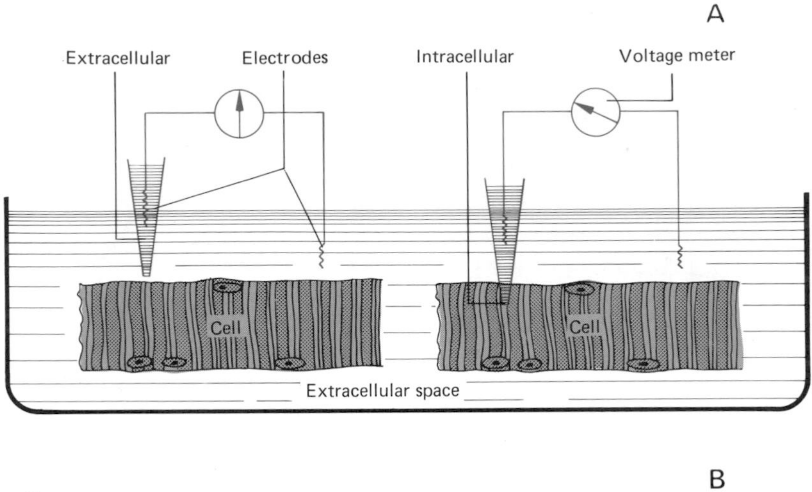

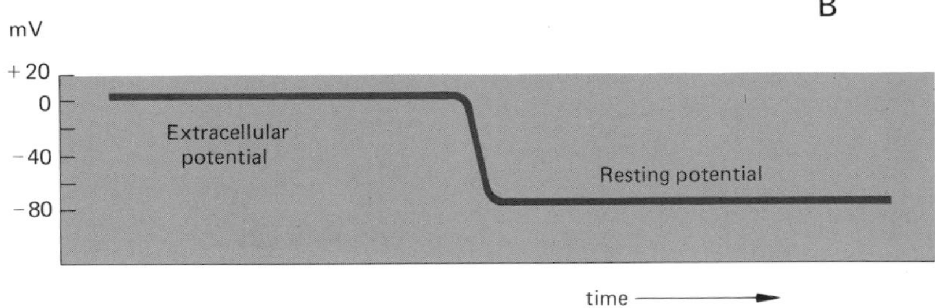

Fig. 2-1. Intracellular measurement of membrane potential. **A.** Diagram of the experimental arrangement. The space around the cell is filled with a saline solution with composition resembling that of blood. On the *left*, both recording and reference electrodes are extracellular, and the voltage meter measuring the potential between the two reads zero. On the *right*, the recording electrode has penetrated the cell while the reference electrode remains in the extracellular space. The meter now indicates the membrane potential. **B.** The potentials recorded before and after insertion of the recording electrode into the cell.

filled with a conducting solution are used as the electrodes, which can be inserted into the interior of the cell. In order not to damage the cells, these glass capillaries have very fine tips (less than 1 μm). At the start of the measurement (left half of Fig. 2-1A) both electrodes are located in the extracellular space, and no potential difference exists between them. The potential of the extracellular space is generally taken to be 0. This zero potential is indicated in Fig. 2-1B (left) as the "extracellular potential." When the tip of the glass capillary is pushed through the membrane of the cell (Fig. 2-1A, right), the potential jumps in a negative direction to approximately −75 mV, as is shown in

Fig. 2-1B. Since this potential difference occurs when the membrane is penetrated, it is referred to as the *membrane potential.*

In most cells, the membrane potential remains constant for long periods, provided that no special influences act on the cell from the outside. When the cell is in such a resting state, the membrane potential is called the *resting potential.* The resting potential is always negative in nerve and muscle cells and has a characteristic constant magnitude for the individual cell types. In the nerve and the skeletal muscle fibers of warm-blooded animals, the resting potentials are between −55 and −100 mV. In smooth muscle fibers, more positive resting potentials between −55 and −30 mV occur.

Origin of the Resting Potential. What physical processes generate the resting potential? If the interior of the cell has a greater negative charge than the surroundings of the cell, then excess negative electric charges will exist within the cell compared with that of the extracellular space. Both the inside of the cell and the extracellular space are filled with aqueous salt solutions. In dilute salt solutions the majority of the molecules dissociate into ions, that is, positively or negatively charged atoms or groups of atoms. Positively charged atoms or molecules are called *cations;* negatively charged ones are called *anions* (because in an electrical field they migrate to the cathode or the anode, respectively). When, for example, sodium chloride (NaCl) is dissolved in water it dissociates into the cation Na^+ and the anion Cl^-. In aqueous solutions the ions are the sole carriers of charge. Consequently, charge disequilibrium, which is expressed by the resting potential, indicates a certain excess of anions (negative charges) inside the cell and a corresponding excess of cations outside the cell.

Since ions can move freely in an aqueous solution, a disequilibrium in charge cannot continue to exist within either the intracellular or the extracellular space; it must be balanced out by the movement of the ions. The charge disequilibrium, which causes the resting potential, therefore must be located at the "solid phase," which demarcates the cell, that is, the cell membrane. The resting potential is thus generated at the cell membrane; on the inner side, within the cell, there is an excess of anions, while on the outer side, there is a corresponding excess of cations.

The *cell membrane* can be regarded as an *electric capacitor* in which two conducting media, the intracellular and the extracellular salt solutions, are separated from one another by a nonconducting layer, the membrane. The insulating membrane is about 6 nm (60 Å) thick. In order to charge a capacitor with this "plate spacing" to the resting potential of −75 mV, approximately 5,000 pairs of ions per square micrometer of cell surface must be displaced. The electric potential across a capacitor is proportional to the number of charges that are held on its "plates."

In order to clarify further the numerical ratios of the ions involved, Fig. 2-2 shows a very small section of the membrane measuring 1 μm × 1/1,000 μm in area and certain adjoining intra- and extracellular volumes, each 1 μm × 1 μm × 1/1,000 μm. Assuming a resting potential of −90 mV, this membrane area is occupied by six anions and six cations. However, there are 220,000 ions in each of the adjoining spaces. The imbalance in the distribution of the charge at the membrane is very slight. Nevertheless, it is the basis of the resting potential and thus of the functioning of the nervous system.

Distribution of the Ion Concentrations. Why is it always a negative resting potential that occurs in nerve and muscle? The source of the resting potential is the *unequal distribution of the types of ions,* particularly the K^+ ions, inside and outside the cell. The distribution of the various types of ions for the intracellular and the extracellular spaces is given in Fig. 2-2. The largest disequilibrium exists in the

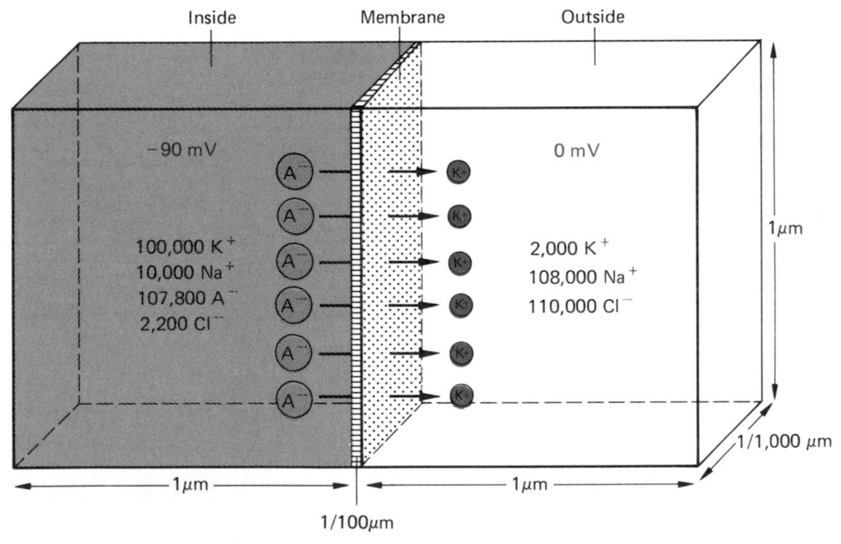

Fig. 2-2. Membrane charge during resting potential. The charge of a small section of the membrane with an area of 1 μm × 1/1,000 μm, six K^+ ions and six anions, is compared with the number of ions in adjacent spaces, each 1 μm × 1 μm × 1/1,000 μm, on either side of the membrane. A^- denotes the intracellular protein anions. The arrows through the membrane indicate that the K^+ ions have diffused through the membrane from the cell but remain fixed at the outside of the membrane because of the charge of the A^- inside the cell.

case of the K^+ ions: these number 100,000 on the intracellular side and only 2,000 on the extracellular side. The figures for Na^+, on the other hand, are 108,000 outside and only 10,000 inside the cell. The distribution of the Cl^- ions is exactly the reverse of the K^+ ions. The majority of the intracellular anions are not chloride but large protein anions, designated as A^-.

Table 2-1 gives the *ionic concentrations inside* and *outside* the *muscle cell* of a mammal (in mM). In these cells the potassium concentration in the cell is about 40 times higher than in the extracellular space, and the sodium concentration is about 12 times higher outside than inside. These ionic distributions are very constant in the individual types of cell. Generally, in nerve and muscle cells the intracellular K^+ concentration is 20 to 100 times greater than the extracellular concentration, the intracellular Na^+ concentration is 5 to 15 times lower than the extracellular concentration, and the intracellular Cl^- concentration is 20 to 100 times lower than the extracellular figure. The concentration distribution of chloride is thus approximately reciprocal to that of K^+. The extracellular salt solution is essentially a sodium chloride solution with an NaCl content of about 9 g/l. A solution of 9 g of sodium chloride in 1 liter of water is also called a "*physiological saline solution.*" This solution tastes just as salty as blood.

The K^+ Ions and the Resting Potential. How is the resting potential generated by the different ionic concentrations in the extracellular and the intracellular spaces? The different ionic concentrations would soon cancel each other by *diffusion* of the mobile particles if this were not prevented or compensated in some way. If the membrane constituted an impenetrable barrier for ions, that is, if it were *impermeable*, the different ionic concentrations on both sides of the membrane could continue to exist without restriction. However, the membrane is not completely impermeable but permits K^+ ions to pass through with

Table 2-1. Ionic Concentrations Inside and Outside the Muscle Cell of a Mammal (mmol/l)

Intracellular		Extracellular	
Na^+	12	Na^+	145
K^+	155	K^+	4
		Other cations	5
Cl^-	4	Cl^-	120
HCO_3	8	HCO_3^-	27
A^-	155		
Resting potential: −90 mV			

relative ease. In other words, it is *permeable to K^+ ions.* We can therefore picture the membrane as being full of pores or channels as indicated in Fig. 2-3. These pores are so narrow that only the relatively small K^+ ions can traverse them, diffusing through the membrane. The "size" of the ions in Fig. 2-3 corresponds to their effective diameter. This is not the same as the "ion radius": in aqueous solution water molecules become attached to the ions, and the latter become hydrated. The hydrated K^+ ions are smaller than the hydrated Na^+ ions; see this relationship in Fig. 2-3.

In order to represent the *diffusion conditions* at the membrane, all the ions except the K^+ ions have been omitted from Fig. 2-4 because it is the K^+ that can pass through the membrane most readily. The K^+ ions will move through the membrane in both directions, that is, from outside to inside and vice versa. But because of their higher concentration within the cell, the K^+ ions will hit and pass through a pore on the inner side approximately 30 times more frequently than on the outer side. A net efflux of K^+ ions results, the motive force being the higher osmotic pressure *of K^+* inside the cell. The osmotic pressure difference for K^+ would very soon lead to an equalization of the K^+

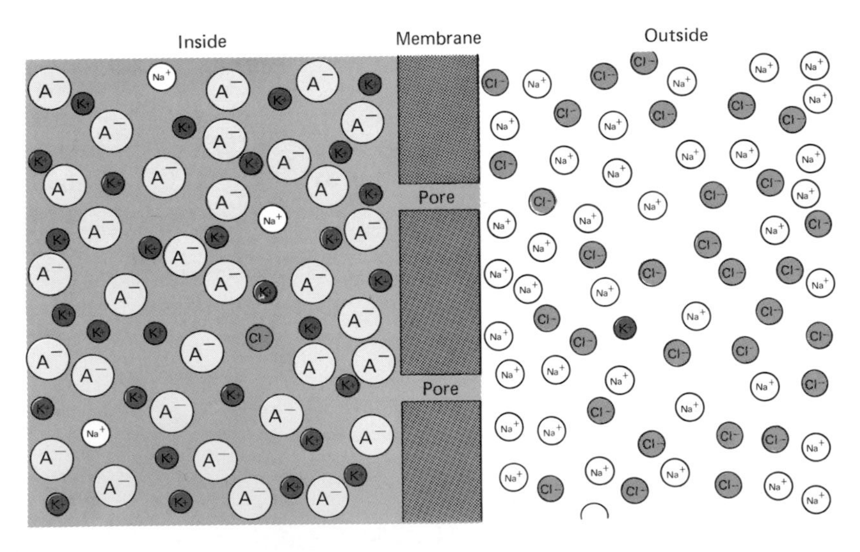

Fig. 2-3. Intra- and extracellular distribution of the ions. On both sides of the membrane, the different ions are indicated by *circles of different diameter,* proportional in each case to the diameter of the (hydrated) ion. A^- designates the large intracellular protein anions. The passages through the membrane, the "pores," are just large enough to permit the K^+ ions to diffuse through.

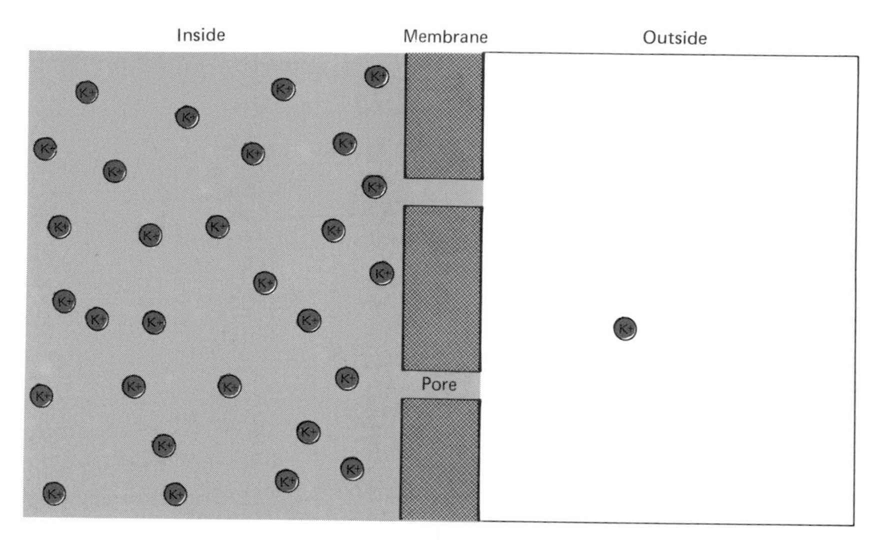

Fig. 2-4. Intra- and extracellular distribution of K^+ ions. This is the same diagram as Fig. 2-3, but all ions apart from K^+ have been omitted, since the latter pass most readily through the membrane.

concentrations as a result of the K^+ efflux were it not for an equally large and opposing force preventing this.

The opposing force is supplied by an electric field, the *membrane potential*, the origin of which will now be explained. We have so far ignored the fact that the K^+ ions carry *positive charges*. As was discussed in connection with Fig. 2-2, the shifting of a cation across the cell membrane results in the *membrane capacitor becoming charged* and gives rise to a membrane potential. When a K^+ ion flows out of the cell, an excess positive charge will appear on the outer side of the capacitor, corresponding to an excess negative charge on the inner side. The membrane potential acts in such a way that it opposes the efflux of more cations. The efflux of positive charges by itself builds up an electric potential that impedes the efflux of more positive charges. The electric potential builds up until the force that it exerts against the efflux of K^+ ions is equal to the osmotic pressure of the K^+ ions. At this potential, the influx and the efflux of K^+ ions are balanced, and we, therefore, refer to it as the *K^+ equilibrium* potential, or E_K.

The potassium equilibrium potential, E_K, is determined by the concentration ratio of the *K^+* ions inside and outside the cell, K^+_i/K^+_o. E_K is proportional to the logarithm of this concentration ratio. The quantitative relationship between concentration ratio and equilibrium potential is called the *Nernst equation*. For K^+ ions, this equation reads:

$$E_K = -61 \text{ mV} \cdot \log (K^+_i/K^+_o)$$

The factor −61 mV is determined by several constants and the temperature*. If $K^+_i/K^+_o = 30$, as in Figs. 2-3 and 2-4, then

$$E_K = -61 \text{ mV} \cdot \log 30 = -61 \text{ mV} \cdot 1.48 = -90 \text{ mV}$$

that is, it is approximately as large as the resting potential. The *resting potential,* as a first approximation, thus corresponds to the *potassium equilibrium potential.* At this potential, the concentration difference of the K^+ ions across the membrane can remain unchanged, despite the good permeability to K^+, because the membrane potential is just large enough to prevent a net efflux of K^+.

Role of Cl^- in the Resting Potential. The representation of the resting potential as a potassium equilibrium potential must be amended to take account of the fact that the chloride ions are also involved. The *membranes* are in fact also *permeable* to *Cl^- ions.* In nerve cells the permeability to Cl^- is less than that to K^+, but in muscle cells it is much greater than that to K^+. The concentration ratio of chloride ions inside and outside the cell, Cl^-_i/Cl^-_o, is usually reciprocal to the corresponding concentration ratio K^+_i/K^+_o (see Table 2-1). For this reciprocal distribution of the chloride, we find, in accordance with the Nernst equation (for anions the sign is reversed), the same potential as for the potassium distribution. The *chloride equilibrium potential* is, then, approximately equal to the *potassium equilibrium potential.*

The reciprocal distribution of K^+ and Cl^- across the cell membrane does not happen by chance. The intracellular chloride concentration can be varied easily by the influx or the efflux of Cl^-, and adjusts itself according to the membrane potential, because when the chloride equilibrium potential deviates from the membrane potential, compensatory currents flow. If the membrane potential adjusts itself at E_K, the result is a Cl^- distribution reciprocal to the K^+ distribution and E_{Cl} becomes equal to E_K.

Unlike the Cl^- concentration, the intracellular K^+ concentration

---

* In general form the Nernst equation is written as follows:

$$E_{\text{ion}} = \frac{R \cdot T}{z \cdot F} \cdot \ln \frac{\text{Extracellular ionic concentration}}{\text{Intracellular ionic concentration}}$$

in which R is the gas constant, F the Faraday constant, z the valence of the ion (positive for cations and negative for anions), and T the absolute temperature.

cannot change very much. The K^+ ions must in fact provide the charge in the cell to balance out the anions. The intracellular anions are mainly large protein molecules (see Table 2-1), which cannot pass through the cell membrane, so their concentration remains constant. These large anions must be opposed in the intracellular solution by an equal number of cations. Since the intracellular Na^+ concentration is kept very low (see Sec. 2.3), neutrality of charge must be guaranteed by K^+ ions. It follows, therefore, that the intracellular K^+ concentration must be about as high as that of the large anions and that the intracellular K^+ concentration, just as that of the large anions, must be very stable. The *high intracellular concentration of* K^+ is indirectly necessitated by the presence of the *intracellular anions*, and, in turn, the negative E_K derives from this high intracellular concentration of K^+. The Cl^- concentration in the cell depends on the membrane potential and is a secondary result of the K^+ distribution. From this standpoint, the negative resting potential is the result of the high concentration of nonpermeable anions retained in the cell.

Q 2.1 Draw a diagram of the test apparatus used for the intracellular measurement of the membrane potential of a cell. At the same time, show at the membrane of the cell the charge of the membrane potential with the cations and the anions that are essential for the resting potential.

Q 2.2 Enter in the following table the ratio of the intracellular to the extracellular ion concentration for Na^+ and Cl^- ions.

Ion	*Internal / External*
K^+	20–100 / 1
Na^+	1 / ______–______
Cl^-	1 / ______–______

The distribution of K^+ and Cl^- ions is ______.

Q 2.3 Which of the following variables are in balance at the equilibrium potential for an ion? (More than one answer may be correct.)
a. The intracellular and the extracellular concentrations of the ion.
b. Osmotic pressure and electric field.
c. Influx and efflux of the ion through the cell membrane.
d. The Na^+ concentration outside the cell and the K^+ concentration inside the cell.

2.2 Resting Potential and Na^+ Influx

The explanation of the resting potential given in the preceding section—that it is determined by the K^+ equilibrium potential—proceeded from the simplifying assumption that the cell membrane is permeable only to K^+ and Cl^- ions. By making this assumption it was possible to invoke an equilibrium of diffusion and potential across the cell membrane. However, to a lesser extent, the membrane is also permeable to Na^+ and other ions. Flows of these ions disturb the equilibrium so that a constant resting potential cannot be maintained by diffusion processes alone.

Dependence of the Resting Potential on the Potassium Concentration. The statement that the *resting potential* is the same as the *potassium equilibrium potential* can be checked experimentally. The extracellular K^+ concentration can be varied within wide limits, and at the same time, the resting potential can be measured. Such resting potentials, measured for various extracellular K^+ concentrations, K^+_o, are plotted as circles in Fig. 2-6*. When the extracellular K^+ concentration is increased as shown, the resting potential decreases in absolute magnitude from −90 to −20 mV. In addition, E_K can be calculated in accordance with the Nernst equation for the various extracellular K^+ concentrations. This calculated dependence of E_K on K^+_o is shown as a straight line in Fig. 2-6. The relationship is linear because K^+_o is plotted on a logarithmic scale along the abscissa. The measurements agree essentially with the calculated straight line and, to a first approximation, confirm the proposal that the resting potential is the same as

* Figure 2-5 appears on p. 323 as an answer to Q 2.1.

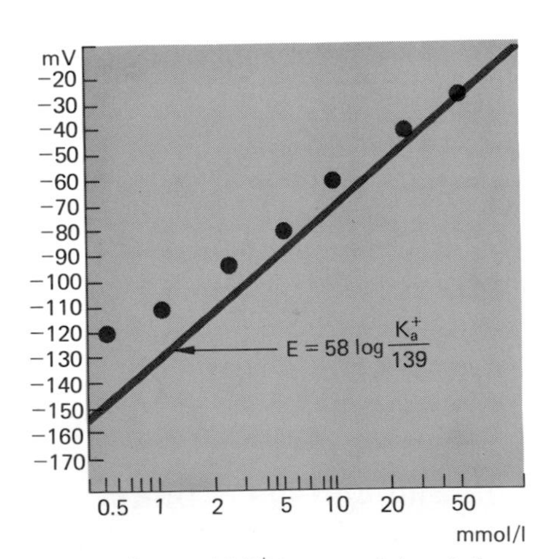

Fig. 2-6. Dependence of the resting potential on the extracellular K_0^+ concentration. The abscissa shows the extracellular potassium concentration K_0^+, plotted on a logarithmic scale; the ordinate gives the intracellularly measured membrane potential (*circles*) and the potassium equilibrium potentials given by the Nernst equation (*red line*) at the different K^+. Modified from Adrian (1956) *J. Physiol.* 133:631.

the potassium equilibrium potential. It will be noticed, however, that only in the upper region of the curve, that is, at high K^+_o values, do the measurements agree well with the theoretical relationship, while at low K^+_o values, they deviate upward to an increasing extent. Thus, at low extracellular K^+ concentrations, the resting potential is less negative than E_K. This is also true for the normal extracellular K^+ concentration of about 4 mmol/l. In muscle fibers, as illustrated here, as well as in other types of cell, the *resting potential is up to 30 mV less negative than the* E_K.

An indication of why the resting potential deviates from E_K at low extracellular K^+ concentrations can be obtained from a variation of the experiment illustrated in Fig. 2-6. If the same experiment is carried out in a bathing solution in which the Na^+ is replaced by a large cation that cannot pass through the membrane (for example, choline), then the measured resting potential and the calculated E_K agree exactly even at low K^+_o. The deviation of the resting potential from E_K in a Na^+-containing solution must, therefore, be caused by a flow of Na^+ ions through the cell membrane. It can be concluded that in the resting condition, the membrane is permeable to not only K^+ ions but also, to a lesser extent, to Na^+ ions. As a result, the Na^+ ions, which are present in very much higher concentrations on the outside, flow slowly into the cell, partly discharge the membrane capacitor, and make the membrane potential less negative.

Membrane Conductance for K^+ and Na^+. The resting potential usually does not quite agree with E_K because the membrane is permeable to not only potassium (and chloride ions) but also to some extent *sodium ions*. The degree to which the resting potential deviates from E_K is determined by the ratio of the membrane permeability values for Na^+ and K^+. In order to give a quantitative explanation of the resting potential, it is, therefore, necessary in some way to measure the ionic permeability of the membrane. For this purpose it is usually the *membrane conductance, g*, that is determined. The conductance is the reciprocal of the electric resistance. The electric resistance is determined by the quotient of voltage/current; consequently, the conductance is determined as g = current/voltage. To obtain the membrane conductance for a certain ion the flux through the membrane of the ions in question must be divided by the potential that provides the driving force. This driving potential is 0 at the equilibrium potential because at this latter potential the net ion flux is equal to zero. The equilibrium potential should be taken as the reference point for the driving potential. As the membrane potential moves further away from the equilibrium potential, the influx and efflux become more and more out of balance, and the net flow increases. Therefore, the difference between the membrane potential and the equilibrium potential is taken as the driving potential for the (net) ion flux. We can then write for the *potassium conductance*

$$g_K = I_K/(E - E_K)$$

in which I_K is the net potassium flux and E is the membrane potential. Experimental quantitative determinations of g_K and g_{Na} have shown that under resting conditions in nerve and muscle cells g_k is *10 to 25 times larger than* g_{Na}.

From the equilibrium potential and the conductances for K^+ and Na^+ ions, it is now possible to explain the magnitude of the resting potential for various $K^+{}_o$ values. The *sodium equilibrium potential* E_{Na} is situated at positive potentials because the Na^+ concentration inside the cell is smaller than outside. The quotient $Na^+{}_i/Na^+{}_o$ is smaller than 1, and thus log $Na^+{}_i/Na^+{}_o$ is negative. When the ratio $Na^+{}_i$ to $Na^+{}_o = 1:12$, we obtain, according to the Nernst equation (see p. 26),

$$E_{Na} = -61 \text{ mV} \cdot \log 1/12 = -61 \text{ mV} \cdot (-1.08) = +65 \text{ mV}$$

If the potential is more negative than E_{Na}, a net sodium current enters the cell. This, in fact, obtains over the entire potential range in Fig. 2.6. If g_{Na} remains constant in this potential range, then the sodium inward current becomes greater as the deviation from E_{Na} increases $[I_{Na} = g_{Na} \cdot (E - E_{Na})]$. Thus, as the membrane potential becomes more negative in Fig. 2-6, the Na^+ inward current increases, causing a larger deviation of the resting potential from E_K.

Even at the normal extracellular K^+ concentration, the resting potential is about 10 mV less negative than E_K. This can be explained quantitatively by the ratio of g_K to g_{Na} and by the deviations of the resting potential from E_K and E_{Na}. At the resting potential, a small potassium outward current must be in equilibrium with a sodium inward current. If these currents are to be equal, the 20 times greater conductance for K^+ must be offset by a 20 times greater driving potential for Na^+. The following relationships must exist:

$$g_K : g_{Na} = 20:1$$

$$-(E - E_{Na}):(E - E_K) = 20:1$$

From the latter expression, it follows that

$$E = E_K + \frac{E_{Na} - E_K}{21}$$

$E_{Na} - E_K = +65 - (-90)$ mV = 155 mV; according to this calculation, the resting potential E would be 7.4 mV more positive than E_K.

Instability of the Resting Potential Given Purely Passive Ionic Currents. The fact that under resting conditions Na^+ ions continuously flow into the cells, while, correspondingly, K^+ ions must flow out, has far-reaching consequences. The system cannot, in fact, be equilibrated under resting conditions by means of diffusion and buildup of membrane charge alone; this would mean that the intracellular ionic concentrations could not be kept constant. If no other processes (see Sec. 2.3) were involved besides the *passive* ion current mentioned above, the cell would slowly pick up more Na^+ and lose K^+. The decline in intracellular K^+ concentration would bring about a decrease in E_K and thus in the resting potential. As the resting potential declined to less negative values, the intracellular Cl^- concentration would be forced to rise because it adjusts itself to the resting potential. The large intracellular protein anions cannot leave the cell, so as the intracellular Cl^- concentration rose, there would be an increase in the total intracellular anion concentration and, consequently, in the total ion concentration. In order to equalize the osmotic pressure, water would flow into the cell, and the cell volume would expand. In turn, the water picked up by the cell would lower the intracellular K^+ concentration and further reduce the membrane potential. Thus, with the takeup of water and the decline in the resting potential, the intracellular ion concentrations would nearly match the extracellular concentrations.

The process that prevents all this from happening in healthy, living cells is discussed in the next section.

Q 2.4 Which of the following statements indicate that, apart from K^+ and Cl^- ions, Na^+ ions also influence the resting potential?
- a. The resting potential is less negative than E_K.
- b. The resting potential changes approximately in proportion to the logarithm of the extracellular K^+ concentration.
- c. In the absence of extracellular Na^+, the resting potential and E_K are the same.
- d. The sodium equilibrium potential is positive, while the potassium equilibrium potential is negative.

Q 2.5 What is the mathematical expression used to define the chloride conductance of the membrane?

Q 2.6 What would the membrane potential be if the membrane were permeable only to Na^+? Give the symbol and the approximate value for this potential.

Q 2.7 Which of the following statements show why the resting membrane potential is made more positive by the inflow of Na^+?

a. There are more Na^+ ions outside the cell than inside.
b. The cell loses K^+ ions because of the inflow of Na^+ ions.
c. The negative charge on the inner side of the membrane is reduced by the influx of Na^+ ions.
d. The cell also picks up chloride as a result of the Na^+ inflow.

2.3 The Sodium Pump

The preceding section showed how the diffusion of Na^+ ions into the cell during resting conditions could upset the equilibrium of the ionic currents to such an extent that the normal concentration gradients and the resting potential would slowly disappear. The Na^+ ions flowing *passively* into the cell must leave the cell again; otherwise, the intracellular Na^+ concentration could not remain at a constant low level. The Na^+ ions that have entered the cell cannot leave again by diffusion against the potential and the concentration gradients, that is, they cannot flow "uphill"*. They must, therefore, be expelled *actively* from the cell, which requires the expenditure of energy. The passive inflow of Na^+ is in fact balanced by *active transport* of Na^+ from the cell. This transport mechanism is also called the *sodium pump*. The sodium pump transports Na^+ ions from the cell against the concentration and the potential gradients with the expenditure of metabolic energy.

Measuring the Active Transport of Na^+. The active transport of Na^+ ions out of the cell can be determined by measuring the Na^+ efflux from the cell. The number of Na^+ ions that can leave the cell passively against the concentration and the potential gradients is negligibly small. The Na^+ efflux from the cell is thus identical to the number of ions actively transported. In order to measure the Na^+ ions that have flowed out of the cell, it is necessary to distinguish them from the many other Na^+ ions already present in the extracellular space. This can be done by first loading the cell intracellularly with a *radioactive isotope of sodium* ($^{24}Na^+$) and then measuring the occurrence of this isotope in the extracellular space.

Figure 2-7 illustrates two such experiments conducted on a nerve. In part *A* during the first measurement period at 18.3°C, the $^{24}Na^+$ efflux slowly declines, because, as a result of the efflux itself, the $^{24}Na^+$ fraction of the intracellular Na^+ concentration drops. If the *tem-*

* Of course, these statements only apply to a net sodium efflux. A very small fraction of the intracellular Na^+, compared with the rate of influx, can diffuse outward.

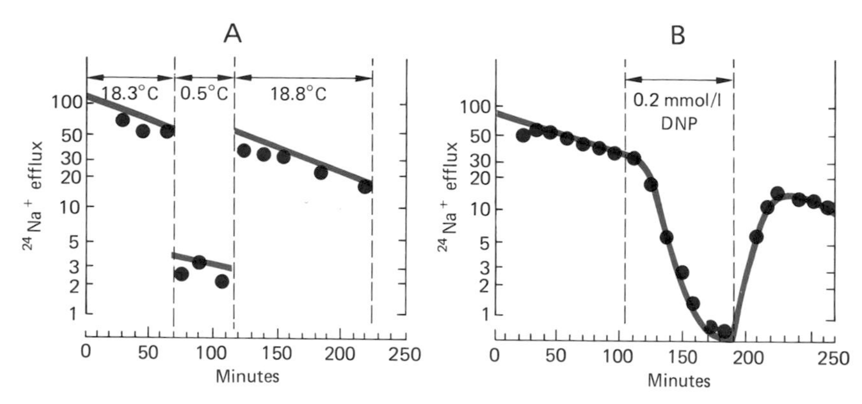

Fig. 2-7. Inhibition of active Na^+ transport as a result of cooling or exposure to DNP. Efflux of radioactive $^{24}Na^+$ from a cell that had been loaded with this isotope before the experiment. Abscissa: time since the beginning of the experiment, in minutes; ordinate: efflux of radioactive $^{24}Na^+$ from the cell. **A.** During the experiment, the cell is cooled from 18.3° to 0.5°C and warmed again. The Na^+ efflux is severely reduced during the cold period. **B.** During the experiment, the cell is exposed for 90 min to 0.2 mmol/l DNP, which causes Na^+ efflux to fall almost to zero. When the DNP has been washed away, sodium ions begin to flow out of the cell again. Modified from Hodgkin and Keynes (1955) *J. Physiol.* 128:28.

perature of the nerve is suddenly *lowered* to 0.5°C, the *$^{24}Na^+$ efflux immediately drops* to about one-tenth its value. When the temperature of the nerve is increased again, the rate of $^{24}Na^+$ efflux returns to the value observed prior to cooling. The strong dependence of the Na^+ efflux on temperature shows that an *active chemical process,* and not passive diffusion, is involved. Diffusion processes would be slowed down only slightly by this drop in temperature. Therefore, active transport must be involved in the Na^+ efflux.

Further proof is provided by a variation of the experiment on which Fig. 2.7B is based. At the start of the experiment, $^{24}Na^+$ flows out of the cell at a fast rate. When dinitrophenol (DNP) is added to the extracellular solution, the $^{24}Na^+$ efflux drops to almost zero within an hour. After the DNP is flushed out, the normal $^{24}Na^+$ efflux picks up again. DNP is a poison that penetrates the cell and blocks energy-supplying metabolic processes. Diffusion processes through the membrane are not affected by DNP. The drop in the Na^+ efflux in the presence of DNP is due to an inadequate supply of metabolic energy. This demonstrates that the *Na^+ efflux is dependent on the availability of energy* and, therefore, that Na^+ is actively transported through the membrane.

Even in the living organism, the supply of adequate amounts of metabolic energy for the cells can break down due to a severe lack of oxygen or to poisoning. When this happens, the Na^+ pump fails and the cells take up Na^+ through passive diffusion. Consequently, as was described in the preceding section, the membrane potential declines, the ionic distributions inside and outside the cell balance each other, and the cells swell. They soon become incapable of functioning and finally suffer irreversible damage. Therefore, an adequately functioning Na^+ pump is vitally important for the existence of the cells.

The Coupled Na^+–K^+ Pump. The extracellular K^+ concentration also has a strong influence on the active Na^+ transport. In the absence of extracellular K^+, the Na^+ efflux falls to about 30% of its normal value. The reason for this dependence of the Na^+ efflux on the K^+ concentration is to be found in an *exchange process:* For each Na^+ ion transported out of the cell, a K^+ ion can be taken into the cell. This exchange process is called a *coupled Na^+–K^+ pump.* The model in Fig. 2-8 was developed to show how the coupled pump could operate. It can be seen that intracellular Na^+ ions bind themselves to a carrier molecule Y on the inner side of the membrane. The NaY complex can diffuse through the membrane. At the outer side of the membrane, the complex breaks down spontaneously so that the external concentration of NaY becomes less than the inner concentration. Consequently, the efflux of the NaY exceeds the influx. Thus, by temporarily combin-

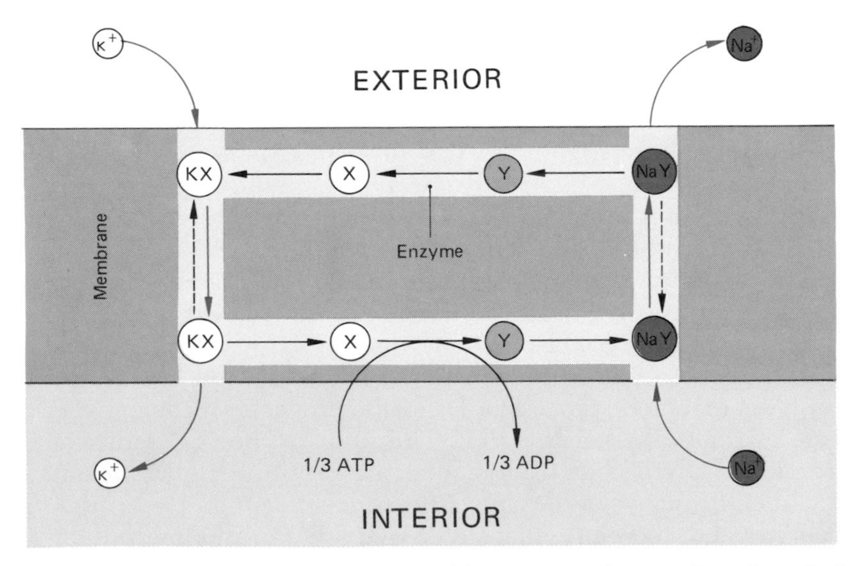

Fig. 2-8. Coupled Na^+–K^+ pump. Diagram of the transport of Na^+ and K^+ through the membrane by the carrier molecules Y and X. See text for further details. Modified from Glynn (1958) *Progr. Biophys.* 8:241.

ing with the carrier molecule Y, the Na^+ diffuses against its concentration and potential gradients.

On the outside of the membrane, carrier molecule Y is converted by an enzyme into carrier molecule X. X combines with extracellular K^+ to form KX and in this form diffuses to the inside. Here, KX, in turn, breaks down, having effected the transport of K^+ as well as the migration of X to the inside of the membrane. Finally, on the inside of the membrane, the carrier molecule X is converted back into carrier molecule Y with the expenditure of energy and is then available once more for the Na^+ transport cycle. In this reaction model, the conversion of carrier molecule X into carrier molecule Y is the active step in the transport process.

The NaY complex is usually electrically neutral. As a result, during the transport process no electric charge is transferred through the membrane, and the membrane potential is unaffected by the transport process itself. This *sodium pump* is therefore also termed *electrically neutral*. The mechanism of the coupled Na^+–K^+ pump probably developed because it *saves metabolic energy*. For the operation of the Na^+ pump, cells require considerable amounts of metabolic energy. It is estimated that *10 to 20% of the resting metabolism* of a muscle cell is *expended for the active transport of* Na^+. This energy requirement would be higher were it not for the fact that the greater part of the Na^+ transport is accomplished by a coupled Na^+–K^+ pump. With a coupled pump, no energy is consumed during the return of the carrier molecule to the inside of the cells, and the energy actually consumed is about half that required for uncoupled Na^+ transport.

Summary of Ionic Currents through the Membrane. Figure 2-9 summarizes the most important ionic currents (with the exception of Cl^-) through the membrane. This diagram of the membrane shows channels for the various ion movements in each direction. The width of the ion channels corresponds to the *magnitude* of the ionic current flowing through them, and their slope corresponds to the *driving potential* for the ionic current in question. The potential between inside and outside, namely, the *resting potential*, is taken to be −80 mV.

Consider first the movement of the *K^+ ions*. The resting potential here is 11 mV less negative than the K^+ equilibrium potential; therefore, the K^+ ions are driven to the outside by this driving potential of 11 mV. This is why the K^+ channels are slightly sloping to the outside. The passive K^+ efflux (uppermost in diagram) therefore outweighs the passive K^+ influx, and more K^+ ions can diffuse "downhill" than "uphill." The difference between the passive K^+ currents is made up by active K^+ transport. Like all active processes in Fig. 2-9, the active K^+

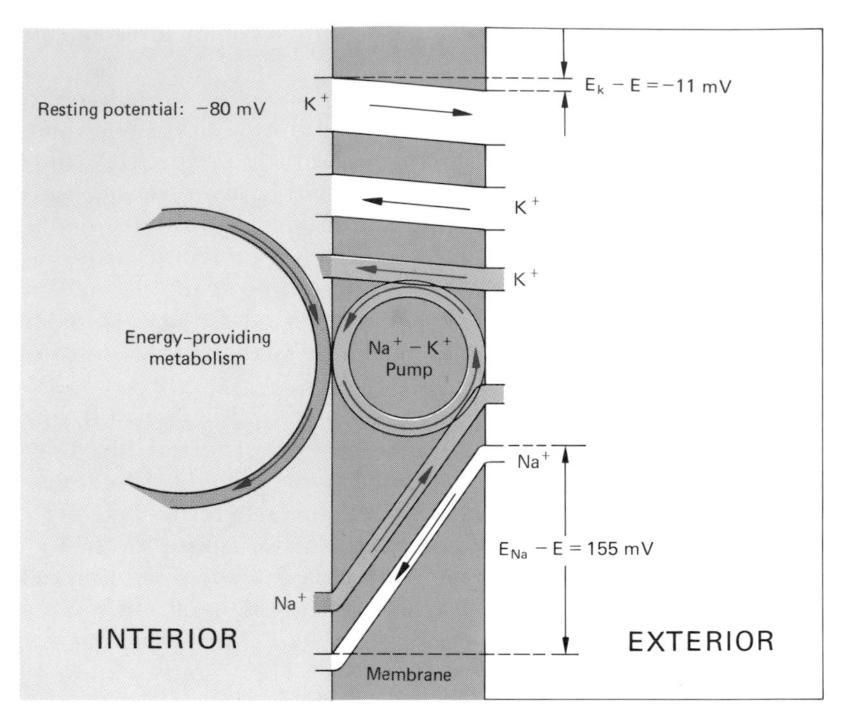

Fig. 2-9. Passive and active movements of ions through the membrane. In the diagram the width of the channels for the various ionic currents indicates the magnitude of the current, and the slope of the channels indicates the driving force behind the ionic current. Because of the Na^+–K^+ pump sodium and potassium currents (*red channels*) occur against the direction of the driving force. Modified from Eccles (1957) *The Physiology of Nerve Cells*, Baltimore, Johns Hopkins Press.

transport into the cell is colored red, and the "active K^+ channel" is also connected to the pump driven by metabolic energy.

Active transport accounts for only a small portion of the K^+ currents. In contrast, practically all the *Na^+ efflux* from the cell is achieved by the *Na^+ pump.* The deviation of the resting potential from the sodium equilibrium potential, and thus the driving potential for the Na^+ ions, is very large (155 mV in Fig. 2-9). The "sodium channels" therefore are very steeply inclined toward the inside. The large driving potential greatly promotes passive sodium influx (bottom channel in Fig. 2-9) and practically blocks passive Na^+ efflux. (The latter is so small that it cannot be indicated in Fig. 2-9.) The passive Na^+ influx is in equilibrium with the active Na^+ efflux. In the "active Na^+ channel," the Na^+ ions are driven "uphill" by the pump. Overall, the Na^+ channels through the membrane in Fig. 2-9 are much narrower than the K^+

channels; therefore, despite large driving potentials, much less Na^+ than K^+ flows through the membrane. This reflects the low membrane conductance for Na^+ ions as compared to K^+.

Q 2.8 The Na^+ efflux from the cell is "active" because
a. the driving potential for the Na^+ efflux is large.
b. no passive net Na^+ efflux can take place against the driving potential.
c. metabolic energy is required for the Na^+ efflux.
d. the sodium conductance of the membrane is much higher than the potassium conductance.
e. the sodium conductance of the membrane is much lower than the potassium conductance.

Q 2.9 Active Na^+ transport can be blocked or greatly reduced by
a. lowering the extracellular K^+ concentration.
b. lowering the intracellular K^+ concentration.
c. increasing the intracellular Na^+ concentration.
d. cooling the cell.
e. poisoning the cell with DNP.

Q 2.10 At constant resting potential the passive sodium inflow is equal to the
a. passive potassium inflow.
b. passive net potassium flow.
c. active sodium outflow.
d. active potassium outflow.

2.4 The Action Potential

The resting potential is the prerequisite for nerve cells and muscle fibers to be able to fulfill their specific functions in the body. It is the task of nerve cells to pick up information, to transmit it throughout the body, and to coordinate and integrate it. Muscle cells contract in accordance with instructions received from nerves. When the cells work and are "active" in this way, brief positive changes in the membrane potential occur. These changes are called "*action potentials*" (also "nerve impulses" and, even more commonly, "spikes"). The generation and the time course of such action potentials are described below.

Time Course of Action Potentials. *Action potentials* can be measured in nerves and muscle cells by *intracellular electrodes*. That is, they can be measured by the apparatus (in Fig. 2-1) used for resting-potential measurement. As will be shown later (see p. 64), the action potentials can also be recorded by extracellular electrodes placed near the cell. However, this procedure usually permits only an approximate determination of the time course of the action potential.

Figure 2-10 shows action potentials measured with intracellular electrodes in vertebrate nerves, muscle cells, and cardiac muscle

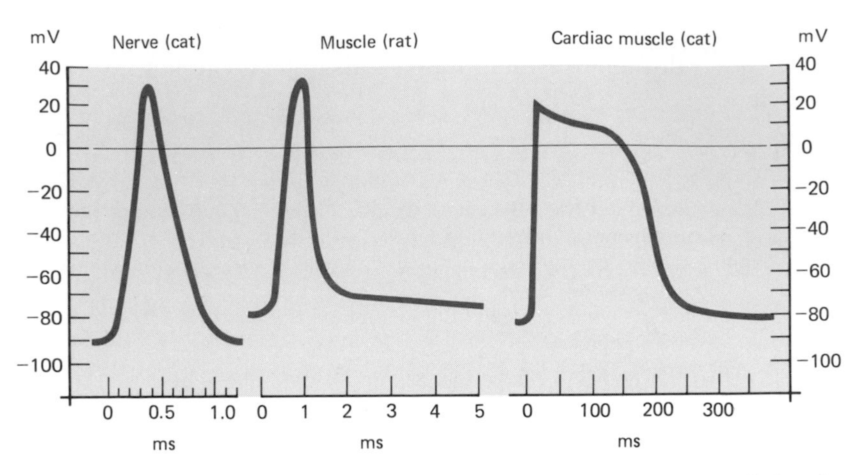

Fig. 2-10. Action potentials of various types of cells measured by intracellular electrodes. Abscissa: time after start of action potential; ordinate: membrane potential. The time scale of the action potentials varies considerably. The nerve action potential of the cat is much shorter than the muscle action potential of the rat, and both are short relative to the action potential of the cardiac muscle.

cells. In all these action potentials, the potential, starting from the resting potential, jumps very rapidly to a positive value and then returns more slowly to the resting potential. The *peak* of the action potential for all the examples shown is located somewhere near +30 mV. On the other hand, the *duration* of the action potential is very *different* in the various types of cell. In the case of the nerve, the action potential lasts approximately 1 ms only, while in the cardiac muscle it lasts more than 200 ms.

The terms for the various *phases of the action potential* are given in Fig. 2-11. The action potential begins with a very rapid positive change in potential, called *the upstroke* or *rising phase*. In nerves and muscle cells of warm-blooded animals, the rising phase lasts only 0.2 to 0.5 ms. During this phase, the cell loses its negative resting charge or polarization. Therefore, the rising phase of the action potential is also called the "*depolarization phase.*"

In most types of cells, the depolarization goes beyond zero to positive potentials. The positive portion of the action potential is referred to as the *overshoot*. Once the peak is reached, the action potential returns to the resting potential. This process is called "*repolarization*" because the normal polarization of the cell membrane is restored.

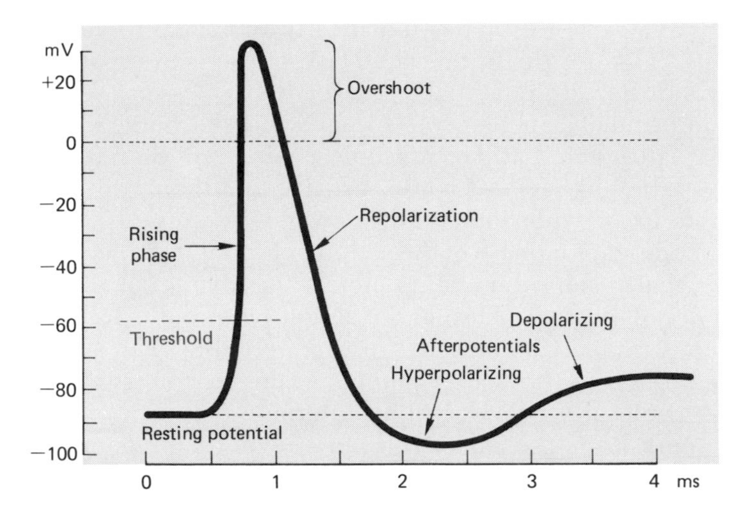

Fig. 2-11. Phases of the action potential. Diagram of the time course of a nerve action potential as shown in Fig. 2-10. The various phases of the action potential named here are discussed in detail in the text.

Toward the end of the action potential, the repolarization slows down in many types of cells, and at the end of repolarization, the potential may exceed the resting value in a negative direction for a certain time. These potential patterns at the end of or after repolarization are called *after-potentials* (see Fig. 2-11). If, at the end of the action potential, the membrane potential remains slightly more positive than the resting potential for a certain time, this is referred to as a *depolarizing after-potential.* If, on the other hand, the membrane potential goes beyond the value of the resting potential for a certain time, this event is referred to as a *hyperpolarizing after-potential.* A well-developed after-potential is seen in Fig. 2-10 in the action potential of the rat muscle.

Triggering the Action Potential and Excitation. According to what has been said so far, the resting potential is a constant and stable state. What event must occur for this stable state to be disrupted, thus triggering an action potential? Action potentials are generated whenever the membrane, starting from the resting potential, is depolarized to approximately −50 mV. The processes that bring about this initial depolarization will be discussed later (see p. 58). The potential from which the action potential starts is called the *threshold* (see Fig. 2-11). The *membrane charge is unstable* at this threshold potential. The charge dissipates rapidly and automatically and usually even reverses its polarity. This results in the rapid rise of the action potential that goes beyond the zero potential (the overshoot phase).

The condition of a spontaneous, progressive discharge of the membrane that is triggered at the threshold is also called *excitation*. The excitation lasts only a short while, usually less than 1 ms. It is thus comparable to an explosion that dies away very quickly.

The depolarization phase of the action potential itself sets processes in motion that restore the resting membrane charge. The excitation-induced depolarization phase of the action potential is followed by a spontaneous repolarization to the resting potential. The stereotyped, cyclic sequence of the action potential can be compared with the operating cycle of a cylinder in an internal combustion engine: an ignition spark heats the gas mixture to a level (corresponding to the threshold of the action potential) at which it explodes (corresponding to excitation). In turn, the explosion sets off mechanisms that restore the system to the state that existed before the explosion (corresponding to repolarization); exhaust gases are expelled, and fresh gas mixture is drawn in and compressed.

Definition of the Action Potential. The action potential in each cell is a stereotyped sequence of depolarization and repolarization of the membrane that occurs spontaneously whenever the membrane is depolarized beyond the threshold potential. Cells in which action potentials can be triggered are called *"excitable."* Excitability is a typical property of nerve and muscle cells.

Action potentials in a particular cell always follow an unvarying sequence. It makes little difference how the threshold is reached by the initial depolarization, nor does it matter whether the initial depolarizing process itself exceeds the threshold to a greater or lesser extent. This unvarying nature of the action potential is also called the *"all-or-none" law of excitation.*

Ion Shifts Occurring During the Action Potential. If the membrane potential undergoes a strong change, even to positive values, during the action potential, then the charge of the membrane capacitor also has to change and ions have to be displaced across the membrane. The type and extent of these ionic shifts will be discussed with the aid of Fig. 2-12. The parts printed in black are a repeat of Fig. 2-2, which showed the ion distribution over a small area of the cell membrane and its surroundings at the resting potential. The *resting potential* was characterized by a *high* K^+ *conductance* of the membrane. Because of the concentration gradient, K^+ ions flowed out of the cell until the resulting membrane charge prevented any further efflux.

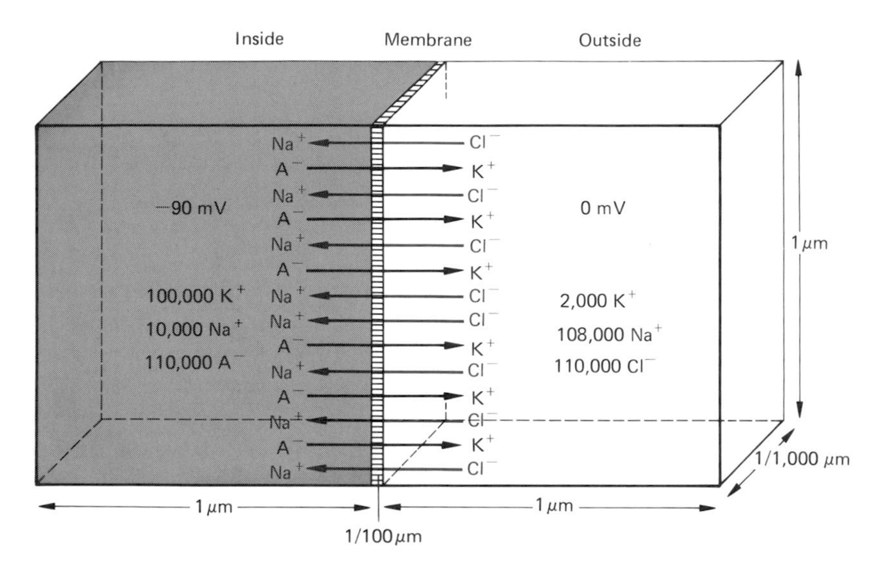

Fig. 2-12. Reversal of membrane charge during excitation. Same diagram as Fig. 2-2, but showing the charge conditions during excitation. The membrane charge is given for the small area of 1 μm × 1/1,000 μm, and the number of ions is given for the two spaces, each 1 μm × 1 μm × 1/1,000 μm, on either side of the membrane. The pairs of ions associated with the membrane during the resting potential are shown in *black*, and the change in membrane charge during excitation, in *red*. During excitation, there is an excess of 2 Na^+ at the inside of the membrane area represented; this corresponds to a membrane potential of +30 mV.

This state of equilibrium was reached in the example in Figs. 2-2 and 2-12 when the membrane area shown became charged with 6 K^+ and the corresponding number of A^-, giving rise to a "resting potential" of −90 mV.

It is a property of the membrane that when it is depolarized to the region of the threshold potential, its *conductance for Na^+ ions*, g_{Na}, *increases.* As a result, Na^+ ions (printed in red in Fig. 2-12) flow into the cell. The Na^+ ions that flow into the cell partly compensate for the resting charge; that is, the potential becomes less negative. As a result of this depolarization, g_{Na} increases still further, and more Na^+ ions flow into the cell. g_{Na} finally attains more than 100 times its resting value; that is to say, *during excitation* g_{Na} *becomes greater than* g_K. If the condition of increased g_{Na} is maintained long enough, the membrane charge is reversed. During excitation, however, the membrane potential can attain no more than the Na^+ equilibrium potential, for at this potential the positive membrane potential balances the inward-directed osmotic pressure arising from the concentration difference for Na^+ ions. The Na^+ equilibrium potential is situated at approximately +60 mV. At this potential an excess of 4 Na^+ should exist at the inner side of the membrane area in the example in Fig. 2-12. Therefore, in order to compensate for the six anions located on the inner side of the membrane at the resting potential, a total of 10 Na^+ should flow into the cell; then the Na^+-equilibrium potential is reached, and Na^+ influx stops.

In accordance with the description just given of the Na^+ inward current during excitation, the peak of the action potential should be situated at the Na^+-equilibrium potential, at approximately +60 mV. As Fig. 2-10 has shown, the peaks of the action potentials are situated at +30 mV and thus do not reach the Na^+-equilibrium potential. There are two reasons for this: first, the *increase in* Na^+ *conductance does not last long enough* to permit the reversal of the membrane charge to go quite as far as E_{Na}. In the diagram given in Fig. 2-12, not 10 but only 8 Na^+ ions have time to flow inward; this creates an excess of only 2 Na^+ ions at the inner side of the membrane, which, in turn, generates a peak potential of +30 mV.

The second reason why E_{Na} is not attained by the peak of the action potential is that the depolarization of the membrane, along with the described increase in g_{Na}, also *gives a powerful boost to the* K^+ *conductance* g_K of the membrane, but *with a lag of not quite 1 msec.* Thus, when the peak of the action potential is reached less than 1 ms after the start of excitation, the K^+ *ions* begin to flow *out of the cell* in increased numbers and rapidly compensate for the influx of positive charges in the form of Na^+ ions. Finally, g_K becomes larger than g_{Na}, the efflux of positive charges outweighs the influx, and the membrane charge becomes more negative. This predominant K^+ *efflux* causes the *repolarization phase* of the action potential. In the nerve fibers of warm-blooded animals, the inner side of the membrane once more acquires a full negative charge, and the resting potential is restored about 1 ms after the start of excitation.

The *ionic movements* during the action potential may be *summarized* as follows: as a result of depolarization beyond the threshold, the Na^+ conductance increases rapidly; the K^+ conductance also increases but with a time lag. Therefore, initially Na^+ ions flow rapidly into the cell, and the membrane potential moves in the direction of the Na^+ equilibrium potential at +60 mV; then K^+ ions flow out of the cell, restoring the resting membrane charge and repolarizing the membrane to the resting potential.

Ionic Shifts During the Action Potential. Despite the large changes in the conductance of the membrane during the action potential the ionic shifts through the membrane are *small* in relation to the quantities of ions surrounding the membrane. In the diagram in Fig. 2-12, only 8 Na^+ ions need to flow in during excitation; similarly, repolarization would be achieved by the efflux of 8 K^+ ions. As a result of these ion shifts, the Na^+ concentration in the very small spaces adjoining the cell (see Fig. 2-12) would vary by less than 1/1,000 during an action potential.

The Na^+ ions that flowed into the cell with the action potential are expelled again in the course of time by the Na^+ pump. The active Na^+ transport compensates for not only the resting sodium influx but also the Na^+ influx during excitation. However, the active Na^+ transport is of no importance for the individual action potentials. If the ion pump is blocked, for example, by poisoning with DNP (see p. 33), then, despite the fact that active transport has been eliminated, thousands of action potentials can occur before the intracellular Na^+ concentration becomes so high that the cell is rendered inexcitable. The action potential thus arises from *passive* movements of the ions along their concentration gradients. Energy-consuming processes such as the Na^+ pump are only necesary insofar as they maintain the concentration gradients.

The Action Potential and Na^+ Deficiency. The role of the Na^+ ions in the excitation process can be demonstrated by a simple experiment. If the extracellular Na^+ concentration is slowly reduced (while balancing out the osmolarity), the resting potential, as discussed above, remains practically unchanged. Usually it becomes about 10 mV more negative (see p. 31). On the other hand, the action potential is clearly affected: the peak potential, that is, the action potential overshoot, becomes less positive, and the rising phase is slowed. If the extracellular Na^+ concentration drops to about one-tenth of the normal, that is, below 20 mmol/l, the cells finally become *inexcitable*. The reason for this is that during excitation under normal conditions a strong Na^+ influx depolarizes the cell, and this influx is now reduced because the extracellular Na^+ concentration is too low. The high intracellular K^+ concentration is a prerequisite for the *resting potential*, while a high extracellular Na^+ concentration is necessary for the *action potential*. In addition, excitability is also dependent on a low intracellular Na^+ concentration so that Na^+ can flow into the cell.

Q 2.11 Draw the action potential of a nerve with amplitude and time scales. Name the various phases.

Q 2.12 Which of the following statements apply to the threshold of the action potential?
a. The membrane potential is positive and close to E_{Na}.

b. The membrane potential is approximately 20 to 30 mV more positive than the resting potential.
c. The membrane charge is unstable and dissipates spontaneously.
d. The potassium efflux is larger than the sodium influx.

Q 2.13 The repolarization of the action potential is brought about by the
a. very small increase in the intracellular Na^+ concentration caused by excitation.
b. potassium efflux, which sets in with a time lag after depolarization.
c. termination of the sodium influx during excitation.
d. removal by the sodium pump of the sodium ions that have entered.

2.5 Kinetics of Excitation

The action potential is caused by a Na^+ current flowing into the cell and a K^+ current then flowing out, both events being caused by depolarization beyond the threshold. These currents depend on the extent of the depolarization as well as on the time that elapses from the initiation of depolarization. The complicated kinetics of these Na^+ and K^+ currents will now be discussed in detail. This knowledge is not only useful for further analysis of the action potential but is, in fact, necessary for a full understanding of how the action potential is propagated and the events through which the threshold of the action potential is reached.

Measuring the Potential Dependence and Time Dependence of the Ionic Currents. The sodium and potassium currents that flow during the action potential are strongly dependent on potential and time. Because the potential changes rapidly throughout the action potential, the potential dependence of the currents cannot be analyzed in more detail during the action potential itself. Such an analysis can be performed, however, if the potential of the cell is held constant artifically after the onset of excitation. An experimental setup that enables this is called a *voltage clamp*.

Figure 2-13 illustrates such a voltage clamp. In this experiment, two intracellular electrodes are used. With one electrode the *membrane potential is measured* in the manner already described in Fig. 2-1. The second intracellular electrode supplies current to the cell. The two electrodes are connected to an electronic regulating device. The apparatus can be programmed in such a way that the membrane potential changes suddenly from one value to another and remains constant there. The regulating device ensures that exactly the right amount of current flows through the current electrode to induce the change in potential, and that once the potential has been changed the

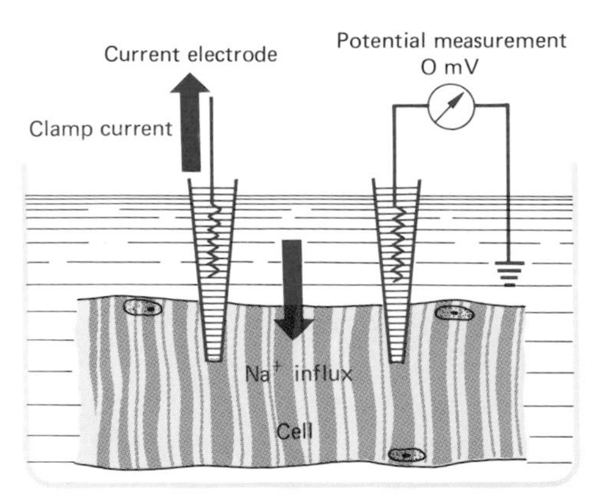

Fig. 2-13. Voltage-clamp currents. The membrane potential, between the inside of the cell and the extracellular fluid, is measured by the electrode on the right. The clamp current (*red arrow pointing upward*), which maintains the potential at 0 mV, flows out of the cell through the left electrode. The clamp current is of the same magnitude, but of opposite polarity, as the Na^+ influx (*red arrow pointing downward*) through the cell membrane at the clamp potential 0 mV.

current through the electrode is adjusted so that the new membrane potential remains constant. If, for example, the change in potential involves a suprathreshold depolarization starting from the resting potential, this will trigger a Na^+ inflow. In this case the regulator will allow just enough current to leave the cell to balance the inflow of Na^+ ions (see Fig. 2-13). As the two currents cancel each other out, the membrane potential remains constant. The clamp current and its time course are measured. Since it is always just big enough to cancel out the membrane current, it is an exact mirror image of the membrane currents. Thus, at a constant *clamp potential*, the *clamp current* indicates the *time course of the ionic currents* through the membrane at this potential.

Membrane Currents After a Depolarization. In 1952 Hodgkin and Huxley published a pioneering analysis of the action potential of the *squid giant axon* using the voltage-clamp technique.* Because of the large fiber diameter of up to 1 mm, this particular preparation is especially suitable for such studies, and since 1952 it has been the subject of continuing research. Figure 2-14 illustrates the clamp currents measured in such a fiber. Positive clamp currents correspond to efflux of positive ions from the cell, and negative clamp currents, to the

---

* (1952) *J. Physiol.* 116:449

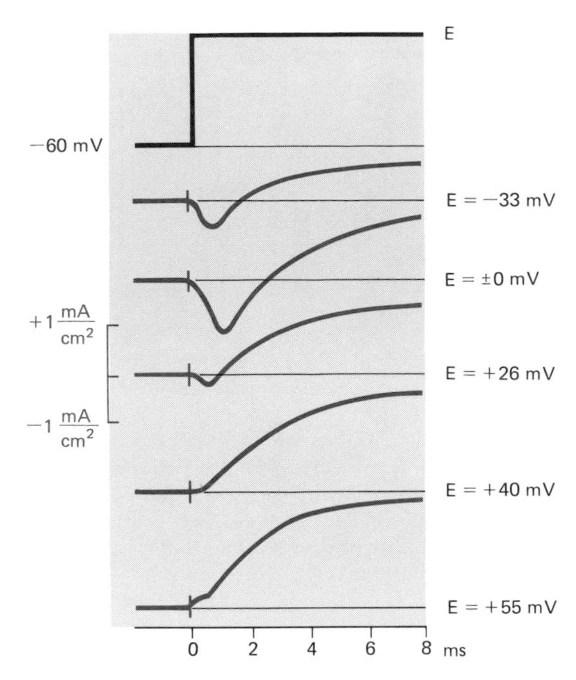

Fig. 2-14. Clamp currents following changes in potential. The *top line* shows the time course of a potential change imposed on a squid giant axon, from a resting potential of −60 mV to the clamp potential *E*. *Below* that are the clamp currents that flow after the change to a potential *E* indicated on the right of each curve. The calibration of the clamp current given for the potential jump to +26 mV is also valid for the other clamp currents. Positive clamp currents correspond to an outflow of positive ions from the cell, and negative clamp currents correspond to an inflow of positive ions. Modified from Hodgkin and Huxley (1952) *J. Physiol.* 116:449.

influx of positive ions. The top trace of the figure represents the programmed jump in potential from the resting potential at −60 mV to the value *E*. The traces below this indicate the clamp currents measured at the different values of potential *E*.

In the case of the smallest jump to $E = -33$ mV, a small negative current flows for about 1 ms after depolarization. This negative current is followed by a steady positive current. If the cell is depolarized to 0 mV, then both the short-duration negative and the subsequent steady positive current components become larger. When the cell is further depolarized to +26 mV, the initial negative current component then becomes smaller, and it disappears entirely at $E = +40$ mV. When still further depolarization occurs to +55 mV, a positive current component takes the place of the hitherto negative one. While the initial current becomes smaller at potentials above 0 mV and finally reverses its direction, the later positive current continues to increase with depolarization.

The reversal of polarity at +40 mV identifies this *initial current as a sodium current,* for in the case of the giant axons of squid, E_{Na} is situated at +40 mV. *Beyond its equilibrium potential, an ionic current must reverse its direction.* At potentials more negative than E_{Na}, sodium ions flow into the cell; at potentials more positive than E_{Na}, sodium ions flow out of the cell. The initial negative clamp current can also be identified as Na^+ current by other experiments. If Na^+ is replaced in the extracellular solution by a nonpermeable ion (see p. 29), this current component disappears completely because there is no Na^+ current in the sodium-free solution. Therefore, in suprathreshold depolarizations to a fixed potential, a *sodium current* flows for 1 to 2 ms.

The positive current following the Na^+ current after a depolarization step is a K^+ current. The time course of the K^+ current is clearly visible at $E = +40$ mV, the Na^+ equilibrium potential E_{Na}. At E_{Na} there is, by definition, no net Na^+ flow; therefore, current measured at that potential must be all K^+ current. In contrast to the sodium current, I_{Na}, which sets in immediately but flows for only a short time, the potassium current, I_K, starts with a *delay,* attains its maximum in 4 to 10 ms, and *does not decline* provided the depolarization is maintained. The amplitude of maximum I_K increases approximately in proportion to the depolarization.

As discussed above, the sodium inflow after depolarization does not take place in a sodium-free solution. The membrane currents, which are then measured after a depolarization step, are (primarily) K^+ currents. Thus, the time course of the current can be determined for each potential. If this K^+ current is subtracted from the current measured in a normal bath solution, we get the sodium current. So it is possible to separate the clamp current at each membrane potential into the Na^+ and the K^+ components. This has been done in Fig. 2-15 for $E = 0$ mV. The figure shows clearly how I_{Na} rises almost without delay after depolarization and drops again after less than 1 ms. I_K, on the other hand, increases with a time lag to a constant final value.

Changes in Membrane Conductance After Depolarization. The changes in *membrane conductance* give a better indication than the membrane currents of the behavior of the membrane following depolarization and during the action potential. At a given potential E, the membrane conductance for an ion is proportional to the ionic current through the membrane. In the case of Na^+, for example, we can write:

$$g_{Na} = I_{Na}/(E - E_{Na})$$

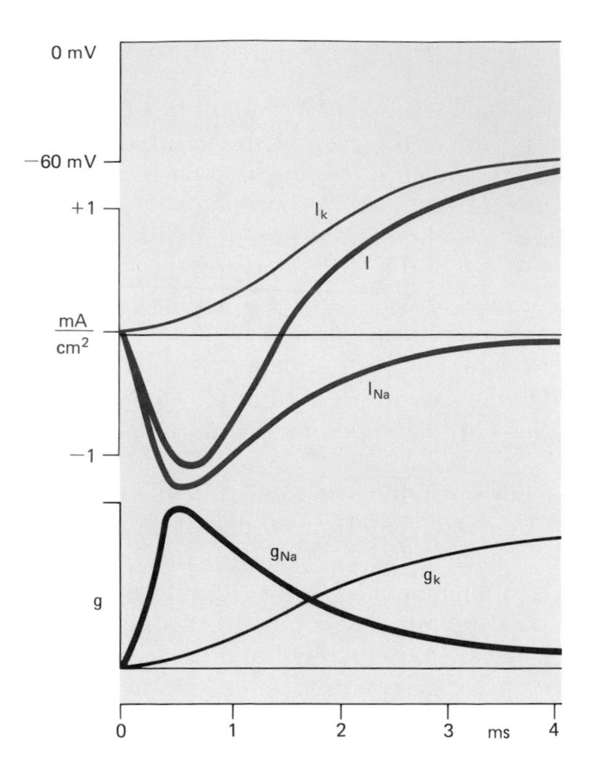

Fig. 2-15. Ion currents and conductance following a change in potential. *Top:* time course of the potential change, −60 to 0 mV, induced by the voltage clamp. *Center:* the clamp current I, with its components I_{Na} and I_K, which flows after the change in potential. *Bottom:* the time courses of membrane conductances g_{Na} and g_K calculated from these currents. Preparation: squid giant axon. Modified from Hodgkin (1958) *Proc. Roy. Soc. B* 148:1.

If the equilibrium potentials E_{Na} and E_K are known, then for a given potential E, one can calculate the time course of g_{Na} and g_K from the time course of I_{Na} and I_K. The former time course is shown for $E = 0$ mV in the bottom trace in Fig. 2-15. g_{Na} reaches its maximum in less than 1 ms after depolarization and has almost disappeared after about 4 ms, although the depolarized state continues. This latter condition is called *inactivation.*

The inactivation of the g_{Na}, which occurs after depolarization, continues for as long as the membrane remains depolarized. Once g_{Na} has been *inactivated* after a depolarization, it is *not available for activation by further depolarization.* The sodium system can recover from inactivation only when the membrane potential returns to the vicinity of the resting potential or to even more negative potentials. In order to permit the inactivated sodium system to become *available for activation* again, the membrane potential must remain more negative than −50 mV for one to several milliseconds. Consequently, the sodium system is available for activation by a depolarization only if the membrane potential has had a sufficiently negative value for at least several milliseconds prior to depolarization. At resting potentials more positive than −50 mV, g_{Na} remains inactivated in the nerves of warm-blooded animals, and, therefore, no excitation can be triggered from this potential range.

The sodium system can exist in any of three different potential-dependent and time-dependent states: (1) available for activation at potentials more negative than −50 mV; (2) activated after suprathreshold depolarizations, but only for a few milliseconds; (3) inactivated after several milliseconds at potentials more positive than −50 mV. The transition from activation to inactivation is dependent on time, and the transition from inactivation to availability for activation is dependent on repolarization and time.

In contrast to g_{Na}, g_K *does not become inactivated during depolarization.* As Fig. 2-15 shows, g_K remains increased after a depolarization for as long as the depolarization lasts. Since the rise in g_K is responsible for the repolarization of the resting potential, the high g_K, which does not decline during depolarization, definitely guarantees the return of the potential to the resting value.

Membrane Conductance During the Action Potential. If the dependence of the membrane conductance on potential and time is known for certain fixed clamp potentials, then it is also possible to calculate the time course of g_{Na} and g_K during the action potential. The result of such a calculation is shown in Fig. 2-16. g_{Na} increases sharply as a function of potential at the start of the action potential and reaches its *maximum before the peak of the action potential.* After attaining its maximum, g_{Na} declines, at first because of the time-dependent inactivation and, when the membrane is nearly repolarized, also because of its potential dependence. At the beginning of the action potential, triggered by depolarization, g_K can *rise only slowly* because of its time dependence, and it reaches its maximum during the steepest section of the repolarization phase. After this it declines slowly because of its *potential dependence.*

Refractory Phases After the Action Potential. At the peak of the action potential, g_{Na} is already partly inactivated, and this inactivation is more or less complete by the time the repolarization intersects the zero line. The inactivation can be reversed and the sodium system can

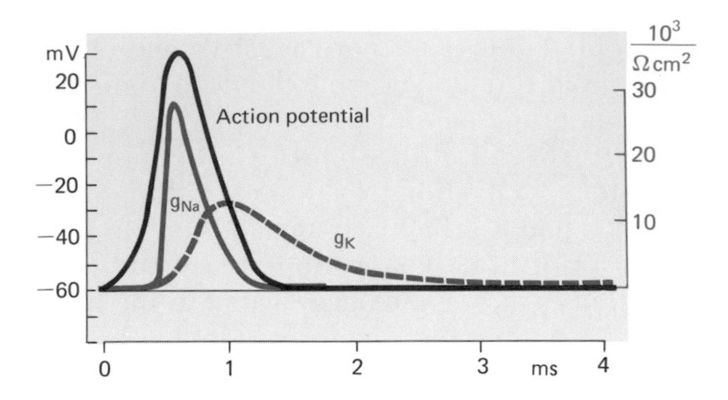

Fig. 2-16. Membrane conductance during the action potential. The time course of the action potential is shown in *black;* the *red curves* show the associated changes in the membrane conductances g_{Na} and g_K. Modified from Noble (1966) *Physiol. Rev.* 46:1.

once again become available for activation only when the potential remains more negative than −50 mV for several ms (see page 49). g_{Na} is thus inactivated during repolarization of the action potential as well as for a short time afterward. During this time, g_{Na} cannot be substantially increased by a new depolarization; that is, the cell is *inexcitable.*

The phase of inexcitability following the action potential can also be detected by depolarizing the membrane to the threshold potential at various times following the action potential, in this way determining the excitability. The result of such an experiment is shown in Fig. 2-17. The induced depolarization of the cell is indicated by the broken curves. The cell is seen to be absolutely *inexcitable* in the first 2 ms after the start of the action potential. No matter how large the depolarizations, the threshold cannot be reached. This phase of complete inexcitability is also called the *absolute refractory period.*

Figure 2-17 also shows that for a few milliseconds after the end of the absolute refractory period the threshold for the triggering of action potentials is located at values more positive than for the first action potential. The period up to the time when the threshold returns to normal is called the *relative refractory period.* The amplitude of the action potential is also reduced during this phase because the sodium system has not fully recovered from the inactivation following the first action potential.

The absolute refractory period limits the maximum frequency at which action potentials can be triggered in the cell. If, as in Fig. 2-17, the absolute refractory period ends 2 ms after the start of the action potential, then the *maximum frequency of the action potentials* in this cell is 500/s. There are cells with even shorter refractory periods, so that in extreme cases frequencies up to 1,000/s can occur in a nerve.

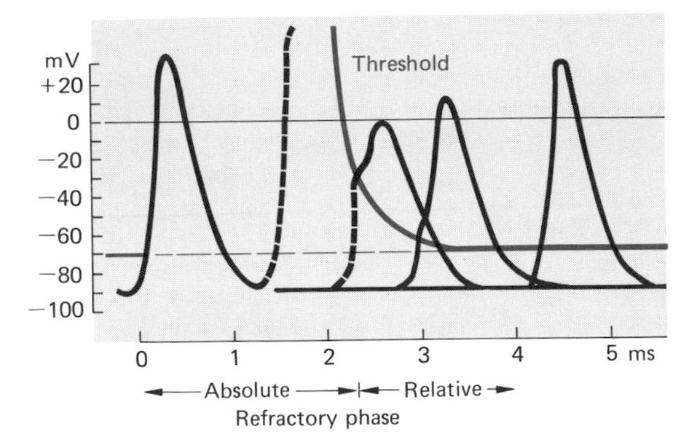

Fig. 2-17. Refractoriness following excitation. An action potential in the nerve of a warm-blooded animal is diagrammed on the *left:* to the *right* of this are shown the time courses of subsequent action potentials triggered at different times after the first. The threshold is indicated by the *red line.* The *dashed black curves* denote depolarization of the fiber to the threshold, while the *solid black curves* represent the spontaneous changes in potential once threshold has been exceeded. During the absolute refractory period the fiber is inexcitable; no matter how large the depolarization, the threshold cannot be reached. In the relative refractory period that follows, the threshold is higher than normal.

In most types of cells, however, the maximum action potential frequencies measured are below 500/s.

The Na^+ Channels in the Membrane. The action potential arises from activation and inactivation of the Na^+ system of the cell membrane. These processes can be described precisely in mathematical terms. But how can we visualize the physical structures involved, the Na^+ *channels* in the membrane through which a rapid and accurately controlled Na^+-ion influx can occur? The diagram in Fig. 2-18 summarizes current knowledge about the Na^+ channels. The channel serves two distinct functions, acting (1) as a *filter* and (2) as a *gate.* The filter, located at the outer end of the Na^+ channel, selects the ions that are to enter; its wall is negatively charged, so that anions are rejected. The diameter of the entrance to the channel, as well as the charge arrangement and water binding within the entrance, permit unrestricted entry of only a few cations, in particular the ions of sodium and lithium. The channel entrance is thus a *selectivity filter,* the particular function of which is to allow Na^+ to flow and to block K^+. Behind the filter is a second obstacle, a *gate.* At the resting potential the gate is closed, but during depolarization it opens. The gate probably consists of a large charged protein molecule that changes its shape or position when the strength of the electric field in the membrane is diminished.

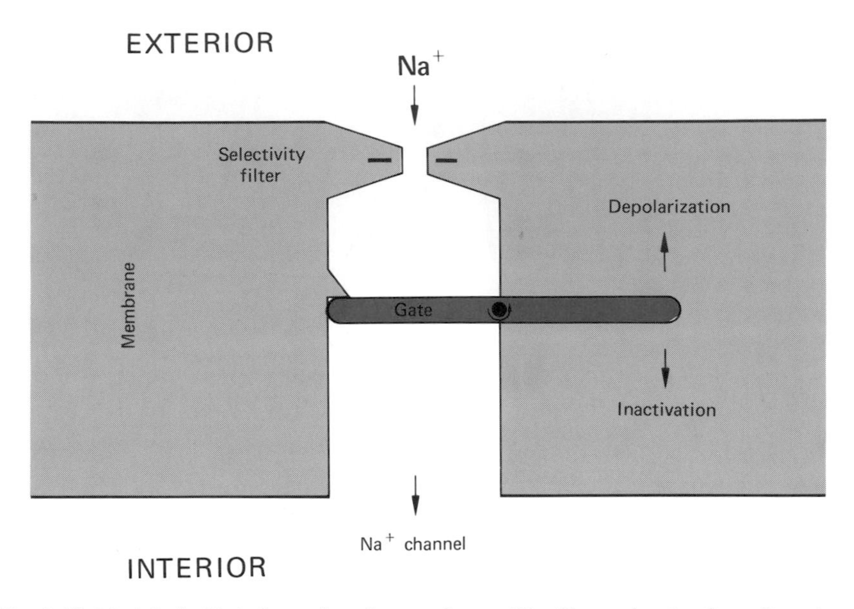

Fig. 2-18. Model of a Na^+ channel in the membrane. The filter selective for sodium ions is formed by a short, narrow pore (0.3 × 0.5 nm) with a negatively charged wall. Within the channel is a gate that swings open during depolarization and is closed again by inactivation.

But potential changes never cause the gate to "open" for very long; after a few milliseconds, *inactivation* occurs. The *filter is at the entrance to the channel* and can be affected only by fairly large molecules approaching the membrane from outside the cell. The *gate is deep within the channel* and can be reached by large molecules only from within the cell. Substances that bind firmly to the entrance of the Na^+ channel (for example, tetrodotoxin) can be used to determine the density of the channels in the membrane. There are about *50 sodium channels per square micrometer of membrane area.* The channels have a diameter of about 0.5 nm and are separated by an average distance of 140 nm. Since the membrane itself is ca. 7 nm thick, we may say that it is penetrated by quite *fine channels, relatively widely spaced.* If the membrane were scaled up to human dimensions and the channels regarded as 1-m-wide doors in a corridor, the next door would be 280 m down the hall.

Q 2.14 At the peak of the action potential, the membrane potential becomes positive because

a. the Na^+ concentration in the cell becomes larger than the K^+ concentration.

b. a Na^+ influx generates a slight excess of positive charges on the inside of the membrane.
c. the membrane potential becomes more positive than the Na^+ equilibrium potential.
d. the membrane potential approaches the Na^+ equilibrium potential.

Q 2.15 During the steep phase of repolarization, what flows through the membrane?
a. Mainly Na^+ current.
b. Mainly K^+ current.
c. Na^+ and K^+ currents of approximately the same magnitude.

Q 2.16 Plot the approximate time course of g_{Na} and g_K during the action potential.

Q 2.17 During the absolute refractory phase after an action potential the
a. cell is inexcitable.
b. sodium inflow is greater than the potassium outflow.
c. sodium conductance is not available for activation.
d. sodium pump is not active.
e. sodium equilibrium potential is negative.

2.6 Electrotonus and Stimulus

Excitation occurs as a result of depolarization of the membrane to the threshold. The actual process of excitation was discussed in detail in the preceding sections, but nothing has been said so far about how the membrane is depolarized to the threshold. This depolarization is also called *stimulation*, and this section will deal with the characteristics of such stimuli.

In cells functioning in the body, the *stimulus* that triggers an action potential is usually an electric current that depolarizes the cell. Normally, this current is not generated in the region of the membrane to be stimulated but comes from somewhere beyond that region. In the case of nerve cells, the stimulating current comes from neighboring sections of the nerve membrane, from synapses, or from receptors. In neurophysiological experiments, the stimulating current is usually supplied by electrodes because this makes it easy to control its magnitude and duration. In this section, therefore, we will first discuss the reaction of the membrane to an applied current and then analyze the conditions under which such a current acts as a stimulus.

Electrotonus in the Case of Homogeneous Current Distribution. Figure 2-19A shows how current can be fed into a cell through an intracellular electrode. The applied current, I, leaves the cell again by crossing the membrane. It flows first via the *membrane capacitance* and second as *ionic current* through the membrane. In the process, the membrane potential, E, is changed. During and shortly after the

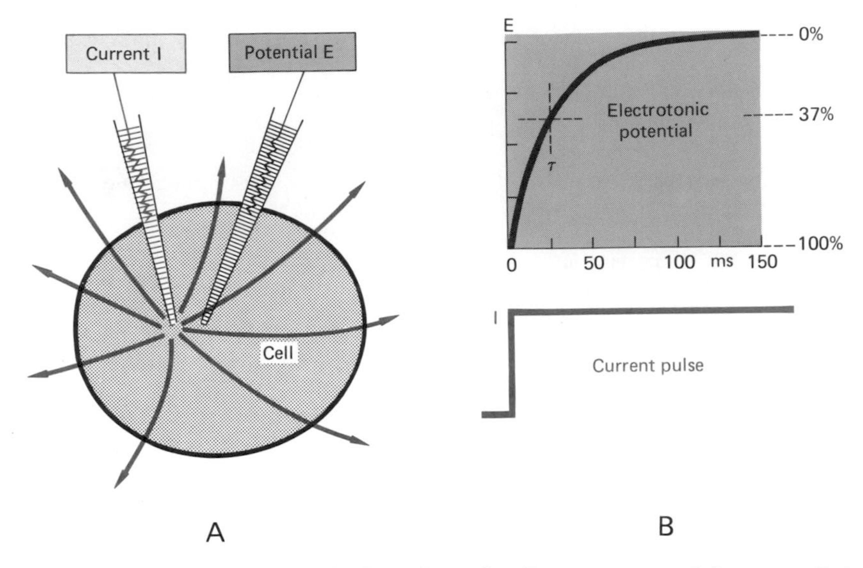

Fig. 2-19. Electrotonic potential of a spherical cell. **A.** Diagram of the intracellular electrodes that measure potential E and supply current I; the current flow through the membrane is indicated by the red lines. **B.** *Bottom:* time course of the current I applied through the current electrode. *Top:* time course of the simultaneously measured membrane potential E, the "electrotonic potential." The exponential rise of the electrotonic potential is characterized by the membrane time constant τ, the time at which the membrane potential has approached to within 37% ($1/e$) of its final value.

end of the current flow, the measuring electrode records an *electrotonic potential.* The term *electrotonic* traditionally refers to the electrical phenomena (voltage, current) associated with the *passive* properties (capacitance, conductance) of the membrane when it is stimulated by currents that do not induce excitation (changes of g_{Na}, g_K).

Let us first consider the current component that flows via the *membrane capacitance.* Depending on their polarity, the excess charges that are introduced into the cell with the applied current can increase or reduce the negative charge on the inner side of the membrane. An inflow of positive charges will reduce the negative charge on the inner side of the membrane (see Fig. 2-2). If the negative charge on the inner side of the membrane is reduced, then the positive charge on the outer side of the membrane declines correspondingly. Thus, the number of positive charges released on the outer side of the membrane is equal to the number used up on the inside to reduce the negative charge, and there has been a current flow through the membrane although no carriers of charge have actually crossed the membrane. Since this current is generated by displacements of charge via the membrane capacitance, it is called a *capacitive current*, I_c.

The membrane potential is proportional to the charge of the membrane capacitor. Given a constant supply of charge, that is, when a constant current is applied, the charge of the membrane capacitor, and as a result the membrane potential, should change at a constant rate. Figure 2-19B illustrates the changes in potential that occur in the cell following application of a constant current. The potential does not, in fact, change at a constant rate; instead, the rate of change decreases with time, and the potential finally attains a constant value despite the fact that current continues to flow. The time course of the change in potential must, therefore, be determined by more than just the flow of a capacitive current.

During the change in potential, an *ionic current* flows in addition to the capacitive current. At the resting potential, the membrane is particularly permeable to K^+ ions, less so to Cl^- ions, and only slightly to Na^+ ions. At constant resting potential, the sum of these ionic currents is zero. If the membrane potential is shifted by charges supplied through an electrode, a net ionic current flows that is proportional to the magnitude of the shift in potential. This is because the ionic currents are proportional to the membrane conductance and vary in proportion to the difference between the membrane potential and the equilibrium potential. If, as in Fig. 2-19B, the membrane charge is reduced by a constant current, then with increasing displacement from the resting potential more ionic current will flow across the membrane. This current is mainly carried by K^+ ions. As depolarization increases, less and less current is available for discharging the membrane capacitor. Consequently, the membrane potential changes more and more slowly with time until it finally becomes constant when all the applied current flows through the membrane as ionic current, I_i.

The exponential time course of the electrotonic potential as shown in Fig. 2-19 is the result. At the start of this electrotonic potential, nothing but capacitive current flows through the membrane, but at the plateau the current is entirely ionic current. This electrotonic potential is characterized not only by the magnitude of the plateau, that is, its amplitude, but also by the steepness of the exponential increase. This, in turn, is characterized by the *membrane time constant*, τ, that is, the time required for the potential to rise to within 37% ($1/e$) of its final amplitude. The value for τ varies for different membranes, ranging from 10 to 50 ms.

Measurements of electrotonic potentials are used often in neurophysiology to determine the resistance and capacity of the membrane.

The *membrane resistance*, r_m, of a cell is the quotient of the final amplitude of the electrotonic potential and the applied current. At the plateau of the electrotonic potential, the entire current flows as ionic current across the membrane resistance, and the latter can be calculated from the change in potential and the current. The time course of the electrotonic potential represents the curve for the charging of the membrane capacitor across the membrane resistance, and the membrane time constant, τ, is the product of resistance and capacitance. The *membrane capacitance*, c_m, can thus be calculated as the quotient of τ and r_m. These simple relationships hold true, however, only for cells in which the applied current is distributed homogeneously.

Electrotonic Potential in Elongated Cells. Almost all nerve fibers and muscle cells are very long in relation to their diameter. A nerve fiber, for example, can be 1 m long with a diameter of 1 μm. In these cells, current applied at one point naturally flows through the membrane at much greater density in the vicinity of this point than in more distant parts of the membrane. Therefore, the electrotonic potentials in such cells must be governed by quite different conditions from those in spherical cells (Fig. 2-19A) where the current is distributed homogeneously.

As shown in Fig. 2-20, the *electrotonic potentials* in an elongated muscle fiber can be measured by inserting intracellular electrodes at various distances—here at 0, 2.5, and 5 mm—from the current electrode. The *time course* measured for the *electrotonic potential* is *no longer simply exponential*, as in the case of the spherical cell shown in Fig. 2-19. At the site of current application, the electrotonic potential (E_0) in Fig. 2-20 rises more steeply than when the current is evenly distributed. This is evident from the fact that at the moment when the membrane time constant, τ, is reached the electrotonic potential is already within 16% instead of 37% of its final value.

This more rapid increase is due to the inhomogeneous distribution of the current: first the membrane capacitor is discharged in a small region close to the current electrode, and only then does current flow through the interior of the cell, which has a considerable longitudinal resistance, to more distant parts of the membrane. Thus, when the current pulse is first applied, the membrane current is concentrated in the direct vicinity of the current electrode, and the potential undergoes a rapid change here. Therefore, the time course of the electrotonic potential becomes slower with increasing distance from the site of current application. As Fig. 2-20 shows, at a distance of 5 mm the electrotonic potential, E_5, starts with a delay, and after 120 ms it has still not quite attained its final value E_{max}.

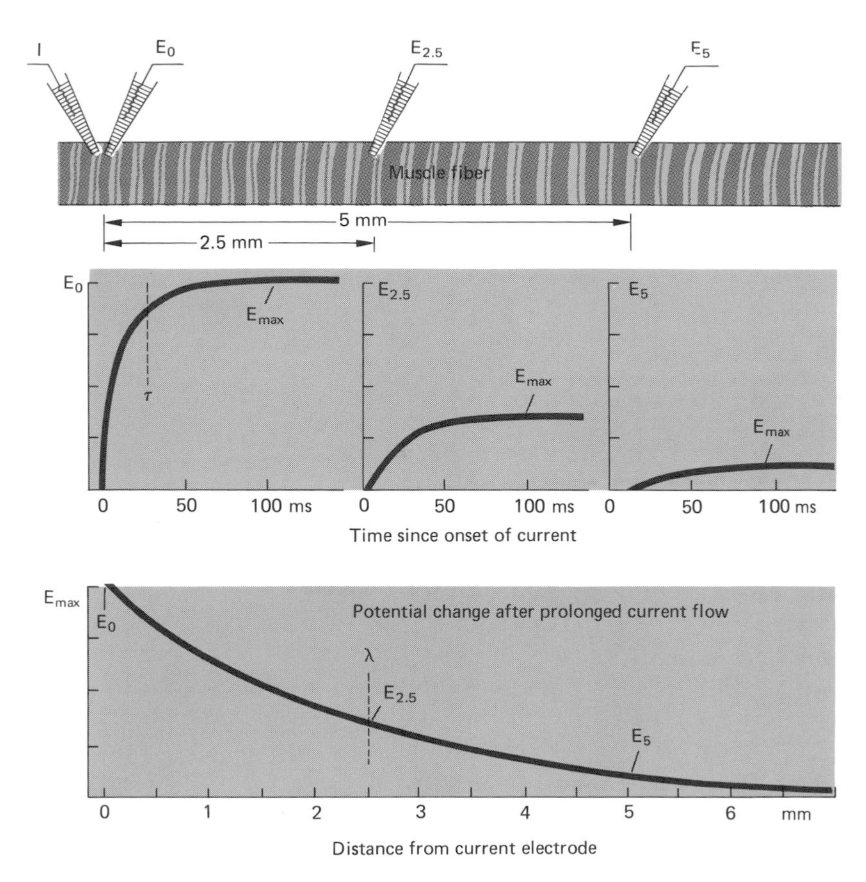

Fig. 2-20. Electrotonic potentials in an elongated cell. *Top:* diagram of the experimental arrangement in which a current is applied (I) to an elongated muscle fiber and the electrotonic potential is measured at distances of 0 mm (E_0), 2.5 mm ($E_{2.5}$) and 5 mm (E_5) from the site of current application. *Center:* time courses of the electrotonic potentials E_0, $E_{2.5}$, and E_5, which reach smaller final values E_{max} with increasing distance. *Bottom:* the final values E_{max} plotted as a function of the distance from the point of current application. The decline of E_{max} with distance is characterized by the membrane length constant, λ, the distance at which the electrotonic potential is only 37% of that recorded next to the current electrode.

Even when the applied current has been flowing for a long time and a new charge distribution is attained, more current still flows through the parts of the membrane close to the point where current is fed in than through more distant parts. The amplitudes, E_{max}, of the final values of the electrotonic potentials are plotted in Fig. 2–20 (bottom) against the distance from the current electrode. It can be seen that the amplitude of E_{max} falls exponentially with distance. The steepness of the exponential decline of E_{max} with distance is characterized by the *membrane length constant,* λ, at which E_{max} has declined to 37% ($1/e$). λ is 2.5 mm in Fig. 2-20. The value of λ varies in different cells from 0.1 to 5 mm. In the length constant, λ, we therefore have a yardstick for judging over what distances electrotonic potentials can spread in elongated cells. At the distance 4λ, for example, the amplitude of the electrotonic potential is about 2% of the value of the potential close to the point of current application. Thus, in nerve fibers electrotonic potentials can really be measured only over distances up to a few centimeters away from their point of origin.

Subthreshold and Suprathreshold Stimuli. The electrotonic potential is a purely passive reaction of the membrane to applied current; that is, the membrane does *not* change its *conductance* (or its capacitance) during the electrotonic potential. The polarity of the current is also, in principle, of no significance to the course of the electrotonic potential. The depolarizing pulses of intensity +1 and +2 in Fig. 2-21 elicit electrotonic potentials; if the current is reversed (−1 and −2), the depolarizing electrotonic potentials produced are exact *mirror images* of the hyperpolarizing potentials.

A depolarizing electrotonic potential is also called catelectrotonus and a hyperpolarizing electrotonic potential is called anelectrotonus. These terms are derived from the procedure of applying current through two extracellular electrodes in contact with the nerve: catelectrotonus is generated near the cathode, and anelectrotonus, near the anode.

Only the first two of the depolarizing pulses in Fig. 2-21 elicit purely electrotonic potentials. When, in response to pulses +3 and +4, the potential exceeds −70 mV, there is progressively more depolarization than would be associated with a pure electrotonic potential. Moreover, these depolarizations are also greater than the corresponding hyperpolarizations produced by pulses of intensity −3 and −4. The amount by which these exceed purely electrotonic responses is indicated by the red areas. The additional depolarization is brought about by the *increase in membrane sodium conductance during depolarization.* In the region of −70 mV, g_{Na} increases slightly above its resting value, and the resulting increase in Na^+ influx brings about additional depolarization. If an even larger current pulse (intensity +5 in Fig. 2-21) is given, so that the depolarization passes the *threshold,* the result is *full excitation*—that is, an action potential. In the subthreshold region, the excitation is not fully developed. The rise of g_{Na} is not sufficient to drive the depolarization over the threshold and elicit an action potential that can be conducted along the axon. The

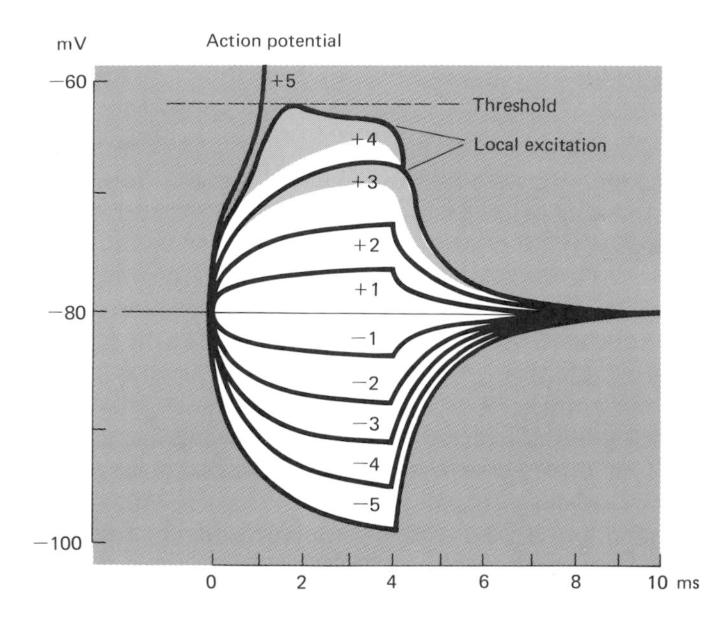

Fig. 2-21. Electrotonic potentials and local responses. Current pulses (4 ms in duration) of relative amplitudes *1*, *2*, *3*, *4*, and *5* in the hyperpolarizing direction produce proportional electrotonic potentials. With depolarizing currents of amplitudes *1* and *2*, the potentials are mirror images of those with hyperpolarizing currents. The depolarizations produced by current amplitudes *3* and *4* exceed those of electrotonic potentials at levels beyond −70 mV, by amounts indicated by the red areas below the curves. The active or nonlinear depolarization, in excess of electronus, is called local excitation. The depolarizing current of amplitude *5* produces a depolarization that passes the threshold and triggers an action potential.

red areas in Fig. 2-21, then, represent an incomplete form of excitation that can be called *local excitation* or local response. During such local excitation, it is quite possible for the flow of sodium into the cell to become greater than the K^+ efflux, so that the excitation process is set off (see the response to current pulse +4 in Fig. 2-21).

Rapid depolarization of the membrane requires that there be much current flowing to discharge the membrane capacitance (cf. p. 54). If the depolarizing current is insufficient, the depolarization rises too slowly, and the sodium system is *inactivated* within about 1 ms (see Fig. 2-15). In this case excitation never reaches threshold and remains *local*. For this reason, very slowly rising stimulus currents can actually depolarize the membrane beyond the "threshold"; during very slow depolarization the sodium system is so strongly inactivated that sodium conductance cannot increase sufficiently even in the above-threshold region. The failure of excitation to occur during very slow depolarization is an example of a phenomenon called *accommodation.* Subthreshold, local excitation plays a considerable part in the activity of cells of the CNS. Often such cells are depolarized to a near-threshold level by the effects of a large number of synapses (see p. 89).

Minimal Stimulus Current and Duration. When a depolarizing electrotonic potential crosses the threshold rapidly enough, an excitation is triggered (see Fig. 2-21); the current pulse producing such a potential change is called the *stimulating current* or simply *stimulus.* A current just sufficient to produce excitation is the *minimal stimulating current.* All currents larger than this minimum act as stimuli. Because of the all-or-none nature of the excitation, they trigger action potentials of uniform amplitude.

In Fig. 2-21 the depolarization brought about by the current pulse of strength +5 reaches threshold about 1 ms after onset of the pulse. During the rest of the rising phase of the action potential, the excitation proceeds independently; the stimulating current could be turned off once the action potential has been triggered without affecting it further. Accordingly, at current strength +5, 1 ms is the *minimal stimulus duration* required to trigger an action potential. If the current were somewhat smaller than +5, depolarization would proceed more slowly and the threshold would be reached later—perhaps after 2 ms. As stimulating current is decreased, then, the minimal stimulus duration increases. Currents *barely above threshold* must flow for a relatively *long time* (several milliseconds). Conversely, an increase in current strength above +5 in Fig. 2-21 would cause the threshold to be reached sooner. With very *large stimulating currents* the minimal durations are very *short* (for example, 0.1 ms). But stimulus duration cannot be reduced indefinitely by increasing the intensity of the stimulus. Extremely short stimuli such as high-frequency alternating current or pulses only microseconds in duration cannot trigger an action potential even if the resulting peak voltage is several thousand volts.

Q 2.18 Without looking at Figs. 2-19 to 20, draw the time course of the electrotonic potential of a spherical cell after application of a constant current to the cell.

Q 2.19 The final amplitude of the electrotonic potential (given homogeneous distribution of the current) is proportional to the
a. membrane capacitance.
b. membrane resistance.
c. reciprocal of the membrane conductance.
d. applied current.
e. duration of current flow.

Q. 2.20 In an elongated cell, how does the final amplitude of the electrotonic potential change with the distance from the site of current application?

a. It remains constant.
b. It increases in proportion to the distance.
c. It decreases in proportion to the distance.
d. It increases in proportion to the square of the distance.
e. It decreases exponentially with the distance.

Q 2.21 A current pulse acts as a stimulus when
a. the sum of stimulating current and sodium inflow is larger than the potassium outflow under resting conditions.
b. it reduces the membrane capacitance.
c. it depolarizes the membrane potential beyond the threshold after 1 ms.
d. it depolarizes the membrane potential beyond the threshold.
e. it reversibly increases the potassium outflow.

2.7 Propagation of the Action Potential

We now come to a discussion of the actual task of the nerve fibers and the membrane of the muscle fibers: the propagation of excitation. Before this could be understood it was first necessary, in the preceding sections, to deal with the mechanism of excitation at the membrane. Then it was shown how changes in potential spread over the length of a fiber as a result of currents flowing through the inside of the fiber. To understand how the action potential is propagated, we must now combine the knowledge we have acquired about excitation and electrotonus.

Velocity of Conduction of the Action Potential. Let us proceed from the simple observation that a nerve propagates action potentials. If the action potential of a nerve fiber is measured at two points not too close together, and if the nerve is stimulated at one end, then an action potential appears first at the measuring point closest to the site of stimulation; a little later an action potential also appears at the second measuring point. This shows that the action potential is *propagated* or *conducted* from the site of stimulation past the first and second electrodes.

The *velocity of conduction* can be determined by dividing the distance between the two measuring points that the action potential passes by the time the action potential takes to travel between these points. (A typical measurement of this sort might involve a distance of 5 cm between the measuring points. If the traveling time between the points is 2.5 ms, then the velocity of conduction is 0.05 m/0.0025 s = 20 m/s.) The conduction velocities measured in nerve fibers range from 1 m/s to more than 100 m/s. The conduction velocity depends on the characteristics of the nerve fiber and is typical for each type of fiber (see p. 67).

Mechanism of Propagation. It is characteristic of action potential conduction that the action potential signal is not weakened by this process. Therefore, conduction cannot take place solely as a result of current flow from an excited to a nonexcited site. Such electrotonic spread would generate potentials that became smaller with increasing distance from the site of current application. Instead, the amplitude of the action potential remains constant along the propagation path because an *excitation* occurs at every point on the membrane. These local responses obey the all-or-none law.

Electrotonic spread and excitation act together in the conduction of the action potential. Current flows from an already excited point on the membrane to a neighboring and as yet unexcited, undepolarized region, where it generates an *electrotonic potential.* This potential reaches the threshold and serves as the *stimulus* initiating excitation in this part of the membrane. The process of excitation now takes place automatically at this point and in turn supplies current for the electrotonic depolarization of other regions of the membrane.

The action potential is conducted like the spark in a blasting fuse. At the point where the fuse is lit, the gunpowder explodes (excitation); as a result, the neighboring section of the fuse is warmed (electrotonic potential) to the point where the gunpowder there also explodes and in turn supplies heat for igniting the next sections.

Membrane Currents During a Propagated Action Potential. Figure 2-22 illustrates in detail the relationships between membrane voltage and membrane currents as an action potential is propagated. The figure represents a "stop-motion photo" of the voltage and current conditions along the fiber. The action potential is being conducted from right to left. The total length of the fiber covered by the action potential depends on the velocity of conduction. Given a fiber conducting at a velocity of 100 m/s and an action potential of 1 ms duration, the length of the abscissa in Fig. 2-22 would correspond to 10 cm. Since the action potential travels past a particular point on the membrane at the conduction velocity, Fig. 2-22 can also be seen as representing the time course of the action potential at one point on the fiber. The abscissa would then, under the above conditions, be 1 ms long.

The lowermost curves in Fig. 2-22 show the shape of the action potential and the associated changes in g_{Na} and g_K. In Section 2.5 the kinetics of excitation in a homogeneous preparation was discussed. In that case excitation occurred simultaneously over the entire membrane. The ion currents flowed uniformly through the membrane at each point, serving only to change the charge on the local membrane capacitance. In conduction, the situation is not homogeneous. Only

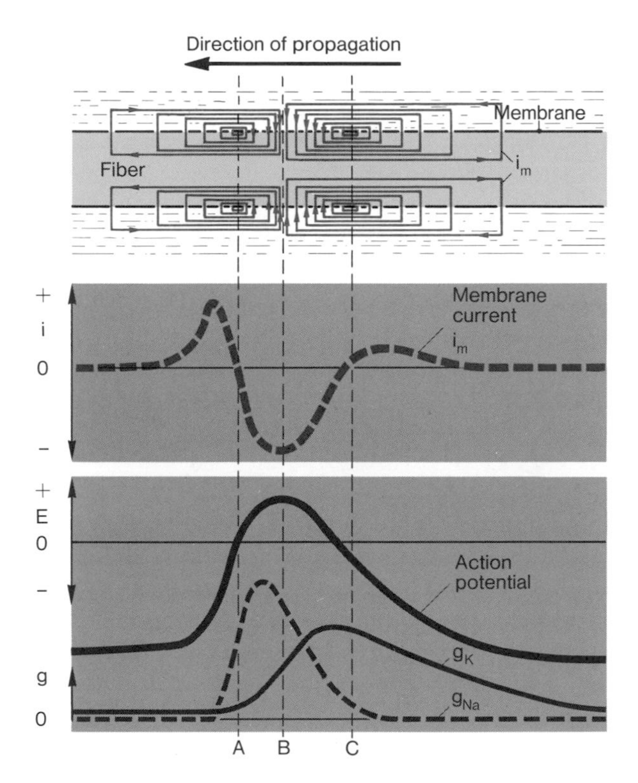

Fig. 2-22. Conduction of the action potential. *Below:* The heavy black curve represents the time course of the action potential at a point, or its spatial distribution along the fiber at a given instant, with the corresponding membrane conductances g_{Na} and g_K shown beneath it. The *red curve* over it shows the membrane current i_m. *Above:* The lines of current flow through the cell membrane and inside and outside the fiber. The vertical dashed lines indicate the times of maximal rate of rise *A*, of the peak *B*, and of the maximal rate of repolarization *C*. Modified from Noble (1966) *Physiol. Rev.* 46:1.

one site on the fiber is excited; ahead of it (left in Fig. 2-22) the fiber is still unexcited, and behind it the fiber is still just barely excited. Because these differently charged regions of the fiber are electrotonically coupled with one another, through the interior of the fiber, equilibrating currents must flow. The directions and magnitudes of these are indicated by the red, dashed curve labeled "membrane current i_m" in the middle of Fig. 2-22. The current loops representing these equilibrating currents are included in the diagram of a nerve fiber above the curve.

The excited region coincides with the peak of the action potential, between lines *A* and *C* in Fig. 2-22. Here, because of the high g_{Na}, current flows into the fiber (i_m negative), not only changing the charge on the local capacitance but also flowing within the fiber to adjacent, less depolarized regions and depolarizing them electrotonically (i_m positive). Accordingly, in the diagram at the top of Fig. 2-22 the current lines in the excited region enter the fiber, turn toward adjacent regions on either side, and there cross the membrane to return to the outside medium. In the part of the membrane that has not yet been excited, to the left of line A, these current loops depolarize the membrane, as a result of which the sodium conductance g_{Na} rises. Hence the Na^+ influx here also increases, the threshold is exceeded, and excitation occurs. By this process the excitation has propagated from the region that had been excited into the previously unexcited region; the excitation has been conducted.

The current loops extend not only toward the left of the excited region, but also into the less depolarized region on the right of line *C*. If this region had not just been occupied by an action potential—that is, if the excitation had not come from the right—the excitation would propagate into it as well. That is, when a nerve fiber is stimulated in the middle the excitation is conducted in both directions. But when the action potential is traveling along the fiber, as shown in Fig. 2-22, the part of the fiber just traversed is in the repolarization phase of the action potential; g_K is high, and the sodium system, having just been activated, is now inactivated. Therefore the current loops running backward from the excited region cannot normally trigger excitation, but only delay repolarization.

The current loops diagrammed for the membrane current at the top of Fig. 2-22 can be recorded with extracellular electrodes. The reason is that they cause small potential changes in the external medium near the excited part of the fiber. The potential changes are large enough (e.g., 100 μV) to be measured with extracellular electrodes relative to ground. Many recordings in the nervous system are obtained with this "unipolar" technique; in this case the recorded potential changes have the triphasic shape of i_m. Given uniform conduction along the fiber, i_m corresponds to the second derivative of the membrane potential with respect to time.

Factors Affecting the Velocity of Conduction; Saltatory Conduction. The velocity of conduction of the action potential can be calculated, with a great deal of effort, from the potential and the time dependence of the ionic currents as well as from the parameters fiber diameter, membrane capacity, and membrane resistance, which determine the spread of electrotonic potentials. At this point, we shall deal only qualitatively with the factors influencing the velocity of propagation.

The velocity of conduction increases with the *amplitude of the Na^+ influx,* because if more current is available during excitation after discharge of the membrane capacitor, then more current can flow into and accelerate the depolarization of adjacent, still unexcited regions. When excitation starts from a normal resting potential, the Na^+ influx into the cell is at its maximum level, and as a result the maximum velocity of conduction is also attained. If the resting potential is lowered (that is, becomes less negative) the Na^+ system is partially inactivated, and the Na^+ influx is reduced during excitation (see p. 49). Therefore, *when the resting potential is lowered, the velocity of conduction is reduced.*

In addition, the *electrotonic flow* of the membrane currents has an important effect on the velocity of conduction. If the electrotonic potentials rise faster and if they decline less with distance, the conduction velocity must increase. The electrotonic potential will rise faster if the membrane capacitance is reduced, and it will fall less with distance if the membrane resistance is increased. These conditions are utilized in the myelinated nerve fibers to increase the velocity of conduction. In these nerve fibers the membrane is thickened by the buildup of insulating myelin layers (see p. 8) that greatly increases its resistance. In the myelinated parts of these nerve fibers, the internodes, an electrotonic potential declines only slightly with distance. The action potential is, therefore, *propagated over the internodes at very high speed.*

If propagated action potentials, as illustrated in Fig. 2-23, are measured simultaneously at many points on a myelinated fiber, it is found that the action potential is not measurably retarded in the individual internodes (thick sections of nerve in Fig. 2-23). On the other hand, propagation is retarded at the *nodes of Ranvier* R_1–R_5. There is no myelin sheath at these nodes, and consequently the membrane capacitance and resistance are normal. At these points, the electrotonic potential increases slowly, and the excitation starts with a delay which is apparent at each node in Fig. 2-23. In myelinated nerve fibers, the excitation jumps from node to node, and the conduction of the excitation is, therefore, called "saltatory." Since the propagation time between the nodes is so short, the overall velocity of propagation is much higher in myelinated fibers than in nonmyelinated fibers of the same thickness. In vertebrates, all the fibers that conduct at velocities of propagation in excess of 3 m/s are myelinated.

Along with the myelination, the *fiber diameter* is the other most important factor determining the propagation velocity. The resistance to the passage of current along the fiber decreases with the square of the inside fiber diameter. When the *resistance along the fiber is low,*

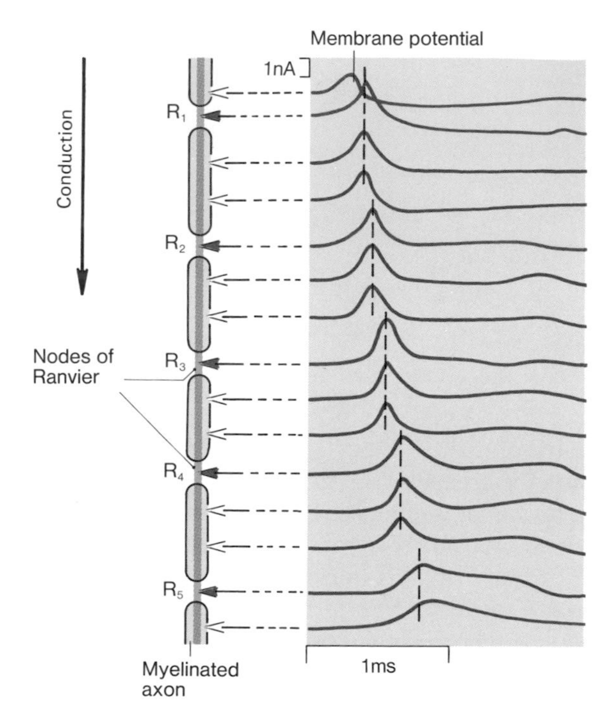

Fig. 2-23. Saltatory conduction of excitation. The curves on the *right* show the time courses of the membrane potential measured at the points on a myelinated axon indicated by the arrows on the *left*. R_1, R_2, R_3, . . . , are nodes of Ranvier. Conduction of the action potential (from *top* to *bottom*) is delayed at each of the nodes. Modified from Huxley and Stämpfli (1949). *J. Physiol.* 108:1.

relatively more current flows from the excited region to the adjacent regions, which are then more rapidly depolarized by electrotonic current. Thus, the velocity of propagation rises with the increase in fiber diameter.

The *relationship between conduction velocity and fiber diameter* is summarized in Table 2-2 a and b. There are two classifications of nerve fibers. One, by *Erlanger and Gasser* (Table 2-2a), uses the letters A, B, and C, while the second, by *Lloyd and Hunt* (Table 2-2b), refers only to afferent fibers and uses the numerals I, II, III, and IV. The groups are subdivided somewhat differently in the two schemes, so that it is not practicable to use only one classification. The fibers of groups A and B, and those of I to III, are *myelinated,* whereas those of C and IV are *unmyelinated* (see Table 1-1). The conduction velocity of myelinated fibers increases more than proportionately to the fiber diameter. The unmyelinated fibers (C and IV) have very low conduc-

Table 2-2a. The Erlanger/Gasser Classification of Nerve Fibers.

Fiber type	Function (examples)	Avg. fiber diameter (μm)	Avg. cond. velocity (m/s)
Aα	primary muscle-spindle afferents, motor to skeletal muscles	15	100
Aβ	cutaneous touch and pressure afferents	8	50
Aγ	motor to muscle spindles	5	20
Aδ	cutaneous temperature and pain afferents	3	15
B	sympathetic preganglionic	3	7
C	cutaneous pain afferents, sympathetic postganglionic	0.5 (unmyelinated)	1

Table 2-2b. The Lloyd/Hunt Classification of Nerve Fibers.

Group	Function (examples)	Avg. fiber diameter (μm)	Avg. cond. velocity (m/s)
I	primary muscle-spindle afferents and afferents from tendon organs	13	75
II	cutaneous mechanoreceptors	9	55
III	deep pressure sensors in muscle	3	11
IV	unmyelinated pain fibers	0.5	1

tion velocities. Measurement of the velocity in unmyelinated fibers of different thickness has shown that here, too, there is an increase of velocity and diameter. Extremely large unmyelinated fibers–for example, the 0.7-mm-thick squid giant fiber–conduct action potentials at about 25 m/s.

Mixed nerves at the periphery of the body–for example, the ischiadic nerve, which serves the musculature and the skin of the leg–contain afferent fibers of Groups I to IV as well as efferent, chiefly motor, fibers. If such a nerve is stimulated at one end, the excitation is propagated at very different speeds in the various groups of fibers. If an electrical recording is made at some distance from the site of stimulation, the first action potentials to arrive will be those of the fastest group of fibers, Group I. Next will come the slower action potentials of the Group II fibers, and then those of the Group III fibers, finally followed by those of the Group IV fibers. Thus, a spectrum of action potentials is generated by a stimulus. For a conduction path of 1 m, and the values listed in Table 2-2, the action potentials of Group I fibers would be recorded after 13 ms and those of Group IV fibers after 1 s.

Q 2.22 Plot the course of the current lines, i_m, in an elongated cell from an excited region to adjacent regions. Beneath this, plot the time course of the action potential.

Q 2.23 Which is the source of the current that depolarizes the membrane to the threshold in a still unexcited area when the action potential is propagated?
a. The driving force for the potassium ions.
b. The sodium inflow of the still unexcited area of the membrane.
c. The sodium inflow of an adjacent, already excited part of the membrane.
d. The axoplasm of the cell.

Q 2.24 Which of the following statements applies to the current flow at the moment when the peak of the propagated action potential is reached? You may use Fig. 2-22 to help you solve this problem.
a. The Na^+ inflow is equal to the K^+ outflow.
b. The Na^+ inflow outweighs the K^+ outflow; the net current reverses the charge of the membrane capacitor.
c. The Na^+ inflow outweighs the K^+ outflow; the net current flows into neighboring regions of the membrane and depolarizes them.
d. The K^+ outflow outweighs the Na^+ inflow; the net current depolarizes neighboring regions of the membrane.
e. At the peak of the action potential, no current flows into the membrane capacitance.

Q 2.25 Which of the following factors reduces the velocity of conduction of a nerve?
a. Reduction in fiber diameter.
b. Decrease in resting potential by 10 mV.
c. Loss of the myelin sheath (in the event of degeneration).
d. Increase in the extracellular Na^+ concentration.
e. 50% reduction in the extracellular K^+ concentration.

3
SYNAPTIC TRANSMISSION

R.F. Schmidt

The junction of an axonal ending with a nerve cell, a muscle cell, or a glandular cell was first called a *synapse* by Sherrington (see also Chapter 1, p. 3). At synapses the propagated action potential is transmitted to the next cell. Originally it was wrongly believed that the axon always formed a "gap junction," that is, was in closest contact with the cell on which it ended so that the propagated impulse could be transmitted without interruption to that cell. However, electrophysiologic and histologic investigations have shown that this form of synapse, which is now called an *electrical synapse*, is rare. Another type of synapse is far more common, particularly in mammals and thus in man. In this type, the axonal ending when stimulated releases a chemical substance that produces an excitatory or an inhibitory effect at the neighboring cell membrane. This type of synapse is called a *chemical synapse*. The structure and the function of excitatory and inhibitory chemical synapses will be explained in this chapter.

The Significance of Synapses. By way of introduction, it should be pointed out that synapses are fundamental to nervous-system function for several reasons. First, at both electrical and chemical synapses, signals are almost always transmitted only from the presynaptic (axonal) to the postsynaptic element. That is, synapses act primarily as one-way relay stations, both between neurons and between neurons and effectors—a feature that would seem essential for the orderly operation of the nervous system. Moreover, the efficiency of synapses can be modified, so that when frequently used they transmit more reliably or with greater probability than when used rarely or not at all.

Because of this *plasticity*, synapses are thought to function in *learning* and *memory*. Finally, synapses are the *sites at which many drugs act.*

3.1 The Neuromuscular Junction: Example of a Chemical Synapse

Structural Elements of Chemical Synapses. Light- and electron-microscope investigations have shown that synaptic junctions vary widely in shape and form. As regards function, however, all the parts of a chemical synapse can be related to the basic elements illustrated in Fig. 3-1 and discussed and described in the following paragraphs.

In Fig. 3-1 the axon ends in the *presynaptic terminal*. Under the microscope, this often stands out as a spherical enlargement of the end of the axon (see Fig. 3-9), called the "terminal button" or "bouton." The presynaptic terminal is separated from the postsynaptic side by a narrow cleft (gap), which is, on the average, 10 to 50 nm (100 to 500 Å)

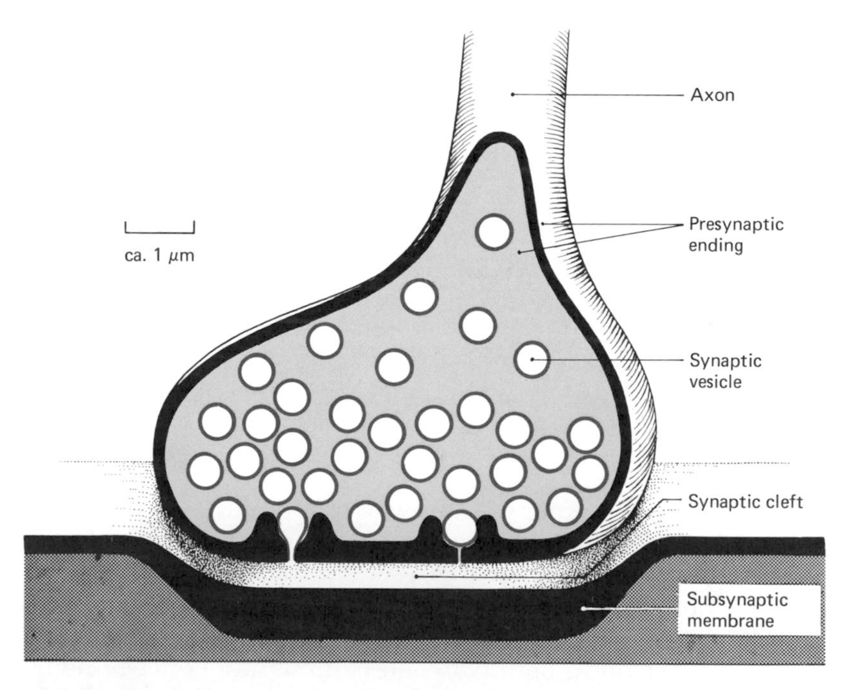

Fig. 3-1. Diagrammatic section through a chemical synapse. All the structural elements important in synaptic transmission are shown. The diameter of the synaptic vesicles and the width of the synaptic cleft are greatly exaggerated with respect to the other elements of the synapse.

wide. This is known as the *synaptic cleft*. It is clearly discernible only under the electron microscope.

The part of the postsynaptic membrane just opposite the presynaptic terminal, forming the postsynaptic boundary of the synaptic cleft, is called the *subsynaptic membrane*. Under the electron microscope, the subsynaptic membrane usually appears somewhat thicker than the other parts of the postsynaptic membrane, which indicates that it has a function different from that of the rest of the postsynaptic membrane.

The presynaptic terminal contains a large number of submicroscopic spherical structures that can be detected only under the electron microscope, namely, the *synaptic vesicles*. They are about 50 nm (500 Å) in diameter. Many experimental findings, the most important of which are discussed in this chapter, indicate that the synaptic vesicles in the presynaptic terminals contain the *transmitter substance* that is released into the synaptic cleft upon excitation and which triggers excitation or inhibition of the subsynaptic membrane.

The End Plate. The motor axons of the motoneurons in the ventral horn of the spinal cord form synapses with striated muscle fibers (skeletal muscle fibers). Because of its appearance, this synapse, and, in particular, the presynaptic section, is called the *neuromuscular end plate* or, alternatively, the *neuromuscular junction* (Fig. 3-2). It possesses all the typical morphological characteristics of a chemical synapse; that is, in addition to the presynaptic terminal with its characteristic synaptic vesicles, it has a synaptic cleft and a subsynaptic membrane on the postsynaptic side. However, in this case, the postsynaptic element is a skeletal muscle fiber rather than a neuron.

It is possible to remove such muscles, together with the associated nerves, from the living organism and to place them in a blood-substitute solution (for example, Ringer or Tyrode solution) where they remain alive and functioning for some time. The frog gastrocnemius and sartorius muscles, with the associated nerves of the same name, and the rat diaphragm with its phrenic nerve, are well-known preparations of this type.

Detection of the End-Plate Potential. Figure 3-3 illustrates the experimental setup for studying synaptic transmission at the neuromuscular junction of an *in vitro* preparation of a skeletal muscle. The figure shows a muscle fiber with its motor axon. A microelectrode is inserted into the muscle fiber to record its membrane potential intracellularly. As soon as this microelectrode is advanced from the bathing solution into the interior of the muscle cell, a *resting potential* of about −70 mV is indicated on a cathode-ray oscilloscope or some other suitable measuring instrument (Fig. 3-4, see also Fig. 2-1).

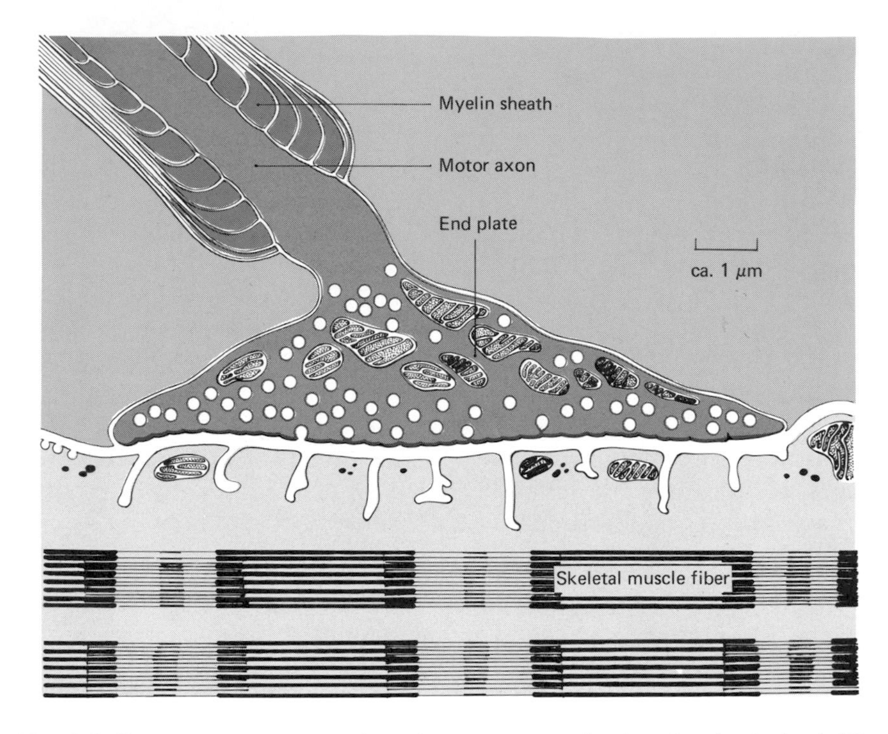

Fig. 3-2. Diagrammatic section through a neuromuscular junction (end plate). The synaptic vesicles are drawn larger than they actually are, in proportion to the other elements of the synapse. Only part of the skeletal muscle fiber is shown.

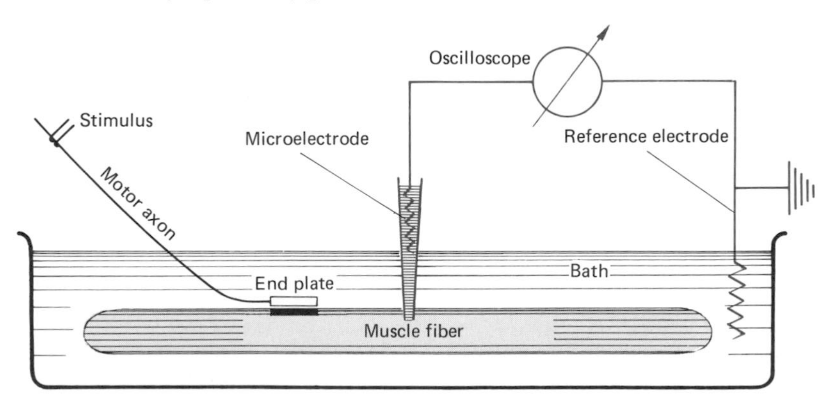

Fig. 3-3. Experimental arrangement for recording end-plate potentials. The bath is filled with Tyrode solution. The intracellular microelectrode records the changes in potential across the muscle-fiber membrane, relative to an external electrode in the Tyrode solution. The motor axon is stimulated electrically. To prevent short circuiting between the stimulating electrodes, the nerve is held in air or in a layer of liquid paraffin during stimulation.

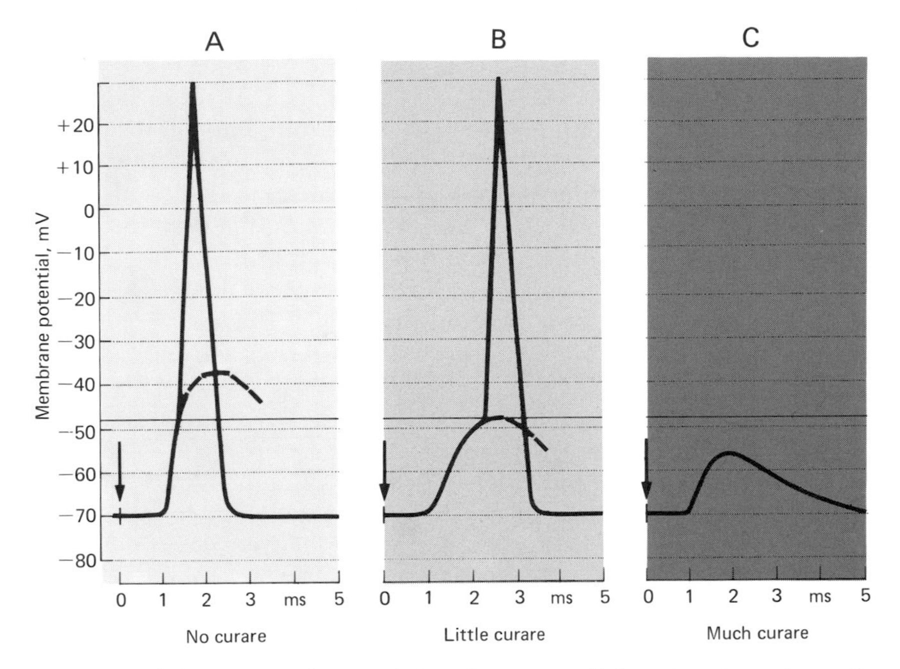

Fig. 3-4. End-plate potentials recorded with an intracellular microelectrode, as diagrammed in Fig. 3-3; the *arrows* indicate stimulation of the motor axon. **A.** Potential change recorded when the muscle fiber is in normal Tyrode solution. **B.** Response to stimulation when the solution contains a small amount of curare. **C.** Response when the curare concentration in the solution has been doubled. The threshold for conducted excitation (action potential) is indicated by the *thin line* at about —48 mV.

If the associated motor axon is now stimulated electrically, an action potential propagates down into the presynaptic terminal of the end plate. On the postsynaptic side (that is, at the membrane of the muscle fiber), it triggers the potential changes indicated in Fig. 3-4A. The arrow in Fig. 3-4A denotes the moment at which the stimulus was applied to the motor axon. After a latent period of about 1 ms (depending on the length of the nerve and the conduction velocity of the motor axon), the muscle membrane potential depolarizes to the threshold, and a typical *action potential* appears (see Figs. 2-10 and 2-11). It can be seen clearly how the initial depolarization induces the steep rising phase of the action potential at about −42 mV. The propagated action potential makes the muscle fiber contract (for details, see Chapter 5).

If a small quantity (10^{-7} to 10^{-6} g/ml) of curare (a substance used by Indians to poison their arrows) is added to the bathing solution, and if the motor axon is again stimulated, the changes in membrane potential shown in Fig. 3-4B are recorded. The initial depolarization is slower, and, therefore, the action potential starts a little later. However, the pattern of the action potential remains unchanged, and a muscular contraction is still observed. But, if still more curare is added to the bathing solution (Fig. 3-4C), the initial depolarization no longer evokes an action potential; instead, it remains subthreshold and returns to the resting potential after a few milliseconds. (The muscle does not contract.) The potential remaining after the action potential disappears is called the *end-plate potential (EPP).* The dotted lines in Figs. 3-4A and B indicate the time courses of EPPs masked by the superimposed action potentials. End-plate potentials can thus vary in amplitude, being *either suprathreshold or subthreshold.*

In healthy muscles the EPPs are always far above threshold. Each presynaptic action potential triggers a contraction in the associated muscle fibers. The amplitude of the normally suprathreshold end-plate potential is severely reduced by poisoning with curare, and if the curare concentration is high enough, the EPP falls below the threshold. That is, it no longer triggers either an action potential in the muscle fiber or a contraction. *Neuromuscular transmission* is, therefore, *blocked* by curare. Thus, a person poisoned by curare will suffocate because neuromuscular transmission in his striated muscles, and these include his respiratory muscles, is blocked.

Mechanism of Neuromuscular Transmission. The experiment just described and illustrated in Figs. 3-3 and 3-4 is not by itself enough to permit conclusions regarding what presynaptic and postsynaptic events lead to the generation of the EPP or how curare, for example, reduces the amplitude of the EPP. Intensive experimental analysis extending over several decades, and still going on in some areas, was necessary to clarify these relationships. The results obtained will be summarized in the following paragraph, and then the most important facts will be dealt with in more detail.

The *action potential* invading the presynaptic terminal causes release of a certain amount of *transmitter substance* into the synaptic cleft. The transmitter substance diffuses to the *subsynaptic membrane* and induces permeability changes that lead to the generation of the *EPP*. The transmitter substance at the neuromuscular junction is *acetylcholine* (*ACh*). It acts on the subsynaptic membrane for only a very short time because it is broken down by an enzyme, *cholinesterase,* into two inactive components, choline and acetic acid.

Even this very brief description shows that there are a number of ways in which the transmission at a chemical synapse may be influenced. A drug, for example, may inhibit synaptic transmission by preventing impulse propagation into the presynaptic ending, by blocking the mechanism that releases transmitter substance when an action potential invades the presynaptic terminal, by inhibiting the production or storage of the transmitter substance, or by quickly breaking down the transmitter substance into inactive components. Finally, a drug may compete with the transmitter substance for the receptor sites on the subsynaptic membrane, or otherwise affect those sites. Examples can be given for almost all of these possibilities. Curare, for instance, *competes* with ACh for its molecular receptors at the subsynaptic membrane.

The Nature of the End-Plate Potential. In the following we will consider which changes in the subsynaptic membrane lead to the generation of the EPP. We will first see that the EPP is generated only at the subsynaptic membrane. Then some experiments will be described that prove that the EPP is the result of a *brief increase in the conductance* of the membrane to small cations (Na^+, K^+, Ca^{++}).

Figure 3-5 shows the intracellular recording of an EPP at various distances from the end plate in a nerve-muscle preparation treated with curare. The points of insertion are spaced about 1 mm apart. It is apparent that as the microelectrode is inserted further away from the end plate, the amplitude of the EPP decreases, and its rise and decay times become longer. This finding is a clear indication of the passive *electrotonic spreading of the EPP* from its site of generation, the subsynaptic membrane (see Sec. 2.6). Thus, it can be stated that excitation

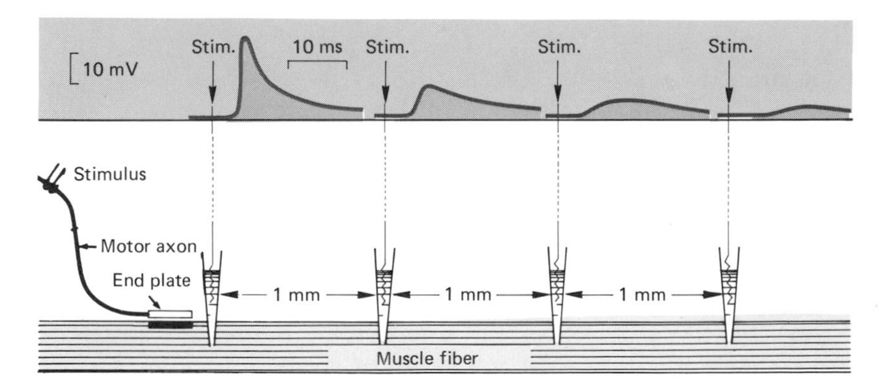

Fig. 3-5. The electrotonic nature of the EPP is shown by recording intracellularly at various distances from the end plate, with an experimental arrangement like that of Fig. 3-3. Sufficient curare has been added to the bath to prevent the triggering of conducted action potentials in the muscle fiber when its motor axon is stimulated.

of the end plate at the subsynaptic membrane gives rise to a depolarization, the EPP. Provided it remains below the threshold (that is, does not trigger a propagated action potential), this EPP spreads electrotonically along the muscle fiber in accordance with the passive electrical characteristics of the muscle fiber membrane.

Voltage-clamp tests and mathematical analysis of the time course and the spatial distribution of the EPP lead to the conclusion that the *initial phase of depolarization*, which occurs while the transmitter substance ACh reacts with the subsynaptic membrane, *lasts only about 1 to 2 ms*. In other words, the change in membrane conductance that leads to a shift in charge of the membrane capacitor takes place in this short time. The subsequent time course of the EPP is determined by the passive electrical characteristics of the muscle fiber membrane—that is, by the membrane capacitance and resistance.

The nature of the *changes in conductance during the initial phase* of the EPP was revealed most clearly by determining the equilibrium potential of the EPP in normal bathing solution and then after systematic variation of the extracellular ionic concentration (in these experiments the EPP amplitude was kept subthreshold by applying curare or by other means). An experimental setup for measuring the equilibrium potential of the EPP—the membrane potential at which no change in potential occurs during the action of ACh—is sketched in Fig. 3-6. In addition to the recording electrode, a second microelectrode is inserted into the muscle fiber. This second electrode is connected to a current source, enabling the membrane potential to be varied at will.

The right half of Fig. 3-6 shows the effect of the stimulus at four different membrane potentials with the muscle fiber held in normal bathing solution. If the EPP is evoked at a membrane potential of −95 mV, its amplitude is approximately 15 mV in the depolarizing direction; at −45 mV, the amplitude is 5 mV in the depolarizing direction; at −15 mV, the amplitude is 0 mV; and at +30 mV, it is 15 mV in the hyperpolarizing direction. This result shows that under normal conditions the *equilibrium potential of the EPP* (E_{EPP}) is located approximately at −15 mV, that is, between the equilibrium potentials for potassium (E_K = −80 mV) and sodium (E_{Na} = +45 mV).

These and other measurements lead to the conclusion that during the time that the ACh acts on the subsynaptic membrane, for about 1 to 2 ms, *the conductance of the membrane to small cations* (Na^+, K^+) *is increased considerably*. Under normal circumstances, therefore, the given ionic distribution (see Table 2-1) will result in predominantly Na^+ ions flowing into the muscle fiber. This reduces the membrane potential, because at a membrane potential of −70 mV the driving

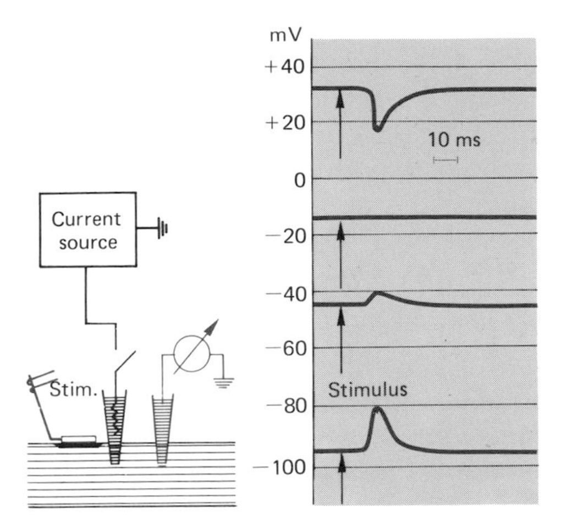

Fig. 3-6. The equilibrium potential of the EPP, measured as diagrammed on the *left*. The arrangement corresponds to that of Fig. 3-3, with an additional microelectrode inserted into the fiber through which current from an external source is applied to change the membrane potential of the muscle fiber. The *arrows* in the right part of the figure indicate suprathreshold stimulation of the motor axon. The EPP is about 10 ms in duration.

force for the Na^+ ions is more powerful than that for the K^+ ions; there is a net inward current carried by Na^+ ions, which is called *end-plate current*. If the change in conductance is large enough, the muscle fiber membrane at the end plate is depolarized to the threshold, and a propagated action potential, which spreads out over the entire fiber membrane, is evoked (see Fig. 3-4).

The Fate of the Acetylcholine. Normally, after its release from the presynaptic terminal, the ACh diffuses across the synaptic cleft to the subsynaptic membrane where it combines with special protein molecules, the *ACh receptors*. The ACh takes only fractions of a millisecond to diffuse across the narrow synaptic cleft. It combines with the subsynaptic receptors and causes an increase in the membrane conductance to small cations.

To use a metaphor, the ACh key is inserted in the receptor lock, and the conductance door for small cations is opened wide. However, the ACh can act for only 1 to 2 ms at the subsynaptic membrane since, as already briefly mentioned, it is broken down by the enzyme cholinesterase into the inactive components choline and acetic acid. (Special staining methods have shown that cholinesterase is present in large quantities at the end plate. In addition, cholinesterase also circulates in the blood so that ACh that diffuses from an end plate into the surrounding extracellular space and, thus, into the circulation is also split into choline and acetic acid.) The decomposition products of ACh, choline and acetic acid, are for the most part reabsorbed by the presynaptic terminal and, with the aid of enzymes, recombined into ACh which is stored in the synaptic vesicles of the presynaptic terminal until it is released once more. This ACh cycle is shown diagrammatically in Fig. 3-7.

Localization of ACh Receptors. If ACh is applied electrophoretically to a muscle fiber with a micropipette, it evokes depolarizations *only at the end plate* and not in other sections of the muscle fiber membrane. It can be concluded that the ACh receptors are located only at the subsynaptic membrane and nowhere else on the postsynaptic membrane. Injection of ACh into the muscle fiber likewise fails to produce

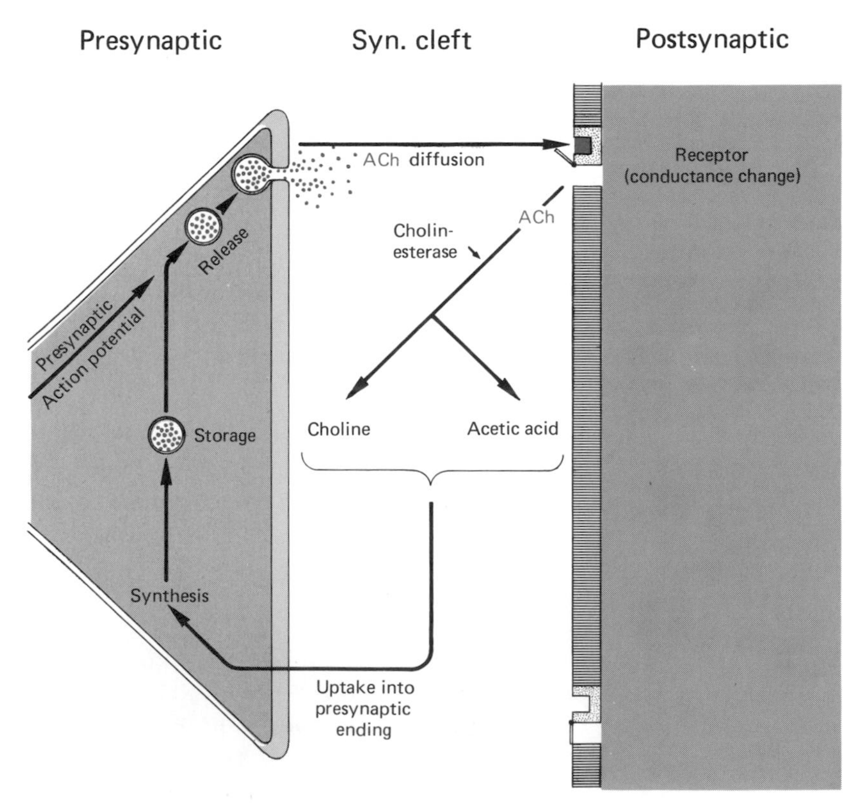

Fig. 3-7. Metabolic cycle of ACh at the neuromuscular junction. For further explanation see text.

depolarization of the membrane because ACh, like all other transmitter substances, acts *only on the outer surface* of the subsynaptic membrane. (It is interesting to note that the sensitivity of the subsynaptic membrane to ACh spreads to the remainder of the muscle fiber in the event of degeneration of the presynaptic axon, such as occurs after its transection; this process is reversed when reinnervation occurs.)

Neuromuscular Block. If a person is injected with a large enough dose of curare, neuromuscular transmission is blocked (see Fig. 3-4), producing a muscular relaxation that can be useful during surgery under anesthesia. A patient treated with this sort of drug requires artificial respiration, but the anesthesia need only be deep enough to render the person unconscious and eliminate the sensation of pain without suppressing motor reflexes. A shallow state of anesthesia has the advantage of being less toxic, easier to control, and quickly reversible. Substances that relax the muscles during anesthesia or in other therapeutic situations are called *muscle relaxants.*

Curare competes with ACh for its subsynaptic receptor, thus effecting a block that is reversible and dependent upon the concentrations of both competing substances. Since such drugs act without changing the membrane conductance, they are called *nondepolarizing muscle relaxants.* A second group of relaxants, which also finds practical application, operates differently. These substances—succinylcholine, for example—have an effect like that of ACh on the subsynaptic membrane, but they are broken down by the cholinesterase only slowly or not at all. They, therefore, cause a prolonged depolarization of the subsynaptic membrane, which blocks neuromuscular transmission. These substances are called *depolarizing muscle relaxants.*

A neuromuscular block can also be produced if cholinesterase activity in the synaptic cleft is prevented by administration of cholinesterase inhibitors. In this case the delay in the breakdown of ACh leads to an enhanced and somewhat prolonged depolarization at the subsynaptic membrane. This method is not clinically useful because of the severe side effects. It is of medical interest, however, since various insecticides and a few nerve gases act as *irreversible* cholinesterase inhibitors. If taken into the body, these produce a transient state of increased excitability of the musculature (convulsions) followed by respiratory paralysis and death.

Under certain circumstances, though, *cholinesterase inhibitors can facilitate neuromuscular transmission.* In cases of poisoning by an overdose of curare, inhibition of cholinesterase can result in an increase of the reduced (and perhaps subthreshold) EPP so that normal neuromuscular transmission again becomes possible. In the disease known as *myasthenia gravis,* resynthesis of ACh in the presynaptic ending is retarded, so that when the synapse is activated repeatedly, progressively less transmitter is released, and neuromuscular transmission is eventually blocked. It is typical of these patients that their neuromuscular transmission is good in the morning and deteriorates during the day (an early symptom is drooping eyelids). Here, too, administration of reversible cholinesterase inhibitors such as neostigmine is useful and provides the most effective treatment yet discovered.

Q 3.1 Release of ACh at the end plate produces a brief increase in subsynaptic membrane conductance of
- a. K^+ ions.
- b. Na^+ ions.
- c. Cl^- ions.
- d. Ca^{++} ions.
- e. Mg^{++} ions.

Q 3.2 Depolarization of a muscle fiber membrane to approximately −30 mV
- a. leaves the EPP unchanged.
- b. shortens the duration of the EPP considerably.
- c. increases the duration of the EPP considerably.
- d. reduces the amplitude of the EPP.
- e. None of the above is correct.

Q 3.3 Inhibition of cholinesterase by poisoning with an irreversible cholinesterase inhibitor blocks neuromuscular transmission because
- a. ACh is competitively driven away from its receptor.
- b. ACh is no longer released presynaptically.
- c. ACh is not broken down, and, consequently, a lasting depolarization of the subsynaptic membrane occurs.
- d. the EPP reverses polarity.
- e. ACh synthesis in the presynaptic ending is blocked.

Q 3.4 Which of the following statements are correct? In the event of poisoning by curare
- a. the presynaptic synthesis of ACh is not seriously affected.
- b. decomposition of the ACh after its release into the synaptic cleft is greatly slowed down.
- c. the ACh is competitively displaced from its subsynaptic receptor.
- d. the equilibrium potential of the EPP shifts toward the resting potential.
- e. the time course of the EPP is considerably slowed down.

3.2 The Quantal Nature of Chemical Transmission

The experimental results summarized in this section indicate that the transmitter substance is not distributed uniformly within the presynaptic ending, but is concentrated in the vesicles to be found there (see Figs. 3-1 and 3-2). The transmitter is released from these vesicles into the synaptic cleft according to specific rules. It is necessary to know these rules in order to devise appropriate procedures for affecting synapses pharmacologically (for example, to produce neuromuscular block as already discussed).

Miniature End-Plate Potentials. If a microelectrode is inserted into a *resting* muscle fiber (see sketch in the upper part of Fig. 3-8), small, brief, and irregularly occurring depolarizations will be recorded (see Figs. 3-8A and B. These spontaneous depolarizations are similar in their time course to normal EPPs, but their amplitude is very much smaller than that of normal EPPs (compare ordinate scale of Fig. 3-8 with that of Fig. 3-4). Because of their similar time courses and very small amplitudes, these spontaneous depolarizations are called *miniature end-plate potentials (MEPPs)*. Experiments of the type illustrated in Fig. 3-5 have shown clearly that the MEPPs, just as the EPPs, are generated only at the subsynaptic membrane and spread out elec-

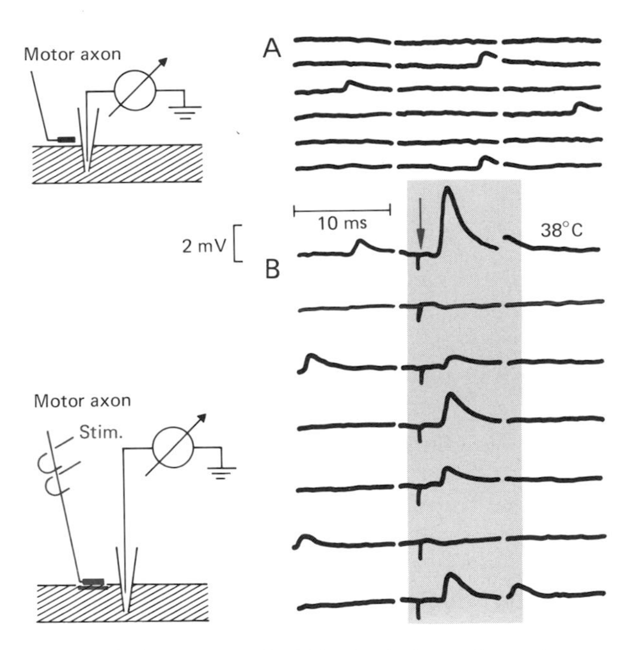

Fig. 3-8. Miniature end-plate potentials, MEPPs. **A.** Recordings from a resting muscle fiber, with the microelectrode immediately adjacent to the end plate (see inset). **B.** EPP produced by electric stimulation of the motor axon (*see arrow*; potentials in *red-shaded* area), with the muscle in saline containing 1 mM Ca^{++} and 6 mM Mg^{++}. Some spontaneous MEPPs also occur in these records. Two stimuli fail to elicit an end-plate potential; in the other cases the amplitude of the end-plate potential corresponds to that of one MEPP or an integral multiple thereof. Modified from Liley, 1956.

trotonically from there over the muscle fiber. (Because of their low amplitude they can be recorded only in the immediate vicinity of the end plate.) The pharmacologic properties of the EPPs and the spontaneous MEPPs are also identical. It is, therefore, justified to assume that the MEPPs are generated by the *spontaneous release of small quantities of ACh.*

Quantal Release of Transmitter. Figure 3-8 also shows that the MEPPs all have approximately the same amplitude. From this it may be concluded that they are triggered by approximately equal quantities of ACh. These nearly equal-sized packets of ACh are called *quanta.* By applying an experimental trick, it is possible to show that the normal EPP is also caused by the release of quanta of transmitter substance. The quantity of transmitter substance released per action potential may be reduced substantially by removing Ca^{++} ions from the bathing solution or by adding Mg^{++} ions. Records of EPPs triggered under these conditions are shown in the red-shaded part of Fig. 3-8B. (A few spontaneous MEPPs appear elsewhere in these recordings.) A total of seven stimuli (arrow) were given. In two cases, the EPP was just about as large as the spontaneous MEPP; in two other cases, it was about twice as large, and once about three times as large. Twice the stimulus failed to elicit any EPP. This result leads one to suspect that the normal EPP also is always composed of integral multiples of the MEPP—that is, that it is brought about by the simultaneous release of a large number of quanta.

Control of Transmitter Release by the Presynaptic Action Potential. At the resting end plate, MEPPs occur at irregular intervals (see Fig. 3-8). That is, the quanta are not released at predictable times, but rather with a certain probability per unit time so that the events can be expressed in statistical terms. This probability is low in the resting state when the MEPPs appear only once per second on the average. But the presynaptic action potential increases it by a factor of several thousand for a short time, less than a millisecond. During this brief period a few hundred quanta are released, and these produce the normal EPP. A single presynaptic action potential has been estimated to release about 200 quanta at the end plate of the frog, while estimates for other synapses are as high as 2,000 quanta.

The elapsed time between arrival of the action potential at the presynaptic ending and the onset of the EPP is called the *synaptic latency.* In most peripheral and central synapses it amounts to about 0.2 ms.

The *temporary increase in the probability of release* by the presynaptic action potential is not an all-or-none phenomenon. It depends at least in part upon the amount of change in the membrane potential, as the following results show. When the presynaptic membrane potential is depolarized by increase in the extracellular K^+ concentration or by externally applied current, MEPPs occur more frequently, and during hyperpolarization they are produced at a lower rate. Experimental variation of the action-potential amplitude and imitation of the presynaptic action potential by externally imposed membrane-potential changes have also shown clearly that the magnitude of the EPP—that is, the number of quanta released per action potential—depends upon the amplitude of that action potential.

The Role of Calcium. As mentioned in the discussion of Fig. 3-8B, a second important factor affecting the amount of transmitter released is the Ca^{++}-ion content of the extracellular fluid. In an *in vitro* experiment, if Ca^{++} is removed from the bathing solution, the presynaptic action potential releases fewer than the normal several hundred quanta; the number of quanta released per impulse fluctuates about a mean determined by the Ca^{++} content of the solution. At the lower Ca^{++} concentrations the EPP amplitudes are low multiples of the MEPP, and occasionally no quantum at all is released (Fig. 3-8B). But the size of the quantum itself is unchanged. These experiments leave no doubt that the *presence of* Ca^{++} *ions is absolutely necessary* for the normal release of quanta by a presynaptic action potential.

Addition of Mg^{++} to the bath has an effect similar to that of removing Ca^{++}. Apparently, the Mg^{++} ions compete with the Ca^{++} ions, crowding them away from the critical sites (of action or of entry) on the presynaptic membrane. Since the number of ACh quanta released varies about as the fourth power of the extracellular Ca^{++} concentration, one hypothesis is that four Ca^{++} ions are required for the release of one quantum.

The role of calcium has been studied even more thoroughly in the case of the synapse of the squid giant fiber than in that of the muscle end plate. The following picture has emerged. Depolarization of the presynaptic membrane, by either an action potential or a pulse of current, opens Ca^{++} pores as well as Na^+ pores in the presynaptic membrane, so that Ca^{++} *ions flow into the presynaptic ending*. This process occurs at a threshold depolarization of 30 to 40 mV. The amount of depolarization beyond this level determines the degree of change in Ca^{++} conductance and thus the amount of Ca^{++} that enters the cell. Accordingly, transmitter release increases with the amplitude and duration of depolarization; the calcium diffusing into and through the membrane participates, in an as yet unknown but decisive way, in the process of emptying the vesicles into the synaptic cleft. As in the case of the end plate, Mg^{++} (as well as Mn^{++}) prevents the influx of Ca^{++} into the presynaptic ending and thus the release of transmitter. Calcium is also necessary for normal release of transmitter substance at other peripheral synapses, so that it may be assumed to play similar roles at all chemical synapses.

Neuromuscular block brought about by removing Ca^{++} or adding Mg^{++} cannot be used in humans, unlike the previously mentioned methods, because the functions of other organ systems, such as heart, kidney, central nervous system, and smooth muscles, are severely disrupted by such changes in the ionic medium. The toxin of botulinus bacteria (in spoiled foods) has an effect on the end plate similar to that of Ca^{++} removal. By inhibiting the release of ACh, botulinus-toxin poisoning causes paralysis of the muscles, often fatal because it prevents respiration. Since botulinus toxin is sensitive to heat, an effective protection against poisoning is to cook the suspected food thoroughly.

Generalization of the Quantum Hypothesis. Since spontaneous miniature potentials associated with vesicles were discovered at the end plate, they have also been found at many other chemical synapses. It has even proved possible to isolate synaptic vesicles by ultracentrifugation and to demonstrate that they contain acetylcholine or other substances believed to function as transmitters. It is a permissible hypothesis, then, that at all these synapses the transmitter is stored in the synaptic vesicles and released in quanta, each of which represents the contents of one vesicle. One of these quanta (not to be confused with the physical concept of a quantum of energy) probably contains several thousand transmitter molecules, which are expelled into the very narrow (ca. 0.1 μm) synaptic cleft in an extremely brief volley and act almost simultaneously at the subsynaptic membrane. A quantum released at the frog end plate contains about 1,000 to 10,000 molecules of ACh; the figure for the rat end plate is 4,000 to 20,000 molecules and for the snake, about 10,000 molecules. There are no reliable estimates for other synapses. Nor are there data to suggest whether the randomly occurring individual miniature potentials play any physiologic role.

Except for the processes just described, little is known about the events intervening between the arrival of a presynaptic action potential and the beginning of the postsynaptic potential—that is, those accounting for the synaptic latency. The chief points to remember are that the number of quanta released per impulse depends upon impulse amplitude, and that this affects transmitter release both directly and via control of the Ca^{++} influx. Other aspects of the control of synaptic transmission will be discussed in connection with synaptic potentiation (p. 111 and presynaptic inhibition (p. 97).

Q 3.5 Which of the following statements regarding MEPPs are correct?
  a. MEPPs are caused by the release of *one* molecule of ACh.
  b. The frequency of MEPPs is independent of the membrane potential of the presynaptic terminal.
  c. The time course of MEPPs is similar to that of normal EPPs.
  d. Curare reduces the amplitude of MEPPs or renders them completely undetectable.
  e. Cholinesterase inhibitors do not affect MEPPs.
  f. MEPPs improve synaptic transmission.

Q 3.6 Which of the following factors *increases* the number of quanta of transmitter substance released per presynaptic action potential?
  a. Decrease in the Ca^{++} concentration in the bathing solution.
  b. Addition of cholinesterase inhibitors to the bathing solution.
  c. Increase in the amplitude of the presynaptic action potential.
  d. Addition of curare to the bathing solution.
  e. Decrease in the Mg^{++} concentration in the bathing solution.

Q 3.7 Which of the following findings at the end plate supports the hypothesis that each presynaptic vesicle contains a quantum of transmitter substance?
  a. The presence of Ca^{++} is necessary for the release of ACh.
  b. MEPPs occur at the end plate.
  c. MEPPs are generated in a random sequence.
  d. The transmitter substance is always released in multiples of a minimum amount.
  e. Decrease in the presynaptic resting potential increases the frequency of the MEPPs.

3.3 Central Excitatory Synapses

With the neuromuscular junction as an example, we have learned the basic events that take place during the activation of a chemical synapse. We are now in a position to examine the somewhat more complex events occurring during excitatory transmission at central neurons. While each muscle fiber possesses only one end plate, and while each EPP is normally far above the threshold, central neurons usually possess many dozen to several thousand synapses, and the excitatory postsynaptic potentials of individual synapses are almost always subthreshold; consequently, only the simultaneous activity of a large number of excitatory synapses can give rise to a propagated action potential. In addition to the excitatory synapses, there are inhibitory synapses on the soma and the dendrites of neurons. Activation of the inhibitory synapses impedes the generation of a propagated action potential.

The motoneurons in the ventral horn of the spinal cord gray matter, whose nerve fibers (motor axons) leave the spinal cord through the ventral roots and innervate skeletal muscle fibers, have proved particularly suitable for the study of neuronal synaptic potentials. They are large (diameter of soma up to 100 μm), are relatively easily accessible, and their excitatory and inhibitory connections are well known. Moreover, the results obtained from the motoneurons may apply without any great reservations to the majority of other central neurons. Consequently, these results will form the basis of our discussion.

Excitatory Postsynaptic Potentials (EPSPs). Figure 3-9 shows diagrammatically that the surface of a neuron, with the exception of the

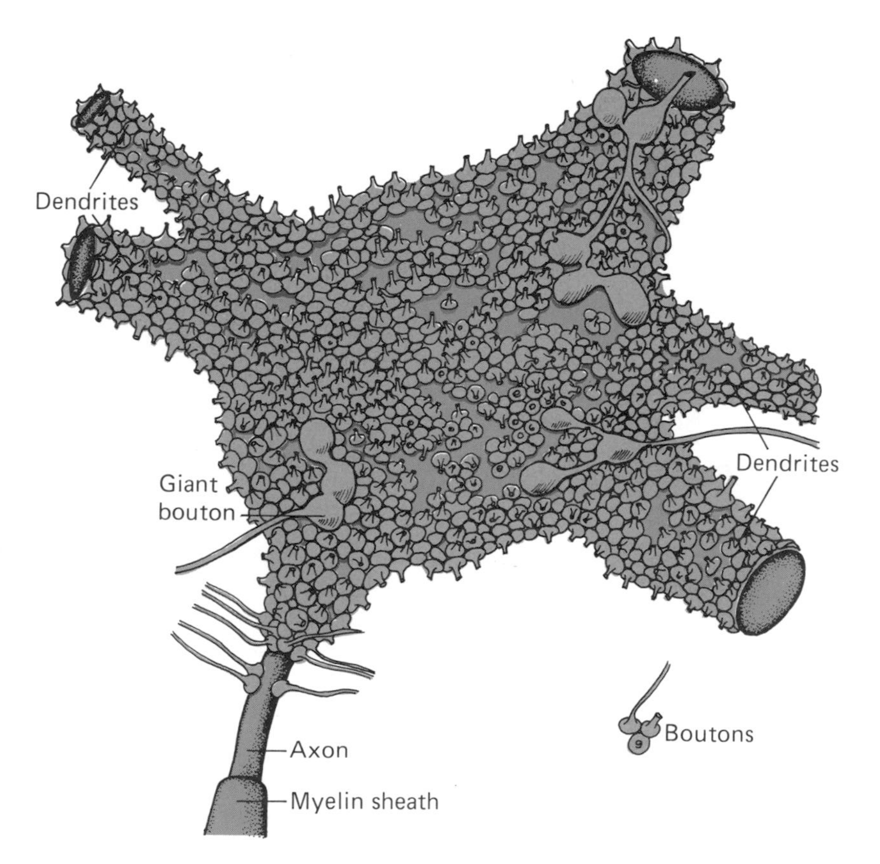

Fig. 3-9. A much simplified, diagrammatic representation of the synapses on motoneurons. Only the bases of the dendrites are shown; on this scale, they would extend far beyond the edges of the page. The soma and dendrites, with the exception of the axon hillock, are almost entirely covered by synapses; some of these, together with their axons, are shown in *red*.

axon hillock and the axon, may have a large number of synapses distributed over it. It is estimated that each motoneuron possesses a total of about 6,000 axosomatic and axodendritic synapses. Some of these synapses are excitatory and some are inhibitory. Their presynaptic axons stem, for the most part, from central neurons. The structure of these synapses resembles that in Fig. 3-1; that is, they are *chemical synapses.*

However, a small fraction of the axons of excitatory synapses come directly from stretch receptors of the skeletal muscles, the muscle spindles; these axons are afferent nerve fibers that enter the spinal cord from the muscle nerves by way of the dorsal roots. As will be shown in detail later (Sec. 4.2), the muscle spindle afferents form direct excitatory synapses only with motoneurons of their own (homonymous) muscle. This situation makes it possible to activate excitatory synapses of a motoneuron by peripheral electric stimulation of the associated muscle nerve. The postsynaptic events can then be observed by means of an intracellular microelectrode.

Such an experimental arrangement is shown on the left in Fig. 3-10. A microelectrode is inserted into the soma of a motoneuron, and the associated muscle spindle afferents are placed on stimulating electrodes at the periphery. The microelectrode records a resting potential of −70 mV. When the afferents are stimulated electrically (arrows in

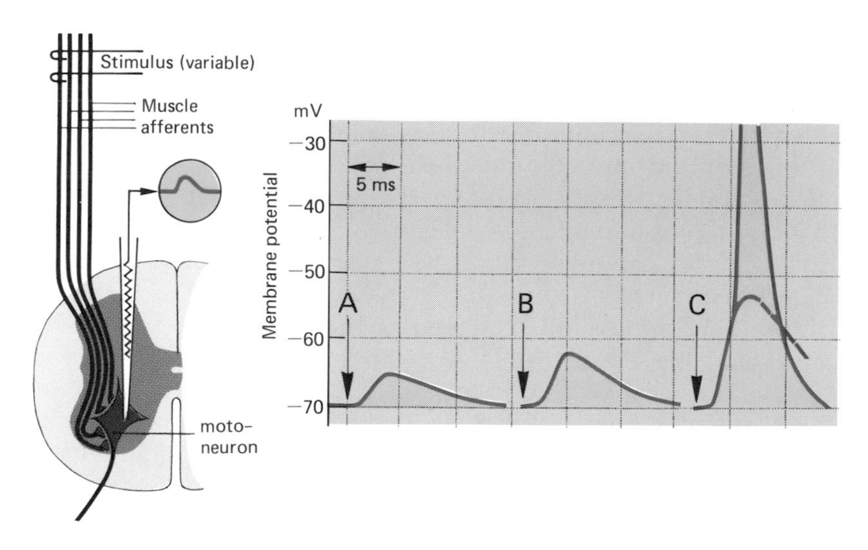

Fig. 3-10. Excitatory postsynaptic potentials, recorded intracellularly from a motoneuron. Afferents in the peripheral nerve from the associated muscle are stimulated electrically at intensities increasing from *A* to *C*.

A,B,C) a depolarization of the membrane potential occurs after a short delay (latency). The time course of the depolarizations in Figs. 3-10 A and B (not C!) is similar to that of the EPP. The amplitudes depend on the number of excited afferents, that is, on the strength of the stimulus in the case of electrical stimulation. The nerve was less strongly stimulated in *A* than in *B*. In *C* the nerve was stimulated even more strongly than in *B*, and the depolarization was so large that a propagated action potential occurred.

Since the depolarizing potentials can excite the motoneuron, they are called *excitatory postsynaptic potentials* (*EPSPs*). The EPSPs are thus analogous to the EPPs at the neuromuscular junction. Whereas the EPP is generated by the activation of an individual synapse, namely, the neuromuscular junction, the EPSPs are usually caused by the simultaneous activation of several or many synapses.

The rising phase of an EPSP lasts approximately 2 ms, and the decline, approximately 10 to 15 ms. As A and B in Fig. 3-10 show, the time course is independent of the amplitude of the EPSP. This means that the EPSPs elicited simultaneously at various synapses summate and do not influence each other's time course. (The unitary EPSPs are independent of each other only within certain limits, which need not be discussed here.)

The Ionic Mechanism of the EPSP. Remember that the EPP is caused by a brief increase in conductance to small cations (Na^+, K^+). Since the EPSPs correspond in many repects to the EPPs, it is assumed that the EPSPs are also generated by brief *increases in conductance to small cations*. (In addition, there also seems to be increased conductance to Cl^- ions.) This assumption is based, among other things, on the fact that the equilibrium potential of the EPSP is at approximately the same membrane potential as that of the EPP, that is, at approximately −15 mV. From the time course of the EPSP and the membrane time constant of the motoneurons, it was possible to calculate that the change in conductance to small cations lasts for about as long as the corresponding permeability change at the activated neuromuscular junction, namely 1 to 2 ms. The unknown (see Sec. 3.5) transmitter substance thus acts for about the same time at the subsynaptic membrane of the motoneuron as the ACh acts at the neuromuscular junction. (Many pharmacologic tests have shown beyond all doubt that the transmitter substance of the motoneuronal EPSP is certainly not ACh.)

The Generation of the Action Potential. When the EPSPs are above the threshold, propagated action potentials are generated (Fig. 3-

10C). It has been found that the membrane of the motoneuron has its *lowest threshold* at the point where the axon leaves the soma, the *axon hillock* (see Fig. 3–9). The threshold of the soma and the dendrites is at least twice as high as that of the axon hillock. Propagated action potentials are, therefore, generated at the axon hillock in motoneurons and probably also in other, if not all, nerve cells. The advantage of the higher threshold of the soma and the dendrites compared with the axon hillock is that no matter which of the synapses are activated at any one time, all the excitatory postsynaptic potentials have a common site of action, namely, the axon hillock. Since the axon hillock merges into the axon, this arrangement also guarantees that once an action potential has been evoked it will continue into the periphery, regardless of the conditions obtaining at the soma and the dendrites. Seen in this way, it is of no importance for the function of the nerve cells whether or not the action potential propagates into the soma and the dendrites.

Since the EPSPs are not actively propagated but spread out electrotonically on the cell membrane, one would expect the axo-somatic synapses in the vicinity of the axon hillock to have a greater influence on the excitability of a motoneuron than axo-somatic and axo-dendritic synapses located further away. To some extent this is true, but the imbalance seems to be partially compensated by the relatively large amplitudes of the EPSPs generated at the dendrites. (This is probably not brought about by an increased release of transmitter substance but by the cable properties of the dendrites; that is, the cause is on the postsynaptic side.) The views of neurophysiologists on the relative importance of axo-somatic and axo-dendritic synapses are, however, still very much at variance.

Other things being equal, the *excitability of a motoneuron is greater, the smaller the neuron,* and this is probably true of other neurons as well. The reason for this correlation is that the membrane of a small cell has a greater total electrical resistance than that of a large cell; the specific membrane resistance (in ohm/cm^2) may be similar for the membranes of both, but a smaller membrane area offers fewer equivalent parallel resistive paths, and thus presents a higher total resistance. As a result, a given ionic current entering through the subsynaptic membrane causes a greater membrane-potential change (proportional to current times resistance) when it flows out through the remaining postsynaptic membrane of a small cell than it does with a large cell, and thus produces a larger EPSP in the small cell. An important consequence is that small motoneurons and the associated muscle fibers are active much more often during the course of a lifetime than large neurons and fibers.

It should be noted, further, that EPSPs of the type just described also occur in other neurons of the CNS. Occasionally, somewhat shorter and longer time courses, both of the rise and the decay of the EPSP, have been observed; taken together, the data currently available suggest that the EPSPs of motoneurons tend to be shorter than most other EPSPs.

Electrical Synapses. Electrical synapses in which presynaptic and postsynaptic membranes are not separated by a wide synaptic cleft but instead are more closely joined together in an electrically conducting manner have been observed occasionally in the nervous systems of invertebrates (crustaceans) and lower vertebrates (goldfish, birds) but so far not in mammals. Therefore, their properties will not be discussed here. It is very likely, however, that electrical synapses exist in mammalian nervous systems as well. There is electron-microscope evidence, at least, in support of this possibility. So-called "*gap junctions*" between neurons have been observed in various parts of the brain; the morphology of these is indicative of an electrical synapse.

Q 3.8 At approximately what membrane potential is the equilibrium potential of the EPSP situated?
a. At −80 mV.
b. At −15 mV.
c. At +40 mV.
d. At +100 mV.
e. The EPSP has no equilibrium potential.

Q 3.9 What area of a motoneuron (motor ventral-horn cell) has the lowest threshold for a propagated action potential?
a. The dendrites.
b. The soma.
c. The axon hillock.
d. The axon.
e. All these areas have the same threshold.

Q 3.10 The total duration of a motoneuronal EPSP is approximately
a. 500 ms.
b. 100 ms.
c. 15 ms.
d. 2 ms.
e. 0.2 ms.

Q 3.11 Which of the following events takes place at the subsynaptic membrane of an excitatory synapse of a motoneuron during the action of the excitatory transmitter substance?
a. Increase in Na^+ permeability.
b. Decrease in K^+ permeability.
c. The permeability to cations remains unchanged.
d. Large anions pass through the membrane.
e. Local hyperpolarization.

3.4 Central Inhibitory Synapses

In addition to the local and propagated excitatory processes in the nervous system that have been described thus far in Chapters 2 and 3, there are phenomena that reduce the activity of the neuronal structures involved. In relatively few cases is this reduction the consequence of preceding excitation such as, for example, the refractory phase that follows an action potential. Much more important are the active processes that lower the excitatory state of the neurons. The former type of process is called *depression*, and the latter is called *inhibition*. The most important form of inhibition is synaptic in nature. That is, activation of the synapse produces not an enhanced but a diminished level of excitation in the postsynaptic membrane. These junctions are called inhibitory synapses.

The importance of inhibitory processes in the normal functioning of the CNS can be illustrated clearly by the following experiment: if several milligrams of strychnine (a drug that blocks inhibitory synapses but leaves excitatory synapses quite unaffected) is injected into an animal, within a few minutes severe convulsions set in, which finally lead to death. It is scarcely possible to demonstrate in a more impressive way that inhibition is a fundamental process of central nervous activity and is just as important as excitation.

We know of two types of inhibition via chemical synapses. In the case of *postsynaptic inhibition*, the excitability of the soma and the dendrites of the neurons is reduced, while in the case of *presynaptic inhibition* the release of transmitter substance at the presynaptic terminals is either reduced or abolished. Postsynaptic inhibition seems to play the more important role in the CNS of vertebrates. Presynaptic inhibition apparently occurs mainly in the presynaptic terminals of somatic and visceral afferents and not so much in the rest of the nervous system.

Inhibitory Postsynaptic Potentials in the Motoneuron. It has long been known from measurements of reflex contractions that stimulation of muscle-spindle afferents not only excites the homonymous motoneurons (Fig. 3-10) but also, at the same time, inhibits the motoneurons of the antagonistic muscle. For example, stimulation of the muscle spindle afferents of the biceps muscle, which flexes the elbow, simultaneously inhibits the antagonist, the triceps, which extends the elbow. Details of this reflex pathway will be given in the next chapter.

Figure 3-11 was recorded with an experimental arrangement similar to that illustrated on the left in Fig. 3-10 and shows the changes in membrane potential recorded with a microelectrode in a motoneuron (red) when antagonistic muscle-spindle afferents are stimulated. The

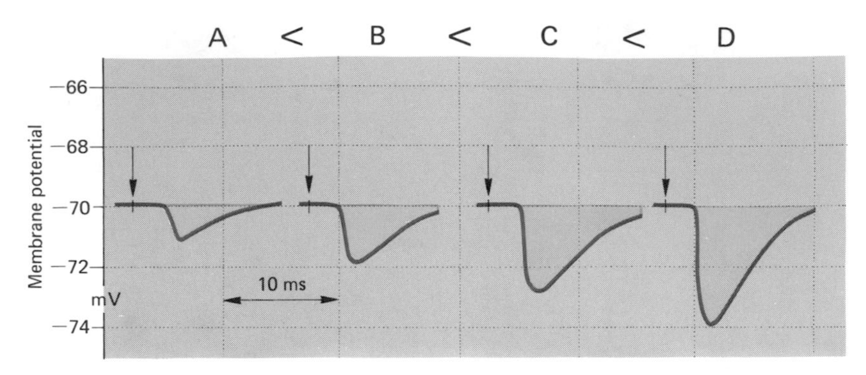

Fig. 3-11. Inhibitory postsynaptic potentials. Experimental arrangement as in Fig. 3-10, except that here an antagonist nerve is stimulated.

resting potential of the motoneuron is −70 mV. At arrows A to D, the antagonistic muscle nerve is stimulated with increasing intensity. Every stimulus triggers a *hyperpolarizing potential change,* and with the selected stimulus strengths, the maximum amplitudes of the hyperpolarizations increase in steps of 1 mV. It can be seen that the time course of the potential change is independent of the amplitude of potential change and is very similar to the time course of the EPSP.

Due to the hyperpolarization, the membrane potential is shifted away from the threshold for initiating propagated action potentials, that is, the motoneuron is inhibited. The hyperpolarizations recorded in Fig. 3-11 are, therefore, termed *inhibitory postsynaptic potentials,* or *IPSPs* for short.

Ionic Mechanism of the IPSP. The time course of an IPSP is practically a mirror image of that of an EPSP, with a rise time of 1 to 2 ms and a decline of 10 to 12 ms. It can be concluded from this result that here, as at the end plate and the excitatory synapse, the subsynaptic change in membrane conductance that leads to the appearance of the IPSPs is short and lasts about 1 to 2 ms. This assumption has been confirmed by further experiments and mathematical analysis of the time course.

The most important experimental method of determining the ionic mechanism of the IPSP is to measure its *equilibrium potential*—that is, the membrane potential at which activation of the inhibitory synapses fails to evoke a potential change. The experimental setup is shown in the left half of Fig. 3-12 and is similar to that in Fig. 3-6. In this case, however, the two single microelectrodes are replaced by a double-barreled microelectrode, for it would otherwise be practically

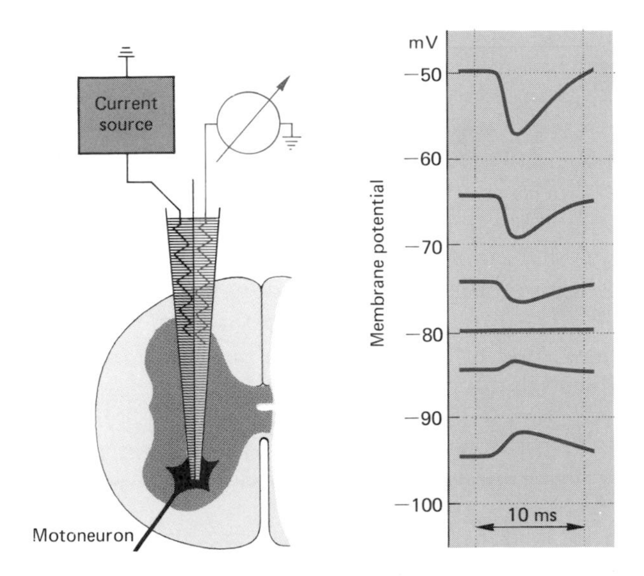

Fig. 3-12. Equilibrium potential of the IPSP, measured in an experiment like that of Fig. 3-10. In this case a double-barreled microelectrode is used; through the second barrel current can be introduced to change the membrane potential of the motoneuron.

impossible to insert both electrodes into the same motoneuron located several millimeters beneath the surface of the spinal cord.

It can be seen that an IPSP triggered at a membrane potential of −65 mV (the resting potential of this cell) has an amplitude of 5 mV in the hyperpolarizing direction. If the membrane potential is lowered by a current applied through a microelectrode, the amplitude of the IPSP (with the same peripheral stimulus) increases greatly. This means that the driving force for the IPSP becomes greater during depolarization of the membrane potential. Conversely, when the membrane potential is increased, the amplitudes of the IPSPs initially become smaller and finally drop to zero at approximately −80 mV. At still higher membrane potentials, the direction of the IPSP is reversed. The equilibrium potential of the IPSP, E_{IPSP}, is thus situated at −80 mV. Since the equilibrium potential for K^+, E_K, is at about −90 mV and since that for Cl^-, E_{Cl}, is at the resting potential, the equilibrium potential of the IPSP, E_{IPSP}, is situated approximately halfway between the two.

The value of the equilibrium potential of the IPSP thus indicates that during the action of the inhibitory transmitter on the subsynaptic membrane, there is a *large increase in the conductance for* K^+ *and*

Cl^- ions. Further experiments confirmed this conclusion. One such experiment involved electrophoretic injection into the cells of large numbers of anions and cations through multibarreled microelectrodes followed by measurements of the resulting changes in the IPSP. It was found that small cations as well as small anions can pass through the activated subsynaptic membrane of inhibitory synapses, while ions whose diameter is larger than that of the hydrated K^+ ions (for example, Na^+ ions) cannot. From these and similar test results at other synapses, the general theory has been developed that the transmitters open *pores* of a certain width at the subsynaptic membranes. All ions with a diameter smaller than that of the pores can pass through. If, in addition, the wall of the pore is electrically charged, this charge acts as a diffusion barrier for ions of the same polarity. Thus, a negatively charged pore would let through only cations, and no anions, and a positively charged pore would let only anions through.

Inhibitory Effects of the IPSP. As stated above, the hyperpolarization occurring during the IPSP shifts the membrane potential away from the threshold for a propagated action potential and thus inhibits the neuron. The inhibitory effects of the IPSP will now be examined in more detail, and particular attention will be paid to the question of whether the hyperpolarization is solely responsible for the inhibition. If this were so, then an IPSP triggered at E_{IPSP}, that is, at -80 mV, would have no inhibitory effect at all on the neuron!

A typical IPSP and EPSP can be seen in Fig. 3-13A. The maximum amplitudes are 2 and 3 mV, respectively. Figure 3-13B shows how the amplitude of the EPSP behaves when it is triggered at different times during the IPSP. If the EPSP is evoked 5 or 3 ms after the start of the IPSP (center and right-hand recordings in *B*), then its maximum amplitude is exactly as large as that of the control EPSP in A. The inhibitory effect of the IPSP is dependent in this case solely on the shift in the membrane potential in a hyperpolarizing direction, that is, away from the threshold for a propagated action potential. If, however, as in the left-hand recording, the EPSP is evoked in the first millisecond after the start of the IPSP, the resulting EPSP is smaller than the control EPSP. Thus, simple addition, such as can be observed at later points in time, does not occur. That is, *during the action of the inhibitory transmitter substance on the subsynaptic membrane, the inhibitory effect of the IPSP is greater than during the passive electrotonic decline of the IPSP to the resting potential.*

The sketches in Fig. 3-13C show the reason for the different effectiveness of the IPSP during and after the active phase. In the left-hand sketch, the excitatory and the inhibitory synapses are activated at ap-

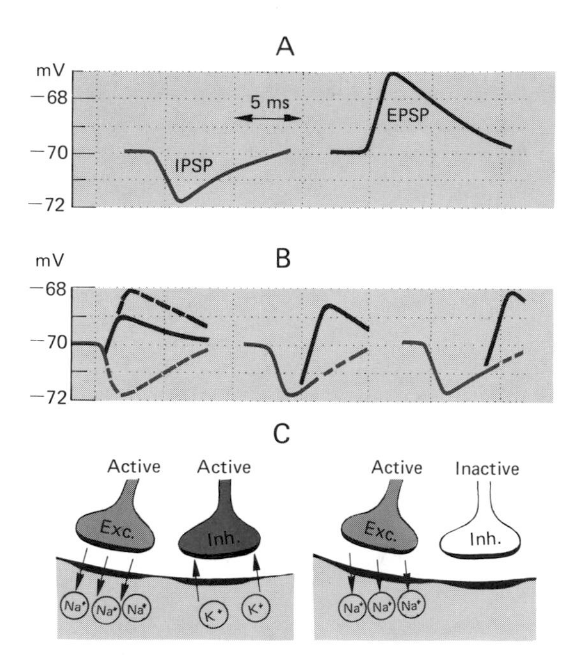

Fig. 3-13. Effect of IPSP on EPSP. Experimental arrangement as in Figs. 3-10 and 3-11. Stimulation of the antagonistic nerve gives the IPSP in **A**; stimulation of the homonymous nerve gives the EPSP. In **B** the EPSP was triggered about 1, 3, and 5 ms after the onset of the IPSP. **C** is a diagram of the subsynaptic conductance changes occurring when excitatory and inhibitory synapses are activated simultaneously (*left*) and when only the excitatory synapse is activated.

proximately the same time, and the inflow of Na^+ ions at the subsynaptic membrane of the excitatory synapse is partially compensated for by the simultaneous outflow of K^+ ions at the inhibitory synapse. The resulting change in potential in a depolarizing direction is, therefore, smaller than at the moment shown at the right in Fig. 3-13C, where the inhibitory synapse is inactive. The increase in conductance beneath the activated inhibitory subsynaptic membrane can also be regarded, conversely, as a decrease in the resistance of the membrane. For a given membrane current (for example, the Na^+ ions in Fig. 3-13C), this reduced resistance leads to a smaller potential change at the membrane.

According to the description given so far, the Cl^- ions play only a small role in the generation of the IPSP. This is correct so long as the IPSP is initiated at the normal resting potential because E_{Cl} is situated at the resting potential. However, if the IPSP starts from a depolarized membrane potential (for example, during an EPSP), the increased Cl^-

permeability will give rise to an increased inflow of Cl^- ions, contributing to the larger IPSPs, which can be seen, for example, in Fig. 3-12.

We can now answer the question posed at the beginning of this discussion, whether the activation of inhibitory synapses fails to produce inhibition when the membrane potential equals the equilibrium potential of the IPSP. By definition, in this case such activation does not in itself lead to a potential change. However, during the increase in subsynaptic conductance the cell is inhibited by the reduced resistance of the membrane. During this time each shift in charge is at least partially compensated by an opposite shift in charge, occurring at the inhibitory subsynaptic membrane (see Fig. 3-13C). When a large number of inhibitory synapses are repetitively and asynchronously activated, the postsynaptic membrane can be effectively short circuited by the large increases in conductance that occur, so that even large excitatory currents produce only small depolarizations.

The excitatory and the inhibitory synaptic events occurring at the membrane of central neurons are summarized in Fig. 3-14. The E_{EPSP} is situated at approximately −15 mV. Activation of the excitatory subsynaptic membrane thus leads to a depolarization that may perhaps reach the threshold (EPSP on left), evoking a propagated action potential at the axon hillock. The equilibrium potential of the IPSP is about

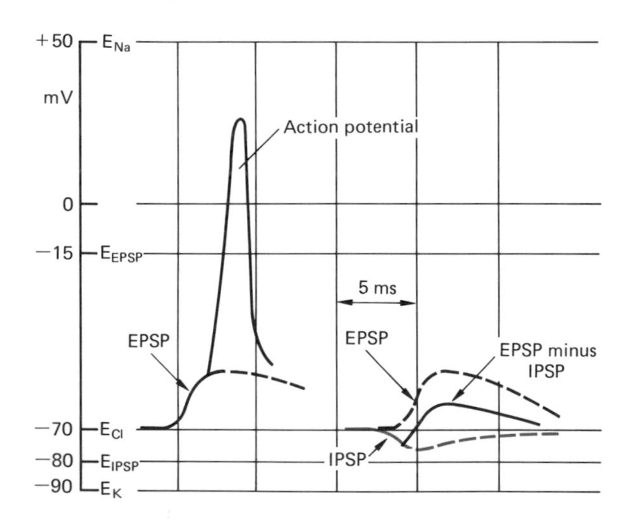

Fig. 3-14. The effect of an IPSP on the action potential; experimental arrangement as in Fig. 3-13. The homonymous nerve is stimulated strongly enough to produce a suprathreshold EPSP (*left*). On the *right*, the antagonist nerve is stimulated about 3 ms before the homonymous nerve. The equilibrium potentials of Na^+, K^+, Cl^-, EPSP, and IPSP are shown.

−80 mV. The conductance to K^+ and Cl^- ions is increased beneath the activated inhibitory subsynaptic membrane, and this results in a hyperpolarization (red curve on right). Because of the IPSP, the EPSP no longer reaches the threshold, and the cell is inhibited.

Presynaptic Inhibition. In the case of presynaptic inhibition, there is no direct inhibitory effect upon the postsynaptic membrane, but instead the inhibitory process brings about a *reduction in the release of transmitter substance* at the presynaptic terminal of the excitatory synapse—an event similar to that with which we are familiar at the end plate when Mg^{++} is added or when poisoning by botulinus toxin has occurred. Presynaptic inhibition is evoked by *activation of axo-axonal synapses.*

Figure 3-15 shows the structure of an axo-axonal synapse and its effect on the postsynaptic EPSP; here axon 1 forms an axo-somatic synapse with neuron 3, while axon 2 forms an axo-axonal synapse with axon 1. As indicated by the arrangement of the synaptic vesicles and the thick portions of the subsynaptic membranes, axon 1 is presynap-

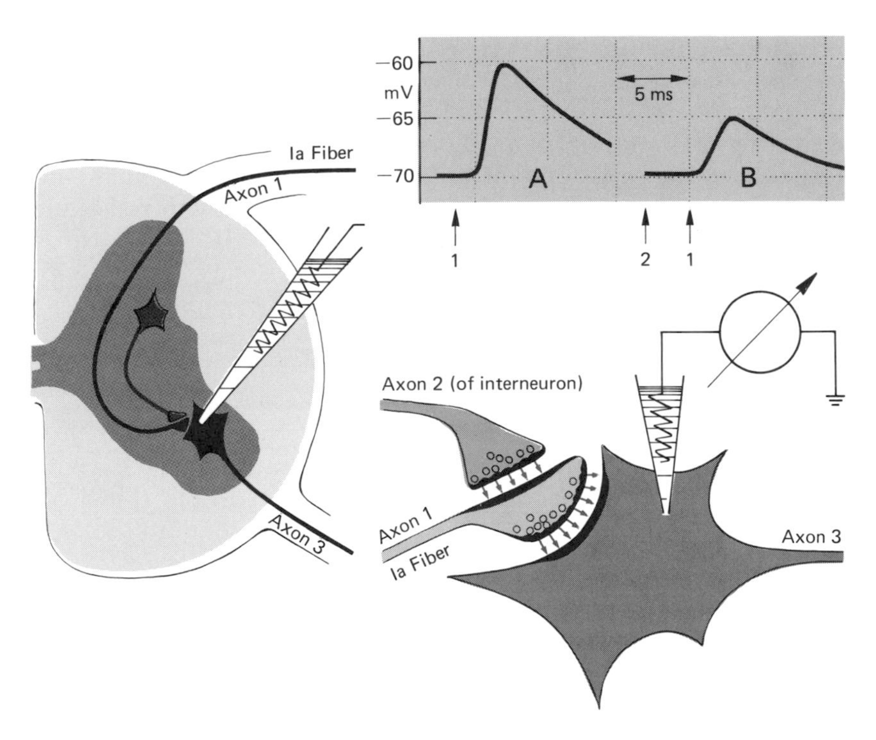

Fig. 3-15. Presynaptic inhibition of a motoneuron. See text for further explanation.

tic to neuron 3, and axon 2 is presynaptic to axon 1. Activation of the synaptic ending 1 (arrow in Fig. 3-15A) evokes an EPSP of approximately 10 mV in neuron 3. The axo-somatic synapse is thus an excitatory synapse. If axon 2 is activated before axon 1 (arrows in Fig. 3-15B), the amplitude of the EPSP is only 5 mV, although no IPSP occurs at the postsynaptic membrane of cell 3. This form of EPSP inhibition *without* any change in the properties of the postsynaptic membrane is called *presynaptic inhibition*. The duration of this inhibition is about 100 to 150 ms—much longer than that of the IPSP.

It is not known for sure what subsynaptic changes in permeability are triggered by activation of the axo-axonal synapse, and what transmitter substance is released at axon 2. There is, however, good evidence that the transmitter substance of axo-axonal synapses in the spinal cord is gamma-aminobutyric acid (GABA, see Sec. 3.5), and that it causes an increase in the conductance for Cl^- ions. The convulsant drug *bicucullin*, for example, a specific antagonist of GABA, inhibits presynaptic inhibition (see the effect of strychnine on postsynaptic inhibition, Sec. 3.5).

With regard to the *mechanism by which excitatory transmitter release is reduced*, at present it seems most likely that it involves reduction of the amplitude of the presynaptic action potential in ending 1. Depolarization of axon 1 can, in fact, be observed during presynaptic inhibition. Since depolarization of the membrane potential leads to a reduced action–potential amplitude, and since the release of transmitter substance is partly dependent on this amplitude, an action potential invading ending 1 during presynaptic inhibition will release less transmitter substance and thus evoke a small EPSP. In an extreme case (for example, repeated activation), ending 1 may become so depolarized following activation of the axo-axonal synapse that it is no longer possible for an action potential to be propagated; therefore, only a very reduced electrotonic image of the action potential reaches the axon terminal, releasing little or no transmitter.

In the mammalian central nervous system (including that of humans), presynaptic inhibition occurs predominantly at the excitatory synapses formed by the endings of afferent fibers in the spinal cord. For example, if axon 3 in Fig. 3-15 comes from a motoneuron, axon 1 is an afferent fiber from a muscle spindle (Ia fiber) and axon 2 comes from an intermediately situated neuron in the gray matter of the spinal cord (an intraspinal interneuron). The *functional significance* of the presynaptic inhibition of primary afferents (that is, fibers from peripheral receptors that enter the spinal cord via the dorsal roots) lies chiefly in control of the sensory signals arriving from the periphery. The ability to inhibit impulses sent by receptors into the nervous system at the earliest possible point—before these impulses have had any excitatory effect on the nervous system—is utilized by the body in various ways. It permits modulation of the sensitivity of the afferent channels, so that undesired information is suppressed and desired information selected (for example, when the attention is being directed toward something); the mechanism is also useful in contrast enhancement (see Fig. 4-4C). Finally, presynaptic inhibition is able to reduce or eliminate selectively some of the inputs of a nerve cell without affecting others.

Q 3.12 At approximately what membrane potential is the equilibrium potential of the IPSP situated?
a. At −90 mV.
b. At −70 mV.
c. At −15 mV.
d. At +40 mV.
e. None of these values is correct.

Q 3.13 The total duration of an IPSP is
a. 1–2 ms.
b. 10–15 ms.
c. 100 ms.
d. 200 ms.
e. 1,500 ms.

Q 3.14 An IPSP inhibits a neuron because
a. it hyperpolarizes the membrane potential.
b. it leads to a reduction in the amount of transmitter substance released at excitatory synapses.
c. it changes the threshold of the neuron.
d. it increases the conductance of the membrane.
e. it shortens the action potential considerably.

Q 3.15 Depolarization of a nerve cell to −50 mV
a. considerably shortens the duration of the EPSP.
b. increases the amplitude of the IPSP.
c. impedes generation of an EPSP.
d. increases the amplitude of the EPSP.
e. leaves EPSP and IPSP unchanged.

Q 3.16 Presynaptic inhibition of a motoneuron
a. is caused by activation of an axo-axonal synapse.
b. involves prolongation of the IPSP.
c. causes no change in excitability of the motoneuronal membrane.
d. cannot be affected by pharmacologic agents.
e. results from a reduced release of transmitter at the primary afferent fiber (Ia fiber).

3.5 The Transmitter Substances at Chemical Synapses

It is known which substances act as transmitters at peripheral synapses of the nervous system—for example, the neuromuscular end plate (see Sec. 3.1) and synapses in sympathetic ganglia (see Chapter

8). An example of the way this knowledge is applied clinically has been given in the discussion of neuromuscular block in Sec. 3.1. By contrast, our information about the transmitters at central synapses (spinal cord, brainstem, cerebrum, cerebellum) is still quite fragmentary. Results are accumulating rapidly, however, in this very active area of research. Since synapses, as *modifiable control points in neuronal networks,* are particularly *susceptible to the action of drugs and chemicals,* results in this field not only are of theoretical interest but have practical clinical applications. There have already been cases (for example, Parkinson's syndrome) in which discovery of the central transmitter involved (dopamine; see below) has suggested directly an appropriate therapy (the administration of the precursor, DOPA).

General Considerations. For all transmitter substances (as described for ACh in Sec. 3.1), the presynaptic endings they occupy have been either shown or assumed to have the capability of synthesizing, storing, releasing, and inactivating them and of taking up their breakdown products. In all cases, of course, transmitters must be able to react with subsynaptic receptor molecules. Most neurons appear to be equipped with only one such system of chemical processes, so that the same transmitter is released at all its endings (no matter how extensively its axon branches). This finding is called *Dale's Principle* after its discoverer. A probable corollary of this generalization is that in the vertebrate CNS each neuron, as a rule, exerts only excitatory or only inhibitory effects. This notion was initially postulated by Eccles, and is known as the *concept of functional specificity.* The concept implies that there are only two basic types of neurons in the nervous system, excitatory and inhibitory neurons.

It is entirely possible for excitatory and inhibitory neurons to use the same transmitter. For example, the ACh liberated by motor axons has an excitatory action on skeletal muscle fibers, whereas the same compound, released by vagus nerve fibers, inhibits the muscle fibers of the heart. It is not the transmitter itself, then, but rather the *properties of the subsynaptic membrane* that determine whether excitation or inhibition results.

ACh as a Transmitter within the Nervous System. Even before leaving the spinal cord, motoneurons send out axon collaterals (branches) that form excitatory synapses in the ventral horn, with interneurons called *Renshaw cells* in honor of their discoverer (see Fig. 4-4 concerning the functional significance of these cells). As Dale's Principle indicates, the transmitter at these synapses is acetylcholine. ACh also acts as the transmitter at certain specifically known synapses of the autonomic nervous system. These will be discussed extensively in

Chapter 8. So far no one has positively identified any other cholinergic synapses in the CNS, although the high content of ACh and its esterase in many parts of the brain suggests their presence.

Adrenergic Transmitters. This group includes adrenalin and noradrenalin, as well as their precursor dopamine. The three transmitters are also categorized as *catecholamines,* and together with serotonin (5-hydroxytryptamine, 5-HT) they are termed *monoamines.* Adrenalin and, to an even greater extent, noradrenalin are important transmitters in the autonomic nervous system, as is described in Chapter 8. In the rest of the CNS as well, more and more synapses are becoming known to release monoamines, and in particular the catecholamines, as transmitters.

The processes involved in biosynthesis, storage, and release of the monoamines are more or less analogous to those for ACh, and will not be discussed further. But the monoamines differ distinctly from ACh with respect to the manner in which *action of the transmitter is terminated* at the subsynaptic membrane; whereas ACh is broken down enzymatically by acetylcholinesterase, the monoamines are removed from their target sites primarily by being *taken back into the presynaptic ending.* This uptake of the transmitter not only puts it rapidly out of operation but also prevents depletion of the presynaptic stores when the neuron is repetitively activated.

An aspect of adrenergic transmission relevant to clinical practice is that it can be affected pharmacologically in a variety of ways at different stages of synthesis, storage, liberation, and inactivation. Examples are given in Chapter 8. Here only one experimental result will be mentioned. It has proved possible by administration of particular drugs to replace the normal physiologic transmitter substances in the presynaptic endings of a few adrenergic synapses with so-called *false* or *substitute* transmitters. The substitute transmitters are relatively or entirely ineffective in transmitting excitation, but the catecholaminergic metabolic systems treat them exactly like the normal substances. As a result, the synapses are completely or partially blocked.

Amino Acids as Transmitters. Certain amino acids occur in particularly high concentrations in the nervous system. This in itself has suggested that they may serve as transmitters. It seems to be emerging from the available data that neutral amino acids such as GABA and glycine act as inhibitory transmitters, while acidic amino acids such as glutamic acid are possibly excitatory transmitters.

As mentioned in the discussion of presynaptic inhibition, there are certain indications that GABA serves as a transmitter at axo-axonal synapses. GABA has been definitely identified as an inhibitory transmitter in crustaceans. *Glycine* is probably the transmitter at certain postsynaptic inhibitory synapses in the spinal cord. Strychnine appears to be a specific antagonist of glycine. Therefore, administration of strychnine produces convulsions as a result of the interruption of postsynaptic inhibition (see the effect of bicucullin at axo-axonal synapses, p. 98).

There may also be *peptidergic transmitters*. Peptides in this provisional category include substance P, enkephalin, angiotensin, vasopressin, and others, which are being found at more and more sites in the nervous system. But a more important function of these peptides may be to act indirectly, by way of the bloodstream and the extracellular fluid as hormones do, so as to change the excitability of single neurons and neuron populations. This type of action is called *neuromodulation*. For example, the administration or release of enkephalin produces an inhibitory modulation of certain neurons, in particular those in the nociceptive ("pain") pathways which are also inhibited by morphine and related compounds. Enkephalin and other substances produced by the body, with a similar action, are therefore also called *endorphins* (endogenous morphines).

Q 3.17 Which of the following statements best describes Dale's Principle?
- a. All neurons of the CNS are either excitatory or inhibitory.
- b. At all motor end plates ACh is released as a transmitter.
- c. A transmitter has either exclusively excitatory or exclusively inhibitory effects everywhere in the nervous system.
- d. Every neuron releases the same transmitter at all its presynaptic endings.
- e. At all catecholaminergic synapses transmitter action is ended chiefly by uptake of the catecholamines into the presynaptic ending.

Q 3.18 Which group of substances constitutes the catecholamines?
- a. Adrenalin and noradrenalin.
- b. Adrenalin, noradrenalin, and serotonin (5-HT).
- c. Adrenalin, noradrenalin, and dopamine.
- d. Adrenalin, noradrenalin, dopamine, and serotonin.
- e. Dopamine and serotonin.

Q 3.19 Strychnine and bicucullin are convulsants. They act by
- a. exciting all the excitatory synapses in the spinal cord.
- b. blocking inhibitory synapses.
- c. being incorporated as substitute transmitters in monoaminergic synapses.
- d. directly exciting the skeletal muscle fibers.
- e. lowering the Ca^{++} level.

4

THE PHYSIOLOGY OF SMALL GROUPS OF NEURONS; REFLEXES

R.F. Schmidt

Axonal impulse conduction and excitatory and inhibitory synaptic transmission are the two fundamental processes of neuronal activity. The complex abilities of the brain are achieved primarily by suitable connections between groups of neurons. In the first part of this chapter, we shall examine some typical neuronal networks that recur often in various regions of the brain. Some of these networks are used to boost weak signals; others serve to suppress overactivity (Sec. 4.1).

Reflex arcs, discussed in Secs. 4.2 and 4.3, are a special basic type of neuronal circuitry. By reflex arcs, we mean complete neuronal networks extending from the peripheral receptor through the CNS to the peripheral effector. They serve to perform reliably and with a minimum of effort the constantly recurring *stereotyped reactions of the organism* to its environment.

4.1 Typical Neuronal Circuits

Divergence. Sooner or later after emerging from the soma, the axons of most neurons divide into a few or many collateral branches that form synapses with several to many other neurons. An example of this *divergence* is given by the afferent fibers of peripheral nerves that enter the spinal cord via the dorsal root and then divide into a large number of collaterals that lead to spinal neurons. A diagram of this process of divergence is given in Fig. 4-1. Divergence serves to make afferent information accessible simultaneously to various sections of the CNS (for example, to the motoneurons, the cerebellum, or the

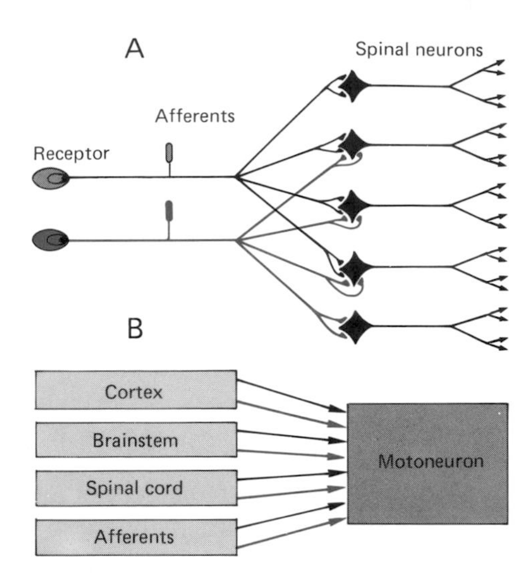

Fig. 4-1. A. Diagram of the divergence of two dorsal-root fibers (afferents) to spinal neurons. The axons of the latter neurons in turn branch into many collaterals. **B.** Diagram of the converging excitatory (*black arrows*) and inhibitory (*red arrows*) inputs to a motoneuron. The motoneuron is the "final common path" for all motor reflexes and commands.

cerebral cortex). The ultimate effectiveness of an action potential coming from a single receptor afferent is considerably magnified as a result of the subdivision of the axon into a large number of collaterals—that is, as a result of divergence. This division of the dorsal root fibers into a large number of collaterals is but one example of divergence, which occurs in nearly all parts of the CNS.

It is extremely difficult to assess divergence quantitatively because histologic and physiologic methods hardly ever permit one to trace all the collaterals of a neuron. However, the motor axons are exceptions worth mentioning here. They emerge from the spinal cord in the ventral roots and lead to the muscles where they divide into collaterals that innervate a varying number of muscle fibers. Since each muscle fiber is innervated by only one collateral, the *average divergence* of each motor axon can be calculated from the number of motor axons and muscle fibers present in a muscle. In humans, values between 1 : 15 (external eye muscles) and 1 : 1,900 (muscles of the limbs) and more are found. (In addition, the motor axons give off collaterals within the spinal cord—that is, before they enter the ventral roots—but we do not know exactly how many.)

Convergence. As might be expected from the prevalence of divergence in the nervous system, most neurons also receive synaptic input from many other neurons. This phenomenon is called *convergence.* Figure 4-1A shows two afferent fibers whose axons diverge in each case to four neurons. As a result, three of the total of five neurons shown are connected to both afferent fibers or, seen the other way round, two afferent fibers *converge* on each neuron. Many dozens to several thousands of axons converge on most neurons in the CNS. Let us take the example of the motoneuron to make this clear. On the average, about 6,000 axon collaterals terminate on the motoneuron. As summarized in Fig. 4-1B, these axon collaterals come not only from the periphery but also from neurons of the spinal cord, the brainstem, and the cerebral cortex. Some form excitatory (black arrows) and others inhibitory (red arrows) synapses.

Since several thousand axon collaterals converge on the motoneurons, it is easy to see that whether or not a motoneuron discharges a propagated action potential depends on the number and the type of the synaptic processes effective at any one time. In this sense, the motoneuron (like many other neurons) processes, or integrates, the excitatory and inhibitory events occurring at its membrane. Long before the discovery of EPSPs and IPSPs the integrative function of the motoneurons was known from studies of muscle contractions after peripheral and central electric stimulation. Around the turn of the century, the English physiologist Sherrington had already described the motoneuron as the *final common path* of the motor system—that is, as the cell that weighs all the excitatory and the inhibitory influences, one against the other. Action potentials are discharged only when the excitatory influences predominate, or in modern terms, for as long as suprathreshold EPSPs occur.

Temporal and Spatial Facilitation. On the left in Fig. 4-2A is shown an experimental arrangement for testing the effect exerted on a neuron by repetitive stimulation of an axon. A single stimulus (arrow, top right) induces a typical EPSP with a total duration of approximately 15 ms. When two stimuli are applied at an interval of approximately 4 ms (arrows, center right) the second EPSP begins before the first has fully decayed. As a result, the cell is more strongly depolarized, and the membrane potential approaches more closely the threshold. A third EPSP occurring 4 ms later (arrows, bottom right in Fig. 4-2A) reaches threshold, and a propagated action potential occurs. EPSPs triggered rapidly one after the other have an additive excitatory effect on a neuron. This type of increase in excitability brought about by succes-

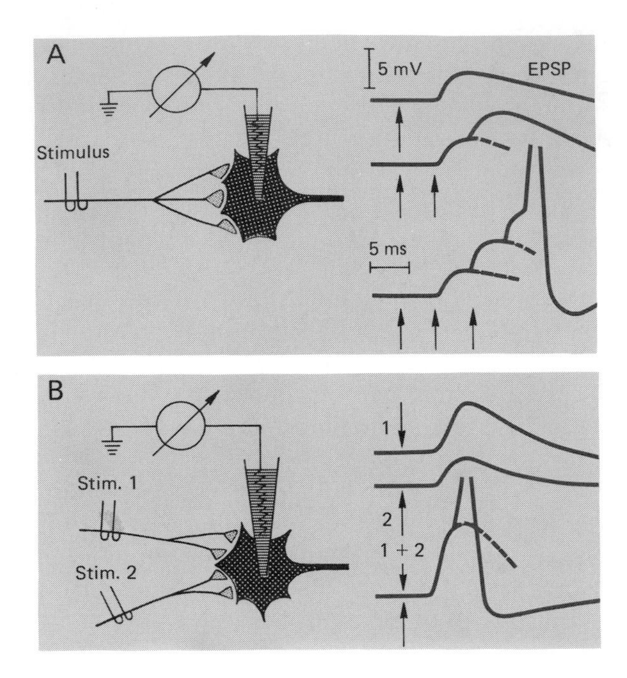

Fig. 4-2. Facilitation in the nervous system. **A.** Temporal facilitation: a single stimulus (*one arrow*) and paired stimuli (*two arrows*, time between stimuli about 4 ms) generate subthreshold EPSPs; the third stimulus (*three arrows*) evokes an action potential. **B.** Spatial facilitation: Stimuli 1 and 2 each generate a subthreshold EPSP, but simultaneous stimulation of both axons evokes an action potential. In the scale of the diagram (1 mm ~ 1 mV; see calibration), the action potential is about 10 cm high, so that only the beginning and end of it are shown.

sive EPSPs is, therefore, termed *temporal facilitation*. That is, the earlier EPSPs facilitate the effectiveness of the action of the later EPSPs in triggering nerve impulses. Temporal facilitation is of great physiologic importance because many nervous processes (for example, receptor discharges) occur repetitively and as a result can summate to produce suprathreshold depolarizations at synapses.

The experimental setup in Fig. 4-2B demonstrates the occurrence of *spatial facilitation:* stimulus 1 alone generates a subthreshold EPSP (top right in B) and stimulus 2 alone (center right) also gives a subthreshold EPSP. But if 1 and 2 are stimulated simultaneously (bottom right), the threshold is reached, and a propagated action potential is elicited. Joint stimulation of 1 and 2 thus leads to a propagated action potential—that is, to a process that could not be evoked by either of the individual EPSPs alone.

The way facilitation operates in a *population of neurons* is shown in Fig. 4-3A to C. The white neurons are unexcited, the light gray neurons receive subthreshold excitation, and the dark gray neurons on a red background are excited above their thresholds. Afferent inputs 1 and 2 each diverge to eight neurons, converging on the four neurons

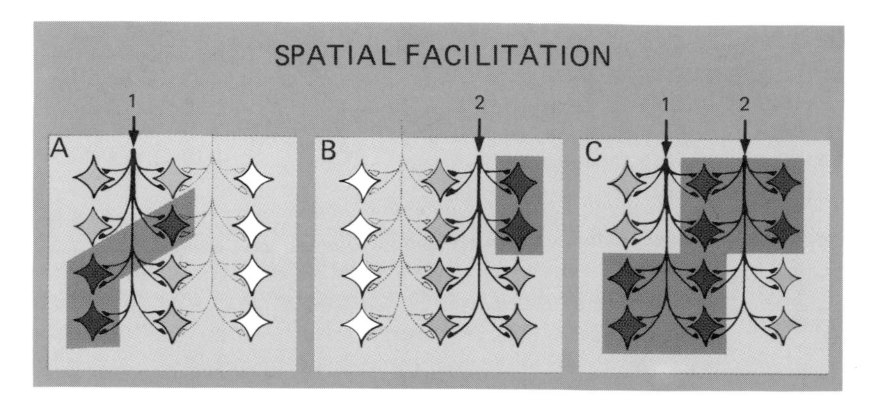

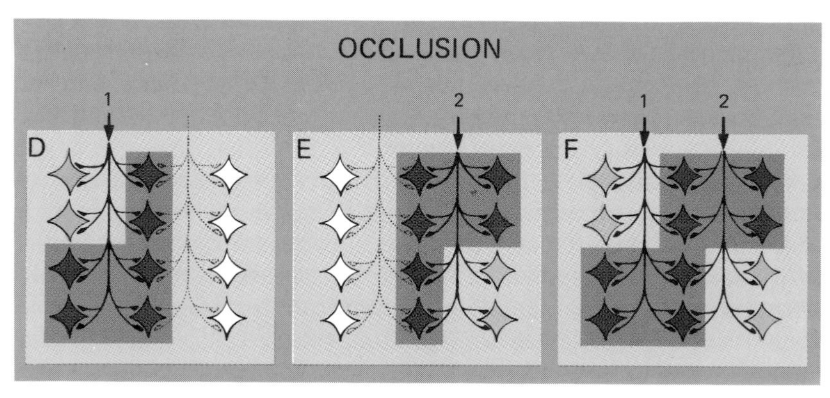

Fig. 4-3. Spatial facilitation. The afferent inputs 1 and 2 each diverge to eight of the total of twelve neurons; both of these inputs converge onto the four neurons in the middle column. **A.** Activation of input 1 produces subthreshold EPSPs in five neurons (light gray) and suprathreshold EPSPs in three neurons (dark gray on red background). **B.** Activation of input 2 results in subthreshold excitation of six neurons and suprathreshold excitation of two. **C.** Simultaneous activation of inputs 1 and 2 provides suprathreshold excitation to all the neurons in the middle column. Thus, the number of neurons excited to above-threshold levels is greater than the sum of those so excited by activation of the inputs singly ($8 > 3 + 2$). **D–F.** Occlusion. If each of the two afferents generates suprathreshold excitation of the neurons with which they both synapse (**D** and **E**), simultaneous activation of both inputs (**F**) produces suprathreshold excitation in a number of neurons smaller than the sum of those excited by activation of one input alone ($8 < 6 + 6$).

in the middle column. Diagram A shows what happens when input 1 is activated; there are subthreshold EPSPs in five neurons and suprathreshold EPSPs in three neurons, one of which is in the middle column. Activation of input 2 produces suprathreshold excitation in only two neurons; the other six, four of them in the middle column, are excited to a subthreshold level. If the two inputs are activated simultaneously (C), the subthreshold influences in the middle column are combined; suprathreshold excitation then appears not only in 3 + 2 = 5 neurons, but in a total of eight neurons. Thus, more neurons undergo suprathreshold excitation than the total number of neurons with suprathreshold excitation in the case of the two single stimuli. As already stated, we call this process *facilitation*, and in this case it is spatial facilitation.

Occlusion. It can also happen, however, that all or nearly all the neurons in a given population receive suprathreshold excitation when only one of two inputs is activated. This situation is illustrated in Figs. 4-3D to F. If input 1 (see D) and input 2 (see E) each produce suprathreshold excitation in six neurons, including the middle four in both cases, simultaneous activation of 1 and 2 (shown in F) excites not 6 + 6 = 12 neurons, but only eight neurons to a suprathreshold level. This phenomenon is referred to as *occlusion*. The process of facilitation illustrated in A to C has thus become occlusion as a result of an increase in the excitability of the neurons involved (for example, via additional excitatory effects acting on the neurons).

The main points are as follows: if the effect of several simultaneous stimuli or of several stimuli occurring in rapid succession is greater than the sum of the effects of the individual stimuli, then we term this *facilitation*. If the response to the stimuli is less than the sum of the responses to the individual stimuli, then the term *occlusion* applies.

One often hears the terms *spatial summation* and *temporal summation* used for the above facilitation phenomena. Strictly, however, the word summation is appropriate only if the response to several stimuli is precisely the sum of the individual responses. This is a rare occurrence in the nervous system, since such inputs often interact with one another (see the discussion of EPSPs and IPSPs in Chapter 3). "Summation" is also used with respect to reflexes (see Sec. 4.3); in this connection, it describes those processes by which, probably as a result of spatial or temporal facilitation, subthreshold stimuli of long duration (for example, a tickle in the nose) eventually elicit an "above-threshold" reflex (sneezing).

Simple Inhibitory Circuits. Pharmacologic interruption of the inhibitory processes of the CNS (by strychnine, tetanus toxin) leads to convulsions and death. Quite clearly, inhibitory circuits serve to suppress superfluous and excessive excitatory effects. This task is taken care of mainly by circuits that act back on the excitation itself: as the response to the original excitation becomes stronger, the excitation is inhibited more strongly. In electronics such circuits are known as "negative feedback circuits." In addition, there are inhibitory circuits that, during an excitatory event, automatically suppress any opposing excitatory events or ensure that an excitation remains unaffected by any neighboring activity. We will now examine various inhibitory circuits characteristic of the CNS.

As mentioned previously, the afferents of the stretch receptors of muscle spindles (called Ia fibers) form excitatory synapses at their homonymous motoneurons and inhibitory synapses with antagonistic motoneurons. In the latter case, an interneuron is interposed between the Ia fiber and the motoneuron. This situation is represented in Fig. 4–4A. The inhibition is termed *antagonist inhibition*. Here (and in B and C) the inhibitory interneurons are shown in red. If, for example, the Ia afferents from the muscle spindles of a flexor muscle are activated (arrows in Fig. 4-4A), they then excite the motoneurons of the homonymous flexor muscle and inhibit the motoneurons of the extensor muscles acting on the same joint. The physiologic implications of this circuitry are discussed later, in connection with Fig. 6-3.

In Fig. 4-4A the antagonist inhibition suppresses excitatory events without influencing the excitatory process by which it was generated.

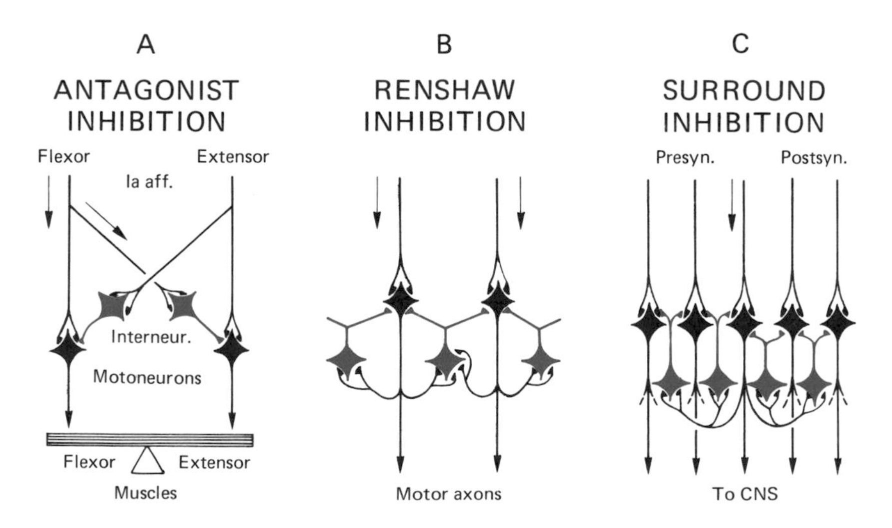

Fig. 4-4. Typical inhibitory circuits. The inhibitory interneurons are shown in *red* in all three neuronal circuits. See text for detailed description.

Since such an arrangement involves no action back on the source of excitation, it is called either *forward lateral inhibition* (if, as in Fig. 4-4A, two or more parallel paths affect one another reciprocally) or *feed-forward inhibition* (if the forward inhibitory path rejoins the original signal pathway "further downstream," after delay or other processing; such circuitry is found, for example, in the cerebellum). Forward inhibition is of common occurrence throughout the nervous system.

On the other hand, the inhibitory interneurons in Fig. 4-4B act back on the cells by which they themselves were activated. In this case, there is negative feedback, and this inhibition is termed *feedback inhibition.* The motoneurons provide a particularly clear example of such a feedback inhibition. As Fig. 4-4B shows, the motor axons give off collaterals to inhibitory interneurons whose axons in turn form inhibitory synapses on motoneurons. This inhibition is called *Renshaw inhibition* after its discoverer, and the inhibitory interneurons are called *Renshaw cells.* As the motoneurons become more excited, the Renshaw cells also become more excited, resulting in greater inhibition of the motoneurons after a short latent period (due to the interneuron). This arrangement guarantees that weak motoneuron activity is transmitted undisturbed to the muscles, while excessive activity is damped to prevent hyperactivity of the muscles or even convulsions.

As with forward inhibition (in which the inhibitory signals arise from the *input* elements), inhibitory feedback effects (in which the inhibitory signals arise from the *output* elements) differ fundamentally depending upon whether the backward inhibition from a cell's output affects the cell itself (feedback inhibition or *self inhibition*) or affects neighboring cells (*backward lateral inhibition*) representing parallel pathways. In Fig. 4-4B, in addition to the self inhibition, lateral effects are shown among "nearest neighbors," but in certain neural arrays the lateral inhibition can be more extensive, as indicated in Fig. 4-4C. The inhibitory interneurons are connected in such a way that they act not only back on the excited cell itself (arrow) but also on many neighboring cells with the same function, in such a manner that these cells are strongly inhibited. This latter type of inhibition is called *backward lateral inhibition;* it ensures that an inhibited zone is generated lateral to the excited zone. Excitation is surrounded on all sides by an inhibited field; therefore, this phenomenon is also referred to as *surround inhibition.* Surround inhibition, or lateral inhibition, plays a particularly important role in afferent systems where it may be mediated by both postsynaptic (Fig. 4-4C, right) and presynaptic (Fig. 4-4C, left) inhibitory synapses. Its important functions are described in detail in *Fundamentals of Sensory Physiology.*

Positive Feedback and Synaptic Potentiation. The great importance of inhibitory circuits for the normal funcioning of the CNS has been demonstrated experimentally on many occasions, and they are a generally accepted feature. On the other hand, the view is often advanced—and is the object of much dissent—that the CNS also contains positive feedback circuits; these, by feeding back excitation to already excited cells, would cause the excitation either to be reinforced or to *reverberate.* An excitatory feedback circuit of this type is illustrated in Fig. 4-5. It could serve to maintain an induced activity for a long time. Many people say that short-term memory is due to the reverberating of excitation in such positive feedback circuits, yet there is almost no experimental evidence to support this (see Sec. 9.4). For the present, therefore, the question must remain unanswered as to whether or not positive feedback circuits exist in the CNS, and if they do, what physiologic importance they might have.

There is another, better known way of facilitating the repetition of neuronal activity. Repeated use of a synapse often results in a considerable increase in the synaptic potentials, a phenomenon referred to as *synaptic potentiation.* This phenomenon often appears during repeated ("tetanic") stimulation and is then called *tetanic potentiation.* The potentiation may outlast the stimulus sequence, or set in only after the end of the tetanus, so that it is then called *posttetanic potentiation.* For example, in the experiment illustrated in Fig. 4-6, single stimuli gave EPSPs of approximately 1 mV amplitude (control values at left). After brief tetanic stimulation (duration 30 s, frequency 640/s = 19,200 stimuli), the amplitude of the EPSP was approximately 2 mV—that is, about twice the initial value. This posttetanic potentia-

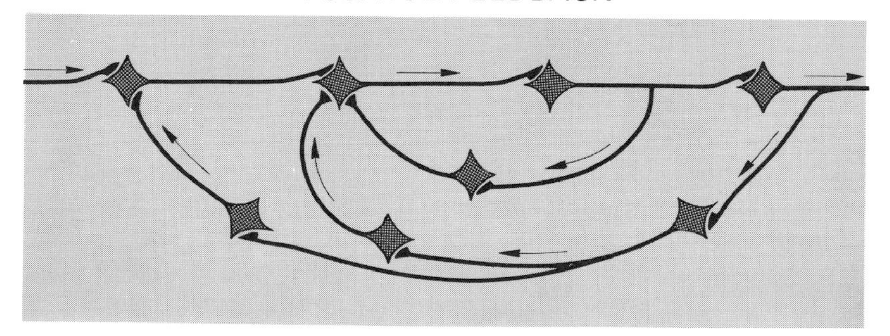

Fig. 4-5. Neuronal circuit giving positive feedback. If suitably dimensioned, this hypothetical circuit could generate reverberating excitation.

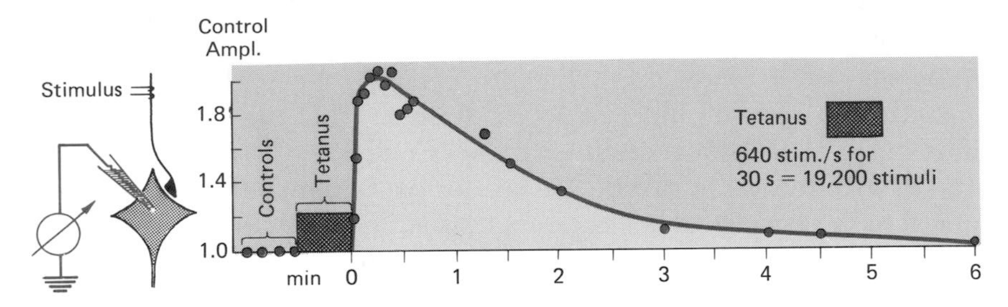

Fig. 4-6. Posttetanic potentiation. The experimental arrangement is diagrammed at the *left*. The presynaptic axon is stimulated at regular intervals except for the period of tetanus, during which the stimulation frequency is increased to 640 Hz for 30 s (a total of 19,200 stimuli). The abscissa of the graph shows the time after the end of tetanic stimulation; the ordinate shows the amplitude of the EPSP as a multiple of the control value. In the text it is assumed, for example, that the control amplitude was 1.0 mV.

tion at first usually declines rapidly, then more slowly, and finally, in the example shown, it disappears after about 5 to 6 min.

The duration and the extent of posttetanic potentiation depend very much on the synapse under investigation and on the duration and the frequency of the repetitive stimulation. The longest known posttetanic potentiations lasted several hours. Functionally, synaptic potentiation amounts to making a central nervous process take place more readily as a result of practice—that is, it could be considered a learning process. It is perhaps significant that particularly long posttetanic potentiations are found in the hippocampus, a central nervous structure that is assumed to play a special role in memory and learning (see Sec. 9.4).

What changes at the synapse are responsible for the improved transmission following repetitive activation? Practically any of the links in the long chain of processes occurring during activation of a chemical synapse (see Chapter 3) could be responsible, but it has been proved experimentally that there are essentially two main *presynaptic mechanisms* to which this improvement may be ascribed.

First, repetitive activation of the presynaptic axon membrane leads to an increase (hyperpolarization) of the resting potential and thus to an increase in the *amplitude of the action potential*. The increased action potential releases more transmitter substance into the synaptic cleft. This process is approximately the reverse of what happens during presynaptic inhibition, where, as we have seen, a reduction in the amplitude of the presynaptic action potential results in the release of less transmitter substance.

Second, repetitive activation gives rise to increased availability of transmitter at the synaptic cleft. This *mobilization* of transmitter substance also causes improved synaptic transmission because a greater portion of the transmitter substance stored in the presynaptic terminal is released per action potential.

Q 4.1 Turn again to Fig. 4-2A. How does a reduction in the stimulus frequency affect the temporal facilitation?
a. It improves it.
b. It makes it weaker.
c. It has no effect.

Q 4.2 In a neuronal population, activation of one nerve leads to suprathreshold excitation of 22 neurons, while activation of another nerve results in suprathreshold excitation of 10 neurons. Combined simultaneous activation of both nerves causes suprathreshold activation of 32 neurons. Would you call this process facilitation or occlusion?

Q 4.3 Which two of the following factors seem chiefly responsible for posttetanic potentiation?
a. Increased synthesis of transmitter.
b. Increased availability of transmitter at the synaptic cleft (mobilization).
c. Slower breakdown of transmitter in the synaptic cleft.
d. Increased sensitivity of the postsynaptic membrane.
e. Larger presynaptic action potentials.

Q 4.4 Which of the following statements is most appropriate? Renshaw inhibition
a. is an example of lateral (surround) inhibition.
b. is produced by activity in Ia fibers.
c. involves an interneuron in its reflex arc.
d. is an example of positive feedback in the nervous system.
e. is exerted entirely as presynaptic inhibition.

4.2 The Monosynaptic Reflex Arc

Definition of a Reflex. Stimuli originating within or outside the body often elicit relatively (though not entirely) invariant responses that have proved, in the course of evolution or individual development, to be especially appropriate. Such stereotyped reactions of the CNS to sensory stimuli are called *reflexes*. You will undoubtedly be able to supply a large number of examples yourself. Touching a hot object makes us withdraw our hand, even before we are aware of any pain and can take any conscious action. Touching the cornea of the eye always triggers a blink (corneal reflex). Foreign bodies in the windpipe make us cough. When food comes into contact with the posterior pharyngeal wall, a swallowing reflex is triggered, and so on.

Most reflexes occur without our being consciously aware of them;

examples include the reflexes that are responsible for the passage or processing of food in the stomach and intestinal tract, or those that continuously adapt circulation and respiration to the required levels. Nor are we normally aware of any of the motor reflexes that day in and day out maintain the upright posture of our bodies in space, keep our balance and, through appropriate regulatory reactions, permit us to carry out voluntary movements successfully. Of the large number of reflex arcs involved in controlling the motor system, we will examine only the simplest, the monosynaptic arc of the stretch reflex—probably one of the most important motor reflexes, despite its simple structural plan.

You are sure to be familiar with a procedure that often forms part of a doctor's examination: a light stroke with a reflex hammer on the tendon just below the kneecap causes a slight jerk of the quadriceps muscle, which extends the lower part of the leg. This test determines whether the reflex arc of the monosynaptic stretch reflex involving the quadriceps is intact. The following discussion of the individual components of the reflex arc begins with the receptor, the muscle spindle.

The Muscle Spindle. Each muscle contains a number of muscle fibers that are thinner and shorter than the ordinary muscle fibers. Usually, several of these fibers are grouped together and, as shown in Fig. 4-7, are encased in a connective tissue capsule. Because of its shape this structure is called a *muscle spindle*. The muscle fibers inside the capsule are called *intrafusal* muscle fibers, while the ordinary muscle fibers, which constitute the major part of the muscle, are called *extrafusal* fibers. To give some idea of the dimensions, we might note at this point that the intrafusal fibers are between approximately 15 and 30 μm in diameter and 4 to 7 mm long. The extrafusal muscle fibers are about 50 to 100 μm in diameter and vary in length from several millimeters to many centimeters and even tens of centimeters (more details are given in Chapter 5).

The *sensory innervation of this stretch receptor* (Fig. 4-7) consists of afferent fibers which wind themselves several times around the center of the intrafusal muscle fibers; this formation is thus called an *annulospiral ending*. The afferent fibers are thick, myelinated fibers (ca. 13 μm in diameter, see also Table 2-2) termed *Group Ia fibers*. In each spindle, the annulospiral endings are formed by one Ia fiber; because of their association with Ia fibers, the annulospiral endings are also called *primary sensory* endings. (The secondary endings are discussed below.)

If the muscle and the muscle spindles it contains are stretched, then action potentials are transmitted to the CNS from the annulospiral

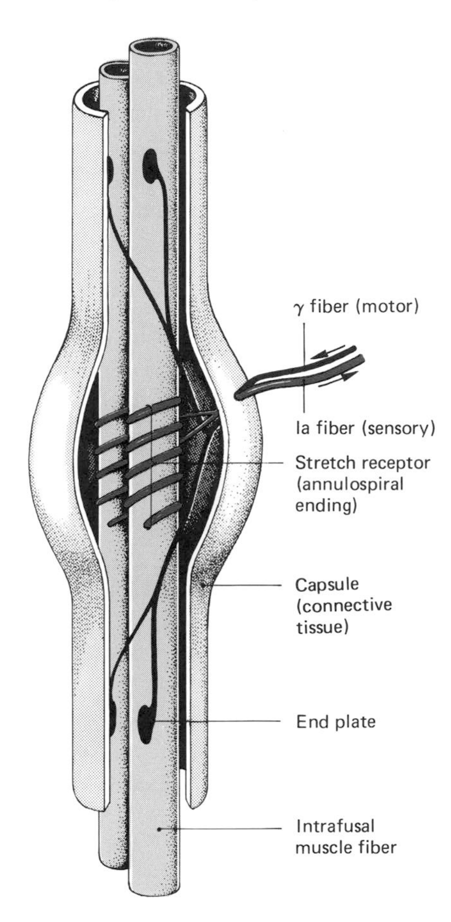

Fig. 4-7. Greatly simplified diagram of a muscle spindle. The intrafusal muscle fibers end a short distance beyond their sites of emergence from the connective-tissue capsule (not shown). So that the innervation can be clearly seen, the diameter of the intrafusal fibers is much increased with respect to their length. See text for further discussion.

endings. The frequency of the action potentials is proportional to the degree of stretching. As the muscle spindle is stretched (that is, as it becomes longer), the impulse frequency in its Ia fiber increases. If the muscle shortens by contraction of the extrafusal muscle fibers, the tension on the muscle spindles is relaxed, and the discharge rate of the Ia fibers is reduced or may even drop to zero. The muscle spindles thus *signal the length of the muscle.*

Many, though not all, muscle spindles have a *secondary sensory innervation,* also with stretch-sensitive endings. The afferent fibers (Group II fibers, diameter 9 μm; see Table 2-2) are thinner than those with annulospiral endings. Just as the latter are called primary muscle-spindle endings because of their Ia innervation, the receptor structures innervated by *Group II fibers* are called *secondary muscle-spindle endings.* Their form resembles that of the primary endings but is much less uniform; they are sometimes referred to as spiral, but also as "flower-spray" endings. (The afferents of the secondary muscle-spindle endings do not participate in the monosynaptic stretch reflex. We shall not discuss their central connections here.)

The intrafusal muscle fibers, just as the extrafusal fibers, possess a *motor innervation.* The motor axons of intrafusal muscle fibers are also thinner than normal motor axons. The latter are usually called Aα fibers, abbreviated to *α fibers* (α = alpha), while the motor axons of the intrafusal musculature are called Aγ fibers, abbreviated to *γ fibers* (γ = gamma). (α fibers are 12 to 21 μm in diameter and γ fibers 2 to 8 μm. The factors governing the relationship "thick muscle fiber = thick axon, thin muscle fiber = thin axon" are not yet fully understood.) The γ motor fibers form synaptic connections, similar to end plates, on the intrafusal muscle fibers, which, as Fig. 4-7 shows, are usually located in each lateral third of the muscle fibers. Morphologically, it is possible to distinguish between two types of intrafusal fibers, two types of γ motor axons and two types of intrafusal end-plate formations. The physiologic importance of these differences is still the subject of debate, and, therefore, we will not go into this matter here. (The function of the motor innervation is discussed below in connection with Fig. 4-10C.)

The Monosynaptic Stretch Reflex. What effect does stretching the muscle spindles have on the homonymous extrafusal musculature? It has already been stated in the discussion of central excitatory synapses, and shown in Fig. 4-4A, that the Ia fibers form excitatory synapses on homonymous motoneurons. *Activation of the primary muscle-spindle endings* by stretching the muscle must lead to *excitation of the homonymous motoneurons.* A sketch of an experiment to test this prediction is given in Fig. 4-8. As the curve (bottom left) shows, brief stretching of the muscle resulting from a light tap of the hammer on the recording arm produces, after a short latent period, a contraction (twitch) of the muscle.

When the knee-jerk reflex is tested during a medical examination, entirely comparable events occur. The hammer stroke on the tendon of the quadriceps muscle below the kneecap briefly stretches the muscle. Stretch of the muscle spindles results, after a short latency, in a slight twitching of the muscle, so that the part of the leg hanging down is raised slightly. (Try it yourself.) The term "patellar tendon reflex"

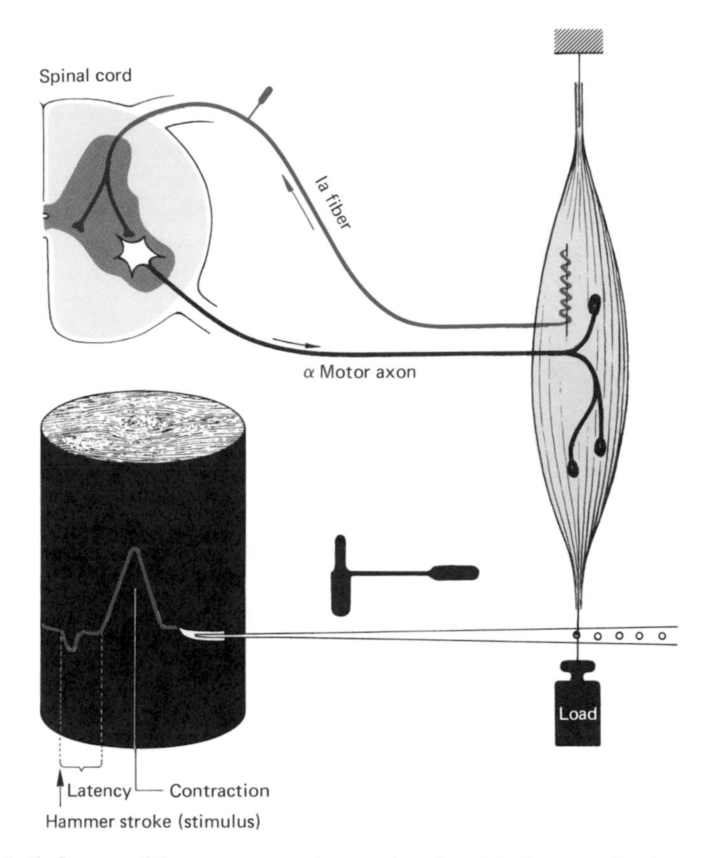

Fig. 4-8. Reflex arc of the monosynaptic stretch reflex. A light tap with a hammer on the stylus recording muscle length (*downward deflection of the trace on the recording paper*) after a brief latency produces contraction of the muscle. The reflex arc underlying this response is diagrammed, from the muscle spindles via the Ia fibers to the motoneurons and back to the muscle.

sometimes used is misleading; it was based on the original assumption that receptors in the tendon produced the knee jerk.

The brief stretch, then, activated primary muscle-spindle endings so that a volley of action potentials in the Ia fibers entered the spinal cord and there directly produced EPSPs in the homonymous motoneurons. Some of these EPSPs were above threshold, and the resulting motor fiber activity produced a slight twitch of the muscle. Since only one synapse within the CNS is involved (that of the Ia fibers with the homonymous motoneurons), the term *monosynaptic stretch reflex* is used.

A complete reflex arc is shown in Fig. 4-9A; it consists of a receptor,

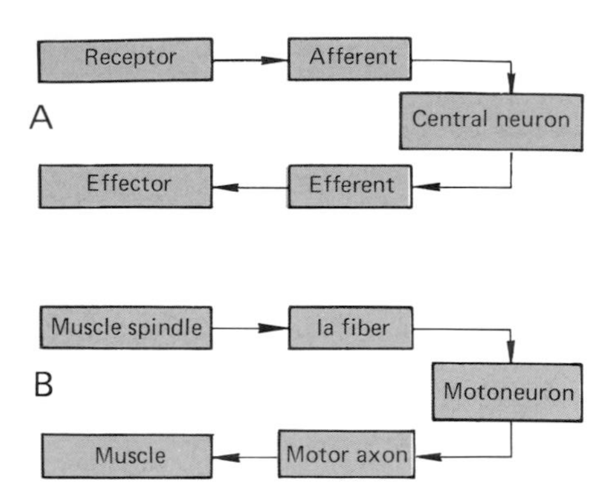

Fig. 4-9.A. General designations of the parts of a complete reflex arc. **B.** Elements in the arc of the monosynaptic stretch reflex.

an afferent pathway, one or more central neurons, an efferent pathway, and an effector. Figure 4-9B shows the corresponding elements in the monosynaptic stretch reflex. Comparison of A and B reveals that *the monosynaptic stretch reflex is the simplest example of a complete reflex arc.* Moreover, in this case the receptors (the muscle spindles) and the effector (the extrafusal musculature) are in the same organ, the muscle. For this reason the stretch reflex is often referred to as an *intrinsic reflex,* but the term stretch reflex is more appropriate.

The time between the start of the stimulus and the action of the effector is called the *reflex time.* In the case of the knee-jerk reflex, the delay between the tap with the hammer and the twitch of the muscle is scarcely noticeable to the observer, which indicates that the reflex time is very short. It is determined mainly by the conduction time of the afferent and the efferent action potentials in the Ia fibers and the α motor axons, respectively. In an adult human, the pathway from the quadriceps muscle to the spinal cord and back measures approximately 80 + 80 = 160 cm. The speed of conduction in Ia fibers and α motor axons is high, approximately 100 m/s. The pathway of 160 cm = 1.6 m is thus traversed in 16/1,000 *s* = 16 ms. In addition to this conduction time, we must allow a few milliseconds for the following: (a) the delay between onset of stretch and the first action-potential discharge by the muscle spindle, (b) transmission in the synapses at the motoneurons (synaptic delay), (c) transmission from the end plates to the muscle fibers, (d) the spread of the muscle action potential along the fiber, and (e) triggering of contraction by the muscle fiber action potential (excitation-contraction coupling). Altogether the *reflex time of the monosynaptic stretch (intrinsic) reflex* is approximately 25 to 30 ms.

The great physiologic significance of stretch reflexes is discussed extensively in Chapters 6 and 7. Here we shall return briefly to their *clinical, diagnostic significance*. Absence, weakening, or overshoot of the responses to tests of the knee-jerk reflex and corresponding reflexes of other muscles indicate a disturbance that should then be examined more closely by other neurologic methods. The trouble could lie in the individual elements of the reflex arc (Fig. 4-9), but it could also be that the other inputs to the motoneurons (Fig. 4-1B) are abnormal, causing subnormal or supranormal excitability of these cells. The knee-jerk and similar reflexes, then, provide a very simple test of the motor system.

Function of the Intrafusal Muscle Fibers. We have seen (Fig. 4-8) that stretching a muscle is one way to excite the receptors in the muscle spindle; passive stretch extends the intrafusal (see A and B in Fig. 4-10) as well as the extrafusal fibers. But there is also a second way of exciting the primary muscle-spindle endings: contraction of the intrafusal muscle fibers, elicited by way of the γ motoneurons (Fig. 4-10C). This contraction of the intrafusal muscle fibers does not change the length and the tension of the entire muscle. It is too weak for that, even if all the intrafusal muscle fibers of a muscle were to contract simultaneously. However, intrafusal contraction is sufficient

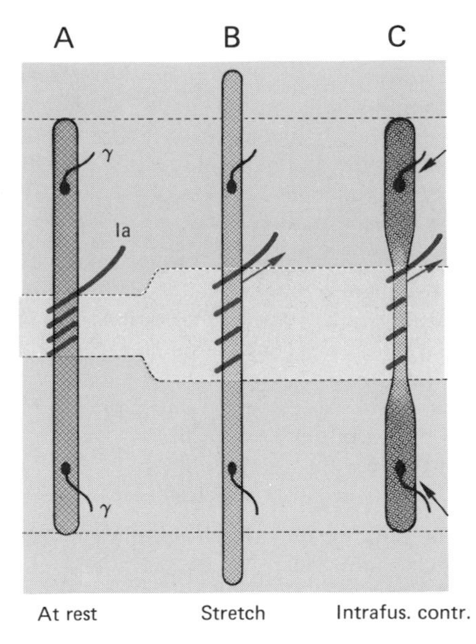

Fig. 4-10. Activation of a primary sensory (annulospiral) ending of a muscle spindle (under resting conditions in **A**) by stretching the whole muscle (**B**) or by contracting the intrafusal fibers (**C**). Compare with Figs. 6-1 and 6-2.

to *stretch the central portion of the intrafusal fibers,* inducing excitation in the primary sensory endings. This, like stretching the whole muscle, produces afferent action potentials in the Ia fibers and hence EPSPs in the homonymous α motoneurons. Consequently, contraction of the extrafusal musculature can be triggered or at least facilitated by the muscle spindles both when the muscle is stretched and when the intrafusal muscle fibers contract, having been activated by the γ motoneurons. The two processes can either be *mutually complementary* or can have a *mutually canceling* effect: intrafusal contraction and simultaneous stretching of the whole muscle will bring about particularly strong activation of the stretch receptors.

Conversely, extrafusal contraction and simultaneous intrafusal relaxation relax the stretch receptor completely. Any desired intermediate value can be obtained between these two extremes of maximum activity and complete inactivation by means of suitable contractions of the intrafusal muscle fibers. In other words, the threshold of a stretch receptor can be varied by intrafusal "pretensioning." When intrafusal contraction occurs, the threshold of the stretch receptor is lowered; in response to stretching of the muscle, the muscle spindle reacts more sensitively (further details will be given in Chapters 6 and 7).

Q 4.5 Which of the following statements about the muscle spindles are correct?

a. The intrafusal muscle fibers are thinner and longer than the extrafusal muscle fibers.
b. The γ motor axons innervate the primary stretch receptor endings.
c. The muscle spindles possess no innervation other than sensory innervation.
d. Afferent volleys in the Ia fibers inhibit the homonymous motoneurons and excite their antagonists.
e. None of these statements is correct.

Q 4.6 Which of the following processes will bring about an increase in tonus of the extrafusal musculature via the monosynaptic stretch reflex?

a. Passive shortening of the muscle (for example, the pendant lower part of the leg is extended by a second person, and this results in passive shortening of the quadriceps).
b. Active contraction of the intrafusal muscle fibers.
c. Stretching of the extrafusal muscle fibers.
d. Relaxation of the intrafusal muscle fibers.
e. Contraction of the extrafusal muscle fibers.

Q 4.7 Let us assume that the afferent and the efferent pathways of a monosynaptic stretch reflex arc are each 120 cm long (those of a foot muscle). The speed of conduction of the afferent Ia fibers is 100 m/s and that of the α motor fibers is 80 m/s. What is the minimum reflex time in milliseconds? Add 13 ms to the conduction times to allow for the synaptic and activation processes mentioned on p. 118.

Q 4.8 Under what conditions are the stretch receptors of the muscle spindles more strongly excited?

a. Contraction of the extrafusal muscle fibers with simultaneous contraction of the intrafusal muscle fibers.

b. Contraction of the extrafusal muscle fibers with simultaneous relaxation of the intrafusal muscle fibers.

4.3 Polysynaptic Motor Reflexes

The motoneurons are the only central neurons involved in the monosynaptic stretch reflex (the cell bodies of the Ia fibers are in the dorsal-root ganglia and thus outside the CNS; see Figs. 1-10, 1-11, and 4-8). In all other motor reflex arcs, several neurons are connected in series, and the motoneuron is always the last link in the chain of central neurons. These reflexes, therefore, are called *polysynaptic* (see Fig. 6-6, p. 167). In addition, in the polysynaptic reflexes, the receptor and the effector are spatially separated in the body, so these reflexes may also be called extrinsic reflexes. (One could refer to reflexes in which receptors in a muscle act upon the same muscle as intrinsic polysynaptic reflexes. There is, for instance, a polysynaptic component in the stretch reflex, which will not be considered here.)

Polysynaptic motor reflexes play an important role in locomotion (*locomotor reflexes*), eating (*nutritional reflexes*), and protecting the body from its environment (*protective reflexes*). Examples of each of these types of reflex will be introduced below. In defining the characteristics of the extrinsic reflexes, we should always bear in mind that even in the case of the monosynaptic stretch reflex the reflex response is not firmly coupled to the stimulus (in automatonlike fashion), but instead it can be modified by other excitatory and inhibitory influences acting simultaneously on the motoneuron. Thanks to the large number of neurons involved, it is much easier in a polysynaptic reflex arc to adapt the reflex response to the prevailing requirements of the body.

Examples of Polysynaptic Motor Reflexes. The simplest example of an extrinsic locomotor reflex was given in Fig. 4-4A, which shows that the Ia fibers coming from the stretch receptors of the muscle spindles form not only monosynaptic excitatory synapses with homonymous motoneurons but also inhibitory synapses with antagonistic motoneurons. The inhibitory connection is made through an interneuron (shown in red in Fig. 4-4A). The inhibitory reflex arc of the Ia fibers and antagonistic motoneurons thus has *two* central synapses, one from the Ia fiber to the interneuron (excitatory synapse) and the other from the axon of the interneuron to the motoneuron (inhibitory synapse). It is the shortest known inhibitory reflex arc. This type of inhibition, therefore, is also called *direct inhibition*. The term *reciprocal antagonist inhibition* is preferable, since it implies that the motoneurons of antagonistic muscles (for example, flexor and extensor muscles of the same joint) can be reciprocally inhibited via this reflex arc (see discussion of function in Sec. 6.1).

Most of the other excitatory and inhibitory afferents converging on the motoneurons from the peripheral receptors (from the muscles, joints, and skin) have more than one, often indeed very many, interneurons in their reflex arc. They are not disynaptic but polysynaptic. Let us consider two examples. Brushing a mother's nipple against the lips of her newborn child automatically triggers a sucking reaction. The same sucking reaction can be provoked by touching the lips with the tip of one's finger or by giving the child a pacifer (dummy). This clearly demonstrates the reflex character of this phenomenon. The *sucking reflex* is a *nutritional reflex*. The receptors of its polysynaptic reflex arc are touch-sensitive structures in the skin of the lips (mechanoreceptors). The effectors are the muscles of the lips, cheeks, tongue, throat, thorax, and diaphragm. The sucking reflex is thus a very complicated extrinsic reflex, and it should also be remembered that the sucking action has to be coordinated with normal respiration.

If a piece of filter paper soaked in acid is laid on the back of a decerebrate frog (with the cerebrum removed under anesthesia, the frog can remain alive for many days and even weeks), then after a short latent period the frog will brush the piece of paper off with the nearest hind limb. This is an example of a *protective reflex*. In the case of this protective reflex, the (pain) receptors are located in the skin of the back, while the musculature of the hind limb is the effector. This reflex is, therefore, also a polysynaptic extrinsic reflex.

Characteristics of Polysynaptic Reflexes. The cough reflex serves to keep the respiratory passages free from blockages preventing inhalation and exhalation of air. The cough reflex is, thus, a typical protective reflex. The receptors are located in the mucous membrane of the windpipe (trachea) and its branches (bronchi). Stimulation of the mucous membrane receptors not only triggers the cough reflex but also generates conscious sensations so that we can compare the intensity of the stimulus and the size of the reflex response. We will, therefore, use this reflex to study the characteristic properties of polysynaptic reflexes.

You will almost certainly have noticed that a slight "tickling" or irritation in the throat does not make you cough immediately but only after a short time has elapsed. We can deduce from this that in the case of polysynaptic reflexes stimuli subthreshold to trigger the reflex can summate to a suprathreshold stimulus, if they last long enough. This *summation* is a central phenomenon; that is, it takes place in the interneurons and the motoneurons of the reflex arc and not in the peripheral receptors. The subjective feelings of discomfort (tickling, irritation) experienced before the reflex is triggered are a clear sign that the receptors responsible for the reflex are already excited.

With increasing stimulus intensity, the time between the stimulus and the triggering of the reflex, that is, the reflex time, becomes shorter and shorter even when the stimuli are already above the threshold. We can generalize even more and say that in the case of the polysynaptic reflex the *reflex time is dependent on the intensity of the stimulus;* stronger stimuli cause the reflex to start sooner. (In contrast, in the monosynaptic stretch reflex, the reflex time is relatively constant.) The shorter reflex time of the polysynaptic reflex with increasing stimulus intensity is a result of the more rapid suprathreshold excitation of the central neurons of the reflex arc, which is brought about by the larger number of more intensely activated receptors. The shorter reflex time is caused mainly by temporal and spatial facilitation.

Coughing can range in intensity from simply clearing the throat to a long fit of choking coughs, again depending on the intensity of the stimulus. This increase in the reflex response with increasing intensity of the stimulus is also a typical characteristic of polysynaptic motor reflexes. In the process, the reflex starts to affect previously uninvolved groups of muscles, a phenomenon called *spreading.* Obviously, with strong stimuli, neurons that were previously excited below the threshold now receive suprathreshold excitation. The spreading of the reflex is particularly easy to demonstrate in the case of coughing. When we simply clear our throats it is chiefly the muscles of the throat that are activated, while in the case of a choking cough the muscles of the chest, shoulders, abdomen, and diaphragm are involved.

Motor Reflexes and Autonomic Polysynaptic Reflexes. In the motor reflexes (for example, the locomotor, the nutritional, and the protective reflexes), the motoaxons form the efferent pathway of the reflex arc, while the receptors are situated primarily in the skin, the muscles, the tendons, and the joints. There are also a large number of polysynaptic reflexes that likewise drive motoneurons but that originate in visceral receptors. The most obvious example is that of the respiratory reflexes triggered by stretch receptors of the lung and by chemoreceptors, with efferent pathways formed by the motor axons of the diaphragm and the respiratory muscles in the thorax. The neurons of the autonomic nervous system that lead to glands and smooth muscles are the efferent pathways for a large number of *autonomic reflexes*. We shall take a closer look at some examples of these in Chapter 8. At this point it is sufficient just to mention the key words "circulatory reflexes," "digestive reflexes," and "sexual reflexes."

The didactically expedient grouping and classifying of the polysynaptic reflexes should not blind us to the fact that there are many mixed types of polysynaptic reflexes and that each of the known "classifications" is in one way or another arbitrary. For example, the cough reflex is certainly a protective reflex, but it is not, in the strict sense, a motor reflex, because its receptors are situated in the mucous membrane of the trachea and the bronchi; they are thus visceroreceptors. Many of the complex reflexes (for example, the sexual reflexes) also simultaneously possess motor and autonomic efferent pathways. A further aspect, which can easily be overlooked if individual reflexes are considered in isolation, is that *most motoneurons and interneurons participate in a large number of reflex arcs*. A motor axon in the musculature of the throat will, for example, be involved in swallowing, sucking, coughing, sneezing, and respiratory reflexes; that is, it will offer the same final common pathway for a large number of reflex arcs.

Congenital and Acquired Reflexes. The reflexes we have considered so far all involve stereotyped reactions of the organism that are predetermined in the structural design of the CNS. They can be observed in practically the same form in all individuals of the same species. The neurons of these preformed reflex arcs are mostly situated in the phylogenetically older parts of the CNS, that is, in the spinal cord and the brainstem, even when the reflexes involved are highly complex (recall, for example, the protective reflex in the decerebrate frog when acid is applied to its back). Each individual in addition has the ability to acquire reflex reactions so that the body can respond better and with less effort to the constantly changing situations in his environment. The reflex arcs of these acquired reflexes usually run in the higher levels of the CNS. The acquired reflexes (which can also be forgotten) are distinguished from the stereotyped congenital reactions of the organism by a wide range of criteria. These criteria and classification systems will not be treated here, however. Examples of well-known learned reflexes that have been studied extensively are the *conditioned reflex* and the behavioral changes produced by *operant* *conditioning,* both of which are described in *Fundamentals of Sensory Physiology.*

Q 4.9 Which of the following are nutritional reflexes and which are protective?
- a. Reflex governing secretion of tears.
- b. Reflex governing secretion of saliva.
- c. Corneal (eyelid closure) reflex.
- d. Sucking reflex.
- e. Sneezing reflex.
- f. Cough reflex.

Q 4.10 Which of the following are due to summation?
- a. The increase in the size of the reflex response with increasing intensity of the stimulus.
- b. The modification of the reflex response by influences simultaneously acting on the interneurons of the reflex arc.
- c. The shortening of the latent period between commencement of the stimulus and onset of the reflex with increasing stimulus intensity.
- d. The simultaneous inhibition of antagonistic motoneurons when homonymous motoneurons are excited by Ia fibers.

Q 4.11 How many central synapses are involved in the reflex arc of direct inhibition?
- a. None.
- b. One.
- c. Two.
- d. Three.
- e. Many.

Q 4.12 Which of the following central neuronal processes are involved in the spreading of polysynaptic motor reflexes?
- a. Direct inhibition.
- b. Temporal facilitation.
- c. Synaptic potentiation.
- d. Occlusion.
- e. Spatial facilitation.

5
MUSCLES

J. Dudel

In terms of sheer size, the most extensively developed organs in the bodies of man and other vertebrates are the musculature, the "flesh." The muscles make up 40 to 50% of the total body weight. Their main function is to develop force and to contract. They are also, among other things, important for the thermal regulation of the body, but the heat-producing role of the musculature will not be discussed here in connection with the neurophysiological features.

Man can work and make an impact on his environment only by utilizing his muscles. This is true not just for physical work but for intellectual activity as well, for both writing and speaking require a finely tuned interplay between muscles. One could regard the nervous system, perhaps a bit simplistically, as an organ that responds to the stimuli acting on the organism by producing corresponding muscle contractions. This means that the muscle is a highly important subject for the neurophysiologist. Furthermore, the modus operandi of the muscle cells is better known than that of most other types of cells. The morphology and the chemical components and reactions as well as the physiological functions of muscle cells have been very extensively researched, and in recent years the various approaches have been combined into a unified theory of muscle contraction. When discussing the function of muscles, we must, therefore, pay particular attention to their structure and chemical composition.

5.1 Contraction of the Muscle

The most important part of the musculature, the *skeletal musculature*, is composed of individual muscles, such as that shown in Fig. 5-3A.

Such a muscle is an elongated "bundle of flesh" terminating at both ends in cordlike *tendons*. The muscle is connected by the tendons to the bones, the "skeleton," and can thus act on the latter. In studies of its functions, the muscle is usually isolated. This is easily achieved by severing the tendons. The muscle can then be suitably attached by way of the tendon stumps and mounted in a bath for experiments (see Fig. 5-1). The reactions of such a muscle preparation to excitation of its fibers will now be discussed.

Isotonic and Isometric Conditions. If a muscle is stimulated by impulses in the motor nerve (see p. 73) or by direct suprathreshold depo-

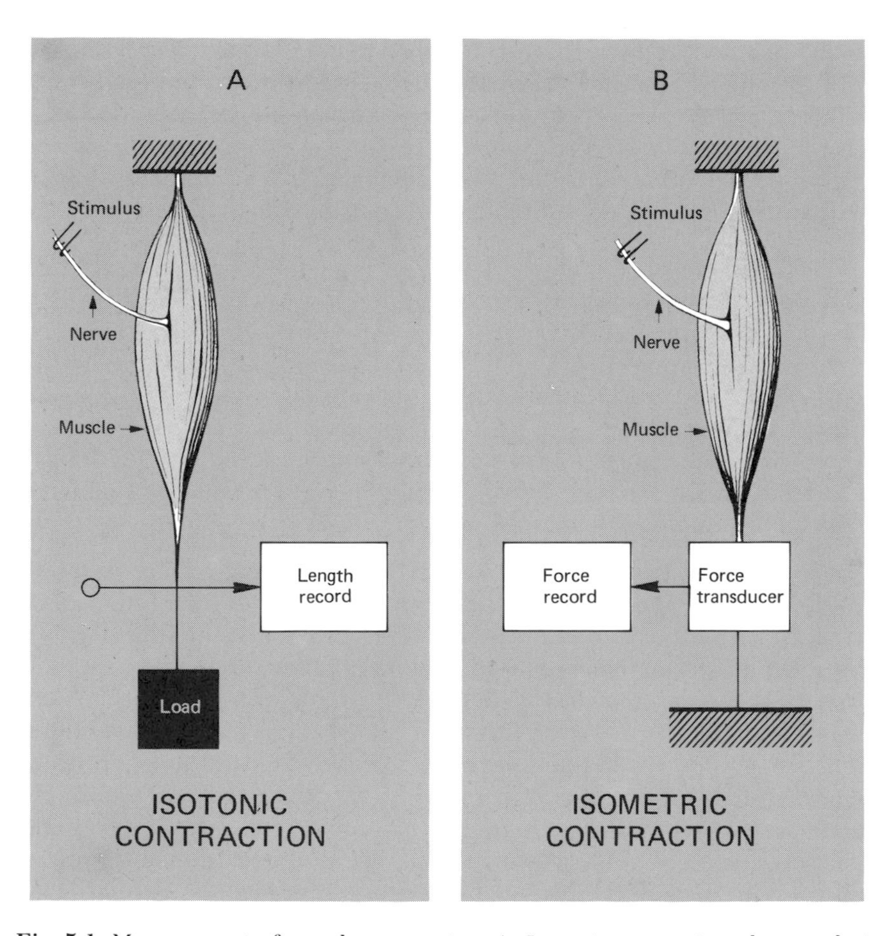

Fig. 5-1. Measurement of muscle contraction. **A.** Isotonic contraction; the muscle is attached at one end only, and as it raises a constant load, its length is recorded. **B.** Isometric contraction; the muscle length is kept constant by attachment at both ends, and the recorded variable is the force with which it pulls on its attachments.

larization of its fibers, then it *contracts;* that is, it exerts tension on its points of attachment that tends to make it shorter. Whether or not shortening of the muscle occurs during this process depends on whether its attachment points are able to move.

Because different properties of the muscle are revealed depending upon whether it can shorten or merely develops tension, it is customary to define and use two standard experimental situations:

1. The contraction can be measured by firmly attaching the muscle at one end and hanging a *movable but constant load* on the other end (Fig. 5-1A). The shortening of the muscle measured under these constant-load conditions is called *isotonic contraction.*
2. The contraction can also be measured by firmly clamping both ends of the muscle so that they cannot move and incorporating a force meter at one end, which does not vary in length under load (Fig. 5-1B). The change in force measured while the *length of the muscle remains constant* is called *isometric contraction* (or, more meaningfully, "tension development under *isometric conditions*").

When the muscles contract naturally, the contraction is rarely fully isotonic or fully isometric. For example, when a bucket of water is raised by the arm, the arm undergoes "isotonic shortening," but, because of the change in the angular configuration of the bones as the arm is bent, the load on the individual muscles in the arm does not remain fully constant. Approximately isometric conditions exist, of course, when the muscles are activated, but the skeleton is stationary; here antagonistic muscles may be opposing one another, or the bucket of water in the above example might be held in one position.

Time Course of Individual Twitches. When a muscle is excited by a single stimulus, it contracts briefly, or twitches. Under isotonic conditions, this single contraction is recorded as a transient shortening of the muscle; under isometric conditions, the event is recorded as a brief increase in force. A typical time course of an *isometric twitch* in a muscle of a warm-blooded animal is shown in Fig. 5-2. The contraction is triggered by the action potential shown in the top half of the figure, and it begins *a few milliseconds* after the action potential. The force exerted reaches its maximum in about 80 ms and returns somewhat more slowly to the resting value. We can distinguish, therefore, between a *rising phase* and a *relaxation phase* of the contraction. In the case of isotonic contraction, the time course of the change in length is quite similar to the time course of force shown in Fig. 5-2. If we compare the time course of contraction with the duration of the action potential, it is obvious that the contraction takes place much

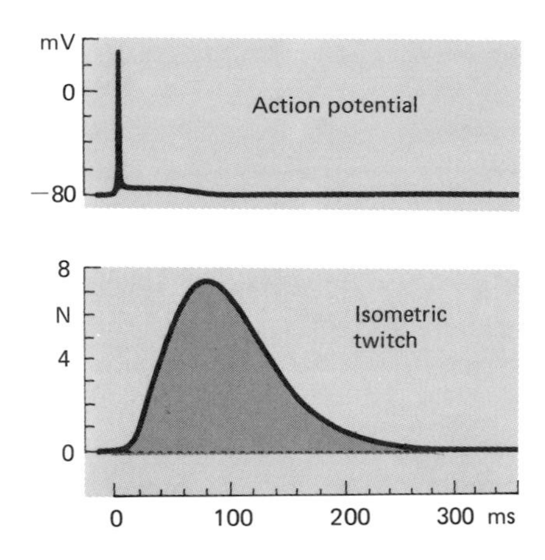

Fig. 5-2. Action potential and muscle contraction. Time course of the action potential and of the subsequent contraction of a human thumb muscle (adductor pollicis). The contraction begins about 2 ms after the rising phase of the action potential and does not reach a maximum until 100 ms have elapsed.*

more slowly. The action potential is almost finished before the start of the contraction it triggers. The contraction of the muscle is thus about 100 times slower than the action potential.

However, the duration of the muscle contraction is less uniform than the duration of the action potential in the various muscles. The contraction curve for a thumb muscle shown in Fig. 5-2 is an example of a "*fast*" muscle as found in warm-blooded animals. These animals also have *slow* muscles that take 200 ms to attain their maximum tension, and in muscles of cold-blooded animals at low temperature the rising phase of contraction can last several seconds.

Fine Structure of the Skeletal Muscles. The mechanism of muscle contraction can be described in greater detail only if the fine structure of the muscles is known. The relevant aspects will be discussed with reference to Fig. 5-3. The skeletal muscle in Fig. 5-3A is composed of bundles of fibers (Fig. 5-3B) that are still easily visible to the naked

* In the lower ordinate of Fig. 5-2, the newton (N) is used as the measure of force. Before this unit was introduced as the standard, the kilogram-weight was customarily used in physiology. The kilogram being the unit of *mass*, its weight is obtained by multiplication by the acceleration due to gravity, 9.81 m/s^2. Thus, 1 kg-wt = 9.81 $kg \cdot m \cdot s^{-2} \equiv 9.81$ N. The newton is a convenient force unit—coincidentally, it is about the weight of an apple!

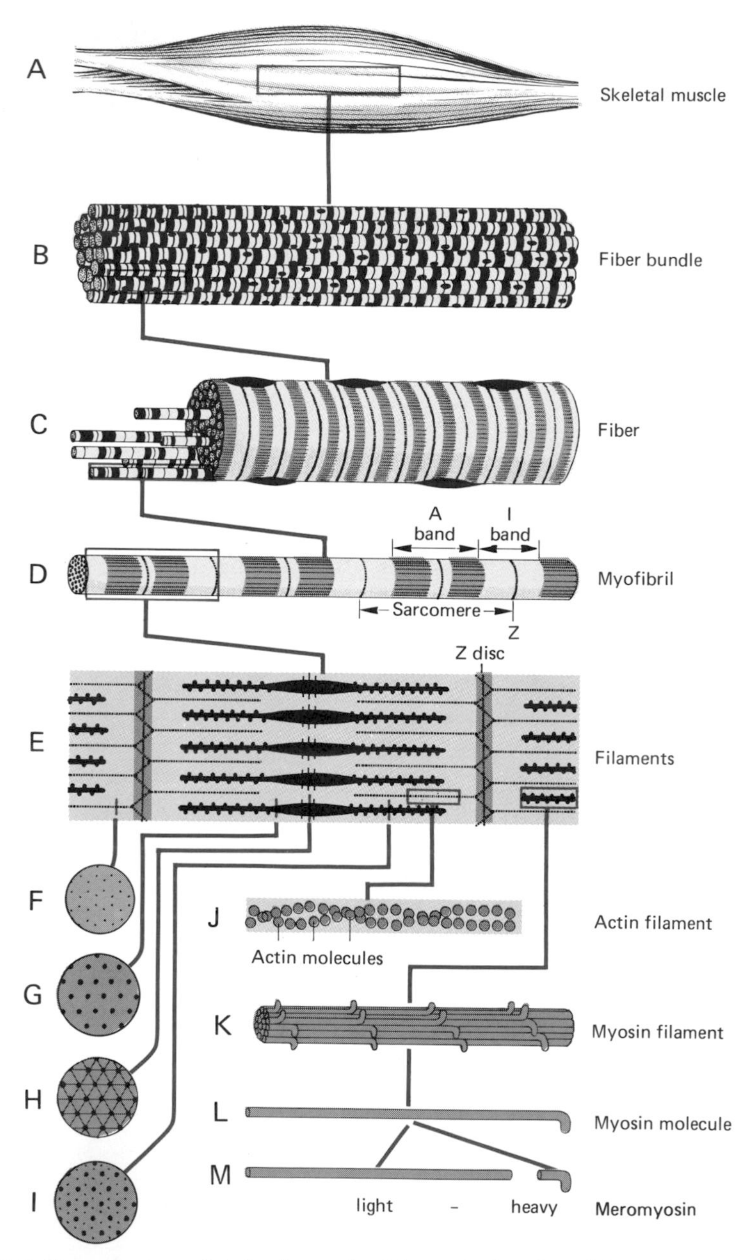

Fig. 5-3. Fine structure of skeletal muscle. Schematic drawing of the structure at different dimensional levels, from the whole muscle through the muscle fiber, muscle fibril, filament, and finally the molecular arrangement of the contractile proteins. Discussed in detail in the text. Modified from Bloom and Fawcett (1969) *A Textbook of Histology*, Saunders, Philadelphia.

eye (think of the "fibers" in boiled beef). The individual *muscle fibers* of the bundle (Fig. 5-3C) are cells measuring several to many centimeters in length and 10 to 100 μm in diameter. The muscle fibers generally run through the entire length of the muscle and terminate at both ends in the connective tissue tendons. The muscle fibers contain closely packed contractile protein structures that are arranged in fiber form in the longitudinal direction of the muscle cell and are called *myofibrils* (Fig. 5-3D).

When viewed under the microscope, the muscle fibers exhibit a characteristic *cross striation*. The striated pattern is produced by the structures running the length of the fibers, the myofibrils, which are themselves cross striated (Fig. 5-3D) and arranged in strict side-by-side configuration in the fiber. The striation of the adjacent myofibrils is caused by the fact that they contain strongly birefringent (anisotropic) and weakly birefringent (isotropic) parts in regularly alternating sequence. In transmitted light, the strongly birefringent striations are darker than the weakly birefringent ones. Accordingly, as shown in Fig. 5-3D, they are termed *A bands* and *I bands*. A thin dark strip is located in the middle of the I band. This is called the Z disc. The region between two Z discs, about 2 μm long, represents the smallest functional unit of the myofibril, the *sarcomere*.

The fine structure of the sarcomere can be resolved further with an electron microscope (Fig. 5-3E to I). As Fig. 5-3E shows, the Z disc links adjacent *thin myofilaments*. In the central section of the sarcomere, *thick myofilaments* are situated between the thin filaments. The cross sections in Figs. 5-3F to I show that the thin and the thick filaments are arranged in strictly ordered arrays in a manner reminiscent of crystal structures. Experiments using chemical methods have shown that the thin filaments consist primarily of the protein *actin* (Fig. 5-3J) and the thick filaments consist of other elongated protein molecules, namely *myosin* (Fig. 5-3K to M). Thus, the I bands of the myofibrils consist mainly of the protein actin, and the central portion of the A bands consists entirely of the protein myosin. In the lateral sections of the A band, both actin and myosin are present.

Displacement of the Actin and the Myosin Filaments During Contraction. The chemical and physical processes on which contraction is based become clear when one observes the behavior of the structural elements of the muscle, as illustrated in Fig. 5-3, during contraction. If the muscle can shorten—for example, in the case of isotonic contraction—the width of the *A bands remains constant*, while the *I bands become narrower*. The birefringence in the A band is caused by the presence of the myosin filaments. If the width of the A bands remains constant during contraction, the length of the myosin filaments must also stay constant.

The I band becomes narrower during isotonic contraction. Nevertheless, electron-microscope images reveal that the actin filaments, which are the sole constituents of the I band, stay the same length during contraction. Since the lengths of the actin and the myosin filaments do not change, the length of the sarcomere can decrease during contraction only as a result of the filaments sliding past each other. This interpretation, now generally accepted, is known as the *sliding filament theory.* The I band is thus narrowed because the myosin filaments slide between the actin filaments, increasing the degree of overlap.

The movement of the myosin and the actin filaments during contraction is illustrated in Fig. 5-4, which shows diagrams of the sarcomere at two different stages of contraction. During contraction (from *A* to *B*) the sarcomere becomes shorter, and the distance between the Z discs becomes smaller because the myosin and the actin filaments slide past each other. The ends of the myosin filaments approach the Z discs, and the I bands, consequently, become narrower.

Molecular Mechanism of Contraction. What forces bring about the displacement of the myofilaments during contraction? In Fig. 5-4 oblique (red) strokes have been drawn between the actin and the myosin filaments. These strokes are meant to symbolize bridges—groups of molecules that constitute chemical links between myosin and actin filaments. When a link is formed between filaments as diagrammed in Fig. 5-4, it *shortens* and thus draws the actin filament in between the myosin filaments. After shortening the link is broken again, and the cycle can begin anew; a new link forms between the myosin and actin filaments that have just been shifted past one another, it then shortens, and so on.

The "links" or "bridges," shortening of which causes displacement of the myosin filaments and hence contraction, originate at the thickened end of the myosin molecule. Fig. 5-3 shows that the myosin filament is composed of myosin molecules (Fig. 5-3K,L) lying next to one another in a staggered array. Within the filament, the thickened ends of the myosin molecules are regularly distributed. This fraction of the myosin is called "heavy meromyosin" (Fig. 5-3M). Heavy meromyosin can, on the one hand, enter into chemical combination with the actin; on the other hand, it is an enzyme that *splits off* one phosphate from the *adenosine triphosphate* (ATP) in the cell. In the process, energy is released, which is used in part for the shortening of the

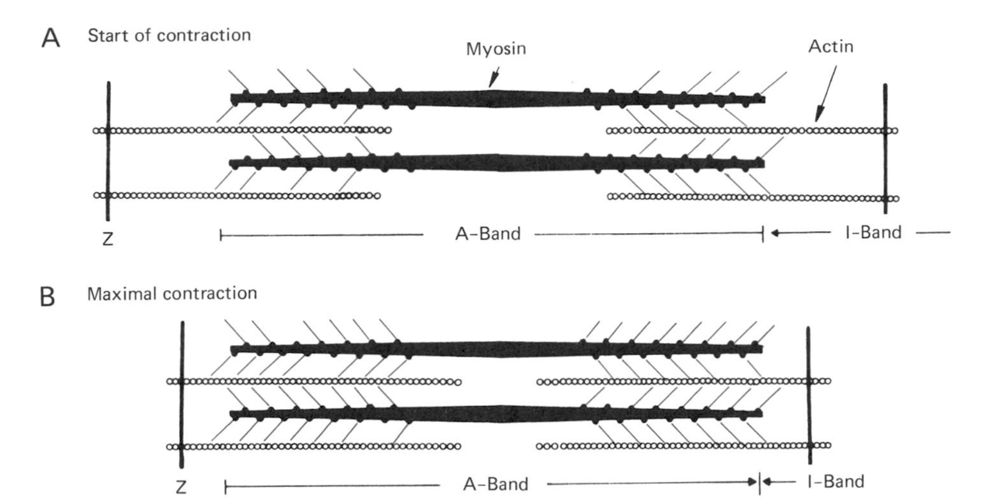

Fig. 5-4. Mechanism of contraction. Diagram of the arrangement of the myosin and actin filaments in a sarcomere (from Z to Z) and of the bridges between these filaments. It is thought that changes in configuration of the bridges produce a shear force that moves the actin filaments between the myosin filaments. **A.** Arrangement at the beginning of contraction. **B.** State of the muscle when contraction is maximal—the sarcomere has become shorter.

bridges between myosin and actin and in part is released as heat. The cause of muscle contraction, then, is the formation of complexes of *myosin, actin,* and *ATP,* in the course of which ATP is split and the bridges between myosin and actin are shortened. Many findings indicate that the "bridges" are composed of two rigid segments set at an angle to one another. During "shortening" of these bridges, then, the angle between the two segments would be reduced.

The reaction of myosin, actin, and ATP, on which muscular contraction is based, can take place only when *Ca^{++} ions* are present in the cell in a free concentration of approximately 10^{-5} mol/l. In addition, contraction requires the presence of Mg^{++} ions. Ca^{++} is present at a concentration of 10^{-5} mol/l only during a contraction. In the relaxed muscle, the Ca^{++} concentration is about 10^{-8} mol/l, and, consequently, the reaction between myosin, actin, and ATP cannot take place. The cell possesses a mechanism by which the Ca^{++} concentration can be raised to 10^{-5} mol/l, permitting the reaction of actin, myosin, and ATP, and thereby initiating contraction. At the molecular level, therefore, contraction is controlled by the Ca^{++} concentration. This process is discussed in detail in Sec. 5.3 (excitation-contraction coupling).

Besides supplying the energy for muscular contraction, ATP has another effect on the contractile system. After contraction, even though the intracellular Ca^{++} concentration drops to 10^{-8} mol/l, the bonds between the myosin bridges and the actin molecules will break down only in the presence of ATP. This action of ATP is known as the *plasticizing effect.* When the ATP concentration in a muscle drops following a reduction in the supply of energy, the muscle cannot relax; it remains "hard" or rigid, and is said to be in the *rigor state.* After death, too, the ATP concentration in the muscle drops, the plasticizing effect of the ATP on the myosin–actin complex is absent, and "rigor mortis" sets in.

Cardiac Muscles and Smooth Muscles. In this section, we have taken a detailed look at the fine structure and the contraction mechanism of skeletal muscle fibers. Apart from this quantitatively predominant type of muscle, there are also the cardiac muscles and the smooth muscles to consider. The latter consist of fibers shorter and thinner than those of the skeletal muscles, and these fibers are interconnected in reticular fashion. The smooth muscle fibers, like those of the skeletal muscle system, contain myofibrils, but the latter are neither as densely packed nor as regularly arranged as the skeletal muscles. Therefore, no cross striation is visible in the *smooth muscles.* The contraction of the myofibrils in the smooth musculature occurs in exactly the same way as in the skeletal muscle fibers. Since the action potentials in smooth muscle are propagated differently from those in skeletal muscle, the time course of contraction in smooth muscles is on the whole slower than in the skeletal muscle system. This fact will be examined in greater detail in Chapter 8.

As regards structure and time course of contraction, the *cardiac muscles* have properties intermediate between those of the skeletal and smooth muscles. However, the contraction mechanism in cardiac muscles is the same as in skeletal muscles.

Q 5.1 Draw a sketch showing how a muscle fibril is composed of thick and thin myofilaments and illustrating how the filaments are connected to the Z discs. Indicate in your sketch the length of the sarcomere and insert the chemical names of the protein elements that make up the myofilaments.

Q 5.2 During isotonic contraction, the
- a. contraction force varies, while the length of the muscle remains constant.
- b. length of the muscle changes, while the load on the muscle remains constant.
- c. length of the sarcomere remains constant.
- d. sarcomere becomes shorter.
- e. anisotropic and the isotropic bands become shorter.

f. isotropic bands become shorter and the anisotropic bands remain constant.
g. anisotropic bands become shorter and the isotropic bands remain constant.
(More than one of the above answers are correct.)

Q 5.3 Indicating the time scale, sketch the time course of isometric contraction in a muscle from a warm-blooded animal.

Q 5.4 Which four substances, apart from Mg^{++} ions, take part in the chemical reaction on which contraction is based or must be present in adequate concentration during the reaction? Underline the substance that supplies the energy for the contraction.
1. . . . .
2. . . . .
3. . . . .
4. . . . .

5.2 Dependence of Force Development on Fiber Length and Velocity of Shortening

In the preceding section, the mechanism of contraction of the muscle fiber was described. During muscle contraction, the actin filaments are drawn between the myosin filaments. However, the extent to which this fundamental process manifests itself as force developed at the tendons and the extent to which the muscle becomes shorter both depend on the circumstances accompanying contraction. Experimentally, it is found that the tension developed by a muscle depends very much upon the length of the fibers at the time of activation—that is, the amount of "prestretching"—as well as upon the velocity of shortening. Since these parameters are also of great importance to the course of the contraction *in vivo,* they will now be discussed in detail.

The Resting Length–Tension Curve. Muscles, or isolated muscle fibers, kept at their "resting length" exert no force at their points of attachment. If one end of the muscle is pulled, the muscle is stretched. If the force necessary to produce stretching is plotted against the muscle length, the *length–tension* curve shown in Fig. 5-5 (resting) is obtained. The force increases exponentially with length starting from zero at the resting length, l_0. The muscle is not damaged by being stretched to about 1.8 times its resting length, but any additional stretching would tear the muscle fibers.

When the muscle is stretched, elongation first occurs in the sarcomere as the actin and the myosin filaments slide past each other (see Fig. 5-4). When this happens, the I bands become broader, and the A bands remain unchanged. Besides the elongation in the contractile

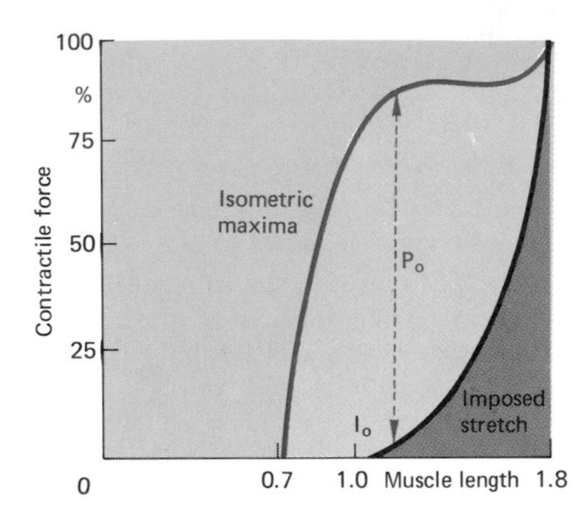

Fig. 5-5. Length–tension curve and curve of isometric maxima. The relationship between force and muscle length in a resting muscle is shown in *black*, and the maximum tension actively generated under isometric conditions is shown in *red*. l_0 is the resting length of the muscle; the abscissa is labeled in multiples of l_0. P_0 is the maximal additional tension that can be generated isometrically.

elements, the sarcomeres, there is also elongation in the passive *elastic elements* during stretching of the muscle. These passive elastic elements are mainly the tendons on which the muscle terminates. As will be discussed below in more detail, the stretching of the passive elastic elements is an important factor in the dynamics of muscular contraction.

Isometric Tension versus Length. The amount of prestretching of the muscle is important not only with regard to the resting tension of the muscle, but it also determines the amount of force developed by a muscle during contraction. For example, if isometric tension is measured by the procedure illustrated in Fig. 5-1, the length of the muscle can be set before each contraction by moving the points of attachment to the desired positions. When the muscle is made to contract from various initial lengths, the force increases in each case above the initial tension to the "*isometric maximum*." The curve of the isometric maxima is drawn in red in Fig. 5-5. The force developed beyond the resting tension for a given muscle length is the difference between the curve of the isometric maxima and the resting length–tension curve.

The muscle develops its maximum force, P_0, at around the resting length, l_0. The force developed is very much reduced if the muscle is shorter than the resting length, l_0, when contraction begins. The point of intersection of the red curve with the abscissa in Fig. 5-5 denotes the shortest length at which the muscle can still just develop force. Thus, in the case of *isotonic* contraction, the muscle could shorten at most to *about 70% of the resting length, l_0*. The force developed during isometric contraction also declines when the muscle is stretched far beyond the resting length. At approximately 1.8 times the resting length, the curves in Fig. 5-5 run together, and at the point where they meet, the muscle develops *no extra force* during contraction. Consequently, the muscle can develop force only in the range from 70% resting length to 180% resting length.

The dependence of the contractile force on the amount of prestretching can be explained by the arrangement of the filaments in the sarcomere. In Fig. 5-6 the isometrically developed force is shown for a sarcomere as a function of the sarcomere length. Below the curve are shown the various arrangements of the myosin and the actin filaments in the sarcomere, as determined by electron microscopy, for various sarcomere lengths (A to E). At maximum prestretching (A), the filaments no longer overlap; no bridges can form between them nor can contractile force be developed. Given less prestretching (between A and B), the overlapping of the filaments increases, and the amount of contractile force developed increases proportionately. The optimum contractile force is attained (between B and C) when bridges can attach to the actin along the entire length of the myosin filament.

At muscle lengths shorter than the resting length (C to E), the contractile force declines rapidly because the actin filaments are drawn in so far between the myosin filaments that they interfere with each other at the center; eventually, the Z discs run up against the myofilaments (E). Moreover, the excitation-contraction coupling (see 5.3) is impaired by distortion of the endoplasmic reticulum. The dependence of contractile force on the amount of prestretching can thus be explained quantitatively by the degree of overlap between the myosin and the actin filaments. The chemical reaction on which contraction is based can take place only when the actin and the myosin filaments are closely adjacent. In the case of partial overlap, the isometric force is proportional to the number of actin and myosin molecules reacting with each other at any instant.

Force and Velocity of Contraction. The dependence of the contractile force on the amount of prestretching has been explained in terms of the structure of the contractile apparatus. Similarly, the influence exerted by the velocity of shortening on the amount of force developed can also be attributed to the characteristics of the contractile apparatus.

Consider some everyday examples of muscular function in our own

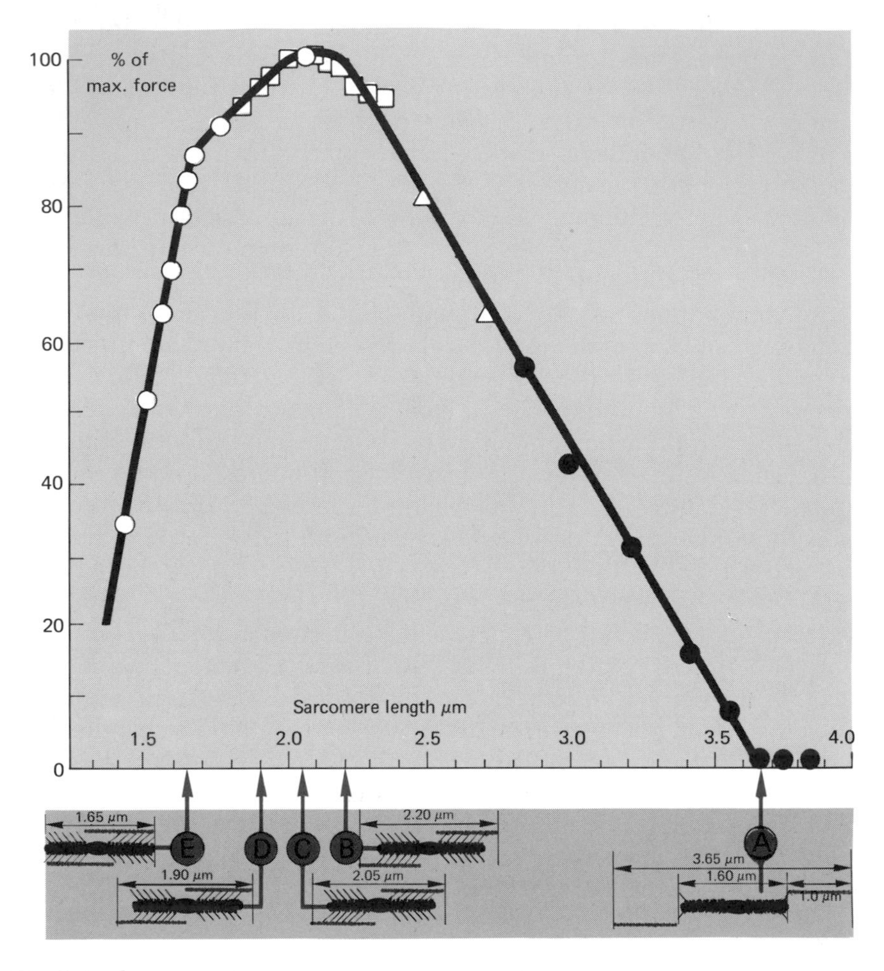

Fig. 5-6. The maximal isometric force developed at various sarcomere lengths. Below the force-versus-length graph are diagrams showing the degree of overlap between actin and myosin filaments at sarcomere lengths indicated by the red arrows (from 3.65 μm in *A* to 1.65 μm in *E*). The relationship between this overlap and the force produced at the corresponding sarcomere length is discussed further in the text. Modified from Ruch and Patton (1965) *Physiology and Biophysics*, Saunders, Philadelphia.

bodies. We can exert maximum force with our muscles only if they do not shorten in the process or if the shortening is only slight—for example, when we press or push against something. In contrast, we can carry out rapid movements only when the muscles are very lightly loaded—for example, when we throw a stone or play the piano.

Figure 5-7 shows the measured relationship between the maximum velocity of shortening and the force developed by an isolated arm

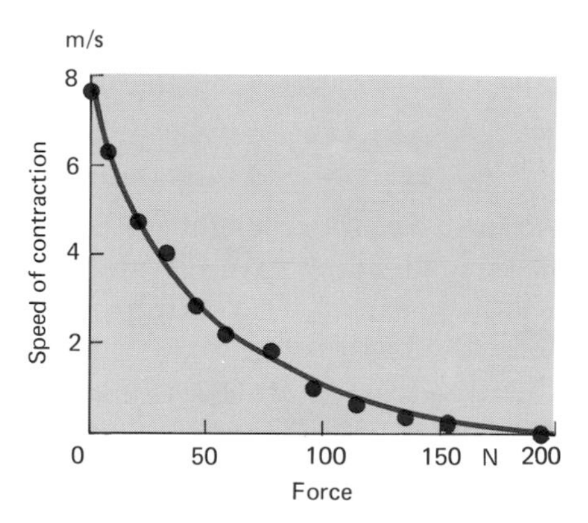

Fig. 5-7. Dependence of the speed of shortening of a muscle on the load. The abscissa gives the load on the muscle during an isotonic contraction; the ordinate indicates the maximal speed of contraction of the muscle. The interpolated curve represents one arm of a hyperbola. Modified from Wilkie (1949) *J. Physiol.* 110:249.

muscle. Without shortening, this muscle can exert a force of 190 N. As the load decreases, however, it can contract at an increasingly rapid rate, and at zero load a velocity of contraction of almost 8 m/s is attained. A similar relationship between force and velocity of shortening is found in all muscles. The decrease in contractile force with increasing velocity of shortening can be derived quantitatively from the contractile mechanism of the actin and myosin filaments sliding past one another. That is, the force development of the muscle is proportional to the number of links formed per unit time between the actin and myosin filaments (see Fig. 5-6). *The more rapidly* the actin and myosin filaments slide past each other during a contraction, *the smaller* is the number of links that can be formed between the filaments in a unit of time. With increasing velocity of contraction, then, the *force* of the contraction must be reduced.

The decrease in the velocity of shortening with increasing load is usually measured in isotonic contractions—that is, the muscle is allowed to shorten under constant load. But contraction velocity also has an effect in isometric contractions, during which there should be no shortening at all. However, at the level of the contractile unit, the sarcomere, there are *no strictly isometric contractions.* During the isometric twitch of the muscle its elastic elements, particularly the tendons, are stretched. Hence the sarcomeres can shorten even during isometric contractions, and the myosin and actin filaments slide past each other. This relative movement of the myofilaments reduces the amount of force development possible during a contraction. Thus even in an isometric twitch the maximal force possible at a sarcomere shortening velocity of zero is never reached. This fact is the precondition for a functionally very important mechanism—the summation of the twitches of muscle fibers. This summation will now be discussed in greater detail.

Summation of Twitches; Tetanus. Figure 5-8A shows a single isometric twitch in a fast muscle of a warm-blooded animal. Tension rises in less than 50 ms to its maximum, and drops again somewhat more slowly (see also Fig. 5-2). If the muscle is stimulated again before it has fully relaxed, then the next contraction starts from a higher initial tension and attains a higher maximum force than the first contraction (Fig. 5-8B). This event is called *summation*.

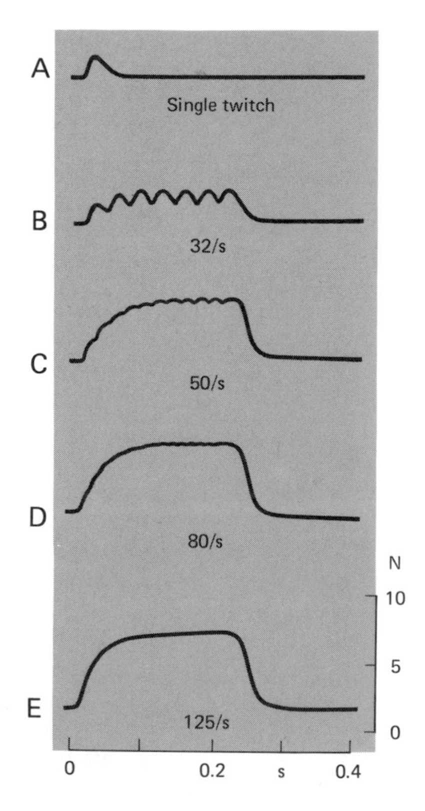

Fig. 5-8. Summation of twitches, leading to complete tetanus at the highest rate of stimulation. The records represent the force generated by a fast muscle of a warm-blooded animal (flexor hallucis longus of the cat) under isometric conditions. **A.** Single twitch. **B–E.** Series of twitches produced by stimuli at frequencies indicated for each record. In **C** and **D** tetanus is incomplete, whereas in **E** fusion is obtained. Modified from Buller and Lewis (1965) *J. Physiol.* 176:337.

In the body, muscles are usually stimulated by a series of action potentials. If the time between the action potentials is shorter than the overall duration of the twitch—that is, if the frequency of the action potentials is greater than 5 to 10/s—then, as shown in Figs. 5-8B to E, summation occurs for each new contraction. However, summation of the individual contractions is *not linear.* Figure 5-8B to D shows clearly that the increase in contractile force per individual contraction decreases with the number of stimuli within the series. If the series of stimuli is long enough, then, after a certain number of stimuli there is no further increase in contractile force during the individual contractions. Tension reaches a plateau about which it fluctuates slightly.

A series of twitches that, through summation, attains a more or less steady and maintained force is called a *tetanus* or tetanic contraction. In Fig. 5-8C to E a tetanus is measured at stimulus frequencies from 50 to 125/s. But an entirely constant force without visible fluctuation occurs only at the maximum stimulus frequency of 125/s. This condition is called *fused (complete) tetanus.* At lower stimulus frequencies, the force fluctuates approximately in time with the stimuli; that is, the tetanus is *incomplete.* The fusion attained at high stimulus frequency by summation of the individual contractions is the *maximum force* the muscle can generate for that particular amount of prestretching. This force is two to five times greater than that attained during a single twitch.

Mechanism of Summation. The maximum possible force of which a muscle is capable is not developed during a single twitch. This might lead one to suppose that the contractile system is not fully activated during a single twitch and that the chemical reactions on which contraction is based do not take place to the full extent. However, this is not the case. Using methods that will not be discussed in further detail here, it has been shown that even during the single twitch the contractile system is fully activated for a short time; that is, the myosin–actin reaction takes place completely with ATP being split. Nevertheless, only part of the tetanic force is developed at the tendons during a twitch because during this phase the sarcomeres become shorter. As discussed above, the same is also true for isometric contractions, for with increasing force the elastic elements become longer. Thus, during the rising phase of isometric twitches the sarcomeres *shorten* at a certain rate, and this *reduces* the force developed (see Fig. 5-7).

If the muscle fiber is restimulated before a twitch is complete, the sarcomeres will still be shortened from the first twitch. Therefore, they can shorten less during the second twitch than during the first, and as a result greater force will be developed during the second twitch, at a smaller velocity of shortening. This mechanism is called *summation.* Summation is thus caused not by a stronger reaction of the contractile system but by prestretching the elastic elements by the preceding twitch. If the muscle is stimulated at a sufficiently high frequency, *tetanic* contractions are generated. In the process, by repetitive activation of the contractile system, the sarcomeres become so short that the tensile force taken up by the elastic elements equals the maximum contractile force. In this state of equilibrium, the sarcomeres no longer change their lengths, and at *zero velocity of shortening the maximum contractile force* is measurable at the tendons.

Q 5.5 Which of the following statements apply to the curve of the isometric maxima? (Check off the correct answers.)
- a. The maximum force is developed at approximately half the resting length.
- b. The maximum force is developed at the muscle length at which the maximum overlapping of myosin and actin filaments occurs during contraction.
- c. The maximum force is developed at the resting length.
- d. The maximum force is developed at the muscle length at which myosin and actin filaments overlap least during contraction.
- e. The overlapping of the myosin and actin filaments is optimum for contraction at the point of intersection of the curve of the isometric maxima and the resting length-tension curve.

Q 5.6 The contractile force of the muscle decreases with the velocity of shortening because
- a. there is an insufficient supply of ATP available at high velocity of shortening.
- b. the work performed increases in proportion to the velocity of shortening.
- c. the number of bridges between the myosin and the actin filaments decreases with the velocity of shortening.
- d. the number of bridges existing per unit time between the myosin and the actin filaments increases with the velocity of shortening.

Q 5.7 Which of the following statements apply in the case of the summation of muscle contraction?
- a. Muscle contractions are summed when the second stimulus occurs in the refractory phase of the first stimulus.
- b. Muscle contractions are summed when the second contraction is triggered before the first contraction has died away.
- c. Through summation of muscle contractions, a force is generated that is proportional to the number of summed contractions.
- d. The greatest force achievable by summation of muscle contractions is the tetanic contraction force of the fiber.

Q 5.8 Muscle contractions of individual fibers can be summed because
- a. not all the myofibrils are activated for each contraction.
- b. the contractile system is not supplied with sufficient ATP during a contraction.

c. in the case of a twitch, the sarcomeres still continue to shorten throughout the state of maximum activation of the contractile system.
d. in the case of a twitch, the state of maximum activation of the contractile system does not last long enough to permit maximum shortening of the sarcomeres.

5.3 Excitation-Contraction Coupling

Humans and animals can move only with the aid of their muscles and can make an impact on their environment only if muscle contractions can be controlled precisely. Control is exerted by the motor nervous system, which, through activation of the motor nerves, elicits EPPs in the muscle fibers (see Chapter 3). These EPPs trigger action potentials that are conducted along the muscle fibers. Excitation of the membrane is followed by contraction of the fiber. A change in the membrane potential thus controls the reaction of the contractile proteins of the muscle. This process is called *excitation-contraction coupling* and will be described in more detail in this section.

Dependence of Contractile Force on Membrane Potential. The phenomenon of excitation-contraction coupling can be studied particularly well by triggering contraction by means of step changes in the membrane potential to any arbitrary value (voltage clamp; see p. 44). The dependence of contractile force on membrane potential as measured in this way is shown in Fig. 5-9. If, starting from the resting

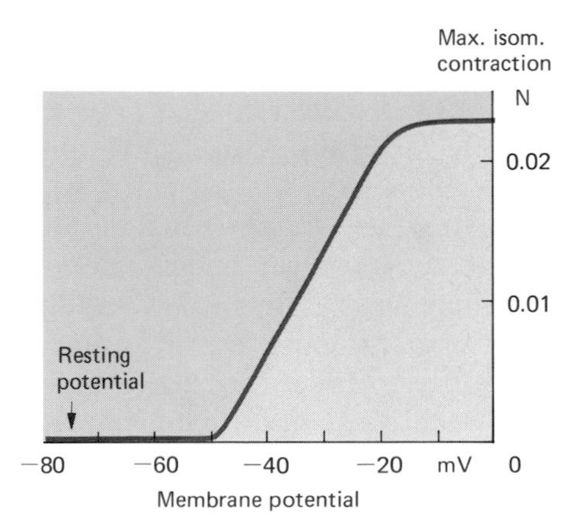

Fig. 5-9. Dependence of contractile force on membrane potential. The muscle develops force only when the potential crosses the "mechanical threshold," at about −50 mV. Between −50 and −20 mV force is about proportional to depolarization, but there is little or no increase at depolarizations beyond −20 mV.

potential, the muscle fiber is depolarized to −55 mV, then no contraction occurs. The *mechanical threshold* is crossed at slightly stronger depolarizations, from −50 mV on, and the muscle contracts. However, unlike the excitation of the membrane, the contraction does not obey the all-or-none law. In the range between −50 and −20 mV, the contractile force increases approximately in proportion to the depolarization. In this potential range the extent of the contraction is *controllable* by the depolarization. Once the depolarization attains −20 mV or more positive values, no further changes occur in the contractile force, even if depolarization continues. In this *saturation range* of depolarization, the maximum force is thus developed independently of the membrane potential.

At its peak, the *action potential* in the muscle fiber reaches a potential more positive than –20 mV (see Fig. 2-10); therefore, it *fully activates* the contractile system for a short time. As already mentioned, this activation of the contractile system by an action potential does not last long enough for the maximum isometric force to be measurable (see p. 139).

Because of the all-or-none character of the action potential, gradation of the contraction by graded depolarization is not utilized in the case of skeletal muscle fibers. Since every action potential in a muscle fiber follows the same time course, the contractile system is always activated to the same extent and for the same length of time. In contrast to the skeletal muscles, however, the smooth musculature can be depolarized to various potentials by synaptic potentials or by stretching (see p. 228). In these muscles, then, the amount of force generated is also controlled by the degree of depolarization of the fiber membrane.

Function of the Endoplasmic Reticulum. The *control* of the development of force by the amount of depolarization is a *very rapid* process. The force starts to increase steeply 1 to 2 ms after the peak of the action potential (see Fig. 5-2), and the contractile system is fully activated within a few milliseconds. This rapid coupling between membrane depolarization and contraction of the intracellular myofibrils cannot be brought about by diffusion of a substance from the membrane to the myofibrils because such diffusion would take more than 1 to 2 ms. Thus, excitation-contraction coupling in the skeletal muscles is not the result of certain substances flowing into the cell during excitation. Instead, the membrane depolarization has to be transmitted to the interior of the cell by special processes.

A special structural complex has been developed in the case of the relatively thick skeletal muscle fibers for the rapid coupling of membrane depolarization and contraction. This is the *endoplasmic reticulum.* It comprises two systems of hollow spaces or "tubes" within the muscle fibers, as shown in Fig. 5-10. On the right in that figure is the cell membrane. At regular intervals thin tubes, the *transverse tubules,* invaginate into this outer cell membrane. These tubes run deeply into the fiber, sometimes crossing it completely, at the level of the Z discs. Since the transverse tubules are invaginations of the outer membrane, their interior is directly connected with the extracellular space.

A second tubular system, the *sarcoplasmic reticulum,* borders on the transverse tubules (see Fig. 5-10). This second system runs at right angles to the transverse tubules and parallel to the filaments of the sarcomere. That is, it is longitudinally aligned. The sarcoplasmic reticulum forms a network of hollow spaces from Z discs to Z discs along the myofibrils of each sarcomere. Where the sarcoplasmic reticulum abuts the transverse tubules, it expands into terminal cisternae. A cross section through this region reveals a characteristic configuration of one thin tube accompanied on two sides by two thick tubes. This configuration is called a triad (see Fig. 5-10, left). It is clear from the figure that the transverse tubules and the sarcoplasmic reticulum do not form a common system of hollow spaces, but instead are only in close contact with one another.

Function of the Transverse Tubules. The transverse tubules are invaginations of the outer cell membrane, and as such they extend the membrane into the interior of the cell and also enclose extracellular space. As a result, excitation of the outer cell membrane can be propagated in the transverse tubules, and electrotonic depolarizations of the outer cell membrane can also be propagated via the transverse tubules into the interior of the cell. That is, the transverse tubules can *rapidly conduct depolarizations of the outer membrane into the interior of the cell.* They permit rapid activation of the contractile system in the interior of the cell even in large diameter fibers. The transverse tubules are, therefore, particularly well developed in thick muscle fibers.

The role of the transverse tubules in excitation-contraction coupling can be demonstrated by an elegant experiment. If, as one applies a fine pipette against the cell membrane, only the mouth of a single tubule is depolarized, the half sarcomeres on either side of the associated Z disc undergo local contraction. The extent of this contraction depends on the magnitude of the depolarizing current. The system of transverse tubules associated with a Z disc thus controls the contraction in the half sarcomeres on either side of that Z disc.

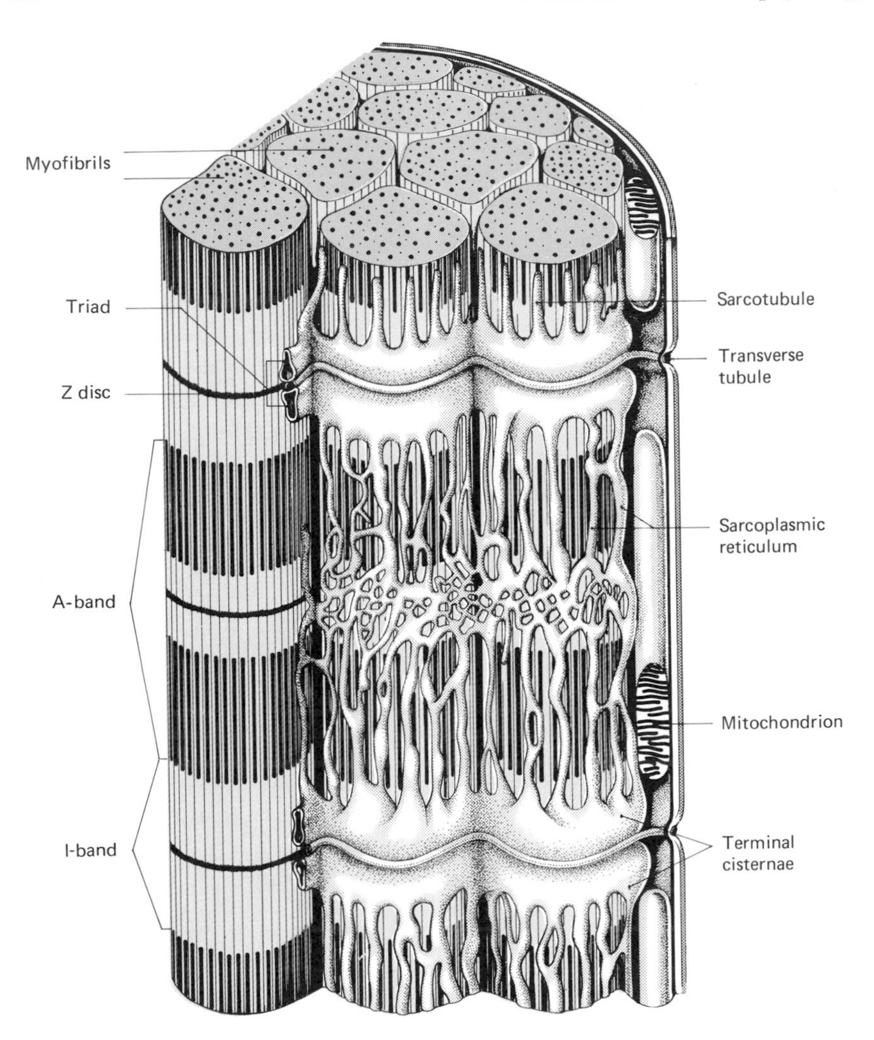

Fig. 5-10. Endoplasmic reticulum of a skeletal muscle fiber. Sketch of the fine structure of part of a muscle fiber based on an electron micrograph. At the *right-hand side* of the sketch is the cell membrane; at the level of each Z disc it invaginates, sending transverse tubules across the interior of the fiber. Between the Z discs and parallel to the myofibrils runs the sarcoplasmic reticulum, sacklike enlargements of which (the terminal cisternae) adjoin the transverse tubules. A cross section through transverse tubule and adjacent terminal cisternae reveals a configuration known as a triad (*upper left*). Modified from Bloom and Fawcett (1969) *A Textbook of Histology*, Saunders, Philadelphia.

Control of Intracellular Ca^{++} Concentration by the Sarcoplasmic Reticulum. But how does the system of transverse tubules control contraction? In the discussion of the molecular mechanism of contraction, it was stated that the Ca^{++} concentration in the intracellular space controls the reactions of the contactile proteins (see p. 133). Myosin and actin can link up, with adenosine triphosphate being broken down, only if the Ca^{++} concentration reaches about 10^{-5} mol/l. The endoplasmic reticulum exercises control on contraction by bringing about rapid changes in the intracellular Ca^{++} concentration.

The *resting muscle fiber* has a very *low* Ca^{++} concentration—about 10^{-8} mol/l. During *contraction* the concentration *increases* to about 10^{-5} mol/l. This change in the intracellular Ca^{++} concentration can be elegantly demonstrated by injecting a luminescent dye such as aequorin into a muscle fiber. The light emission of this dye is strongly dependent on the Ca^{++} concentration. When such a muscle fiber is depolarized, the light emission increases rapidly to a value corresponding to 10^{-5} mol/l Ca^{++}. Then the light emission and thus also the free Ca^{++} concentration rapidly decline again.

What is the source of the Ca^{++} ions that increase the free Ca^{++} concentration in the intracellular space rapidly to 1,000 times the resting value after depolarization? A very high concentration of Ca^{++} ions has been detected in the relaxed muscle in the terminal cisternae of the *sarcoplasmic reticulum,* close to the transverse tubules (see Fig. 5-10). It is asssumed that once the transverse tubules have been depolarized they act in some as yet unknown way to cause an increase in membrane conductance for Ca^{++} ions in the *sarcoplasmic reticulum.* Because of the high concentration gradient, Ca^{++} *ions then flow out* of the sarcoplasmic reticulum into the intracellular space. This raises the Ca^{++} concentration and permits the contraction-inducing reaction of myosin, actin, and adenosine triphosphate to take place. Hence excitation-contraction coupling is effected by controlling the Ca^{++} concentration. The various steps in the coupling process are shown once more in diagrammatic form in Fig. 5-11A. The depolarization of the cell membrane is conducted by the transverse tubules to the interior of the cell. This depolarization causes the sarcoplasmic reticulum to release Ca^{++} ions. The Ca^{++} ions diffuse to the adjacent myofilaments where they make possible the chemical reactions that lead to contraction.

Once depolarization has come to an end, the intracellular Ca^{++} concentration drops very quickly, and *relaxation* commences. The rapid drop in the intracellular Ca^{++} concentration is not due merely to the fact that the sarcoplasmic reticulum stops releasing Ca^{++}. Instead, the sarcoplasmic reticulum picks up Ca^{++} ions actively from the inte-

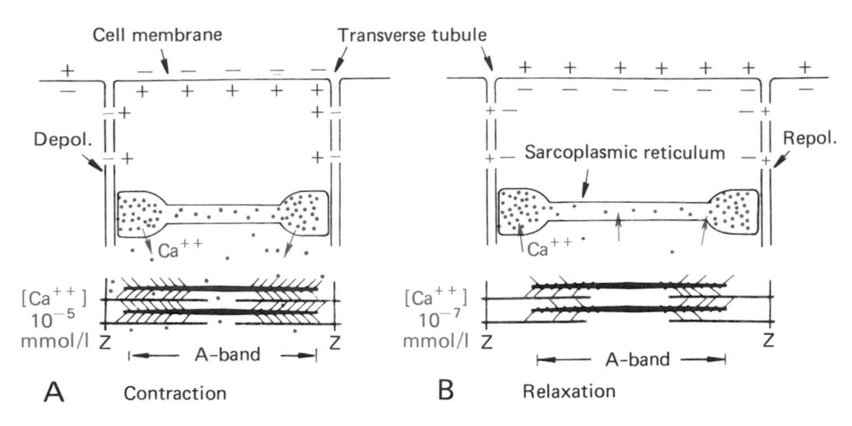

Fig. 5-11. Excitation-contraction coupling. Diagram of the cell membrane showing the transverse tubules with a portion of the sarcoplasmic reticulum between them and a sarcomere. The red dots denote Ca^{++} ions, present in high concentration in the sarcoplasmic reticulum. During contraction (**A**) the membrane of the sarcoplasmic reticulum becomes permeable to Ca^{++} ions after depolarization of the transverse tubules; the Ca^{++} ions flow out and enable the filaments to contract. During relaxation of the muscle (**B**), the transverse tubules are repolarized, the Ca^{++} efflux from the sarcoplasmic reticulum therefore ceases, and the reticulum actively pumps Ca^{++} out of the intracellular space. Contraction ceases when the intracellular concentration of free Ca^{++} ions is sufficiently low.

rior of the cell. An ion pump is situated in the membrane of the sarcoplasmic reticulum. By using metabolic energy, this pump transports Ca^{++} against the concentration gradient into the sarcoplasmic reticulum. The Ca^{++} pump of the *sarcoplasmic reticulum* operates much like the Na^{+} pump in the membrane of nerve and muscle fibers (see p. 32) and other cells. By means of this Ca^{++} pump the free Ca^{++} concentration in the intracellular space is maintained at a very low level (10^{-8} mol/l) under resting conditions, and the sarcoplasmic reticulum can store Ca^{++} in high concentrations. The pump also brings about the rapid drop in Ca^{++} concentration at the end of depolarization. Most of the Ca^{++} ions present at this time in the intracellular space are rapidly taken up into the sarcoplasmic reticulum by the pump.

Figure 5-11B summarizes in diagrammatic form the manner in which muscular relaxation is initiated by the sarcoplasmic reticulum. The cell membrane is repolarized and this repolarization obtains also in the transverse tubules. Consequently, the membrane of the sarcoplasmic reticulum becomes impermeable to diffusing Ca^{++} ions, and the Ca^{++} efflux is terminated. With the aid of the pump the Ca^{++} ions are again transported into the sarcoplasmic reticulum, and the contraction is brought to a stop due to the insufficient concentration of intracellular Ca^{++}.

Q 5.9 The transverse tubules
  a. are open to the extracellular space.
  b. run transversely through the muscle fiber.
  c. are also depolarized when the outer cell membrane is depolarized.
  d. are openly connected with the sarcoplasmic reticulum.
  e. release Ca^{++} ions into the intracellular space to initiate contraction.

Q 5.10 The sarcoplasmic reticulum
  a. is open to the extracellular space.
  b. can store Ca^{++} ions.
  c. runs in the longitudinal direction of the muscle fibers.
  d. can actively take up Ca^{++} ions.
  e. regulates the intracellular K^{+} concentration.

Q 5.11 Contraction of the skeletal muscle fiber is triggered by
  a. depolarization of the membrane during the action potential.
  b. an increase in the intracellular Na^{+} concentration during the action potential.
  c. an increase in the intracellular concentration of free Ca^{++}.
  d. an increase in the intracellular level of free ATP during membrane depolarization.
  e. inhibition of the Na^{+} pump during membrane depolarization.

Q 5.12 The free Ca^{++} concentration in the relaxed muscle is, at 10^{-8} mol/l, very low because
  a. the outer cell membrane is impermeable to Ca^{++}.
  b. a Ca^{++} pump transports Ca^{++} ions into the sarcoplasmic reticulum and thus reduces the free intracellular Ca^{++} concentration.
  c. the contractile proteins bind Ca^{++} ions to themselves and thus reduce the intracellular concentration of free Ca^{++}.
  d. Ca^{++} ions are consumed during contraction so that at the end of the contraction the free Ca^{++} concentration drops to a very low value.

5.4 Regulation of Muscle Contraction

So far in our discussion of muscle contraction we have concentrated on the individual muscle fiber and its myofibrils. However, individual fibers contract by themselves in the body very rarely, if at all; instead, contraction involves varying numbers of the fibers contained in a muscle. When a muscle contracts, many individual fibers act together, and to control the muscular force the nervous system must coordinate the activity of the fibers. We will now examine the way in which the contraction of the entire muscle is regulated.

Summation of the Contraction of Several Fibers. The contractile force of a *single muscle fiber* can be regulated, within certain limits, by the frequency with which it is excited. At very low frequencies (for example, 2/s), the greatest force developed is that of a single twitch. If the frequency of excitation is increased, the maximum (that is, the tetanic) force can be developed by the muscle by means of summation. The tetanic force can be about five times greater than the maximum force developed in a twitch. The range in which the contractile force of a *single* fiber can be regulated by means of the frequency of excitation is thus relatively narrow.

Besides the limited summation of the contractions of the individual fibers, *summation of the twitches of parallel fibers* also takes place in the intact muscle. The parallel fibers of the muscle all end at the same tendons, so that these combine the force they develop. This summation of the contractions of parallel fibers is illustrated by Fig. 5-12. The figure shows equal twitches of three individual fibers that are not summated because the stimulus frequencies (2 to 4/s) are too low. The

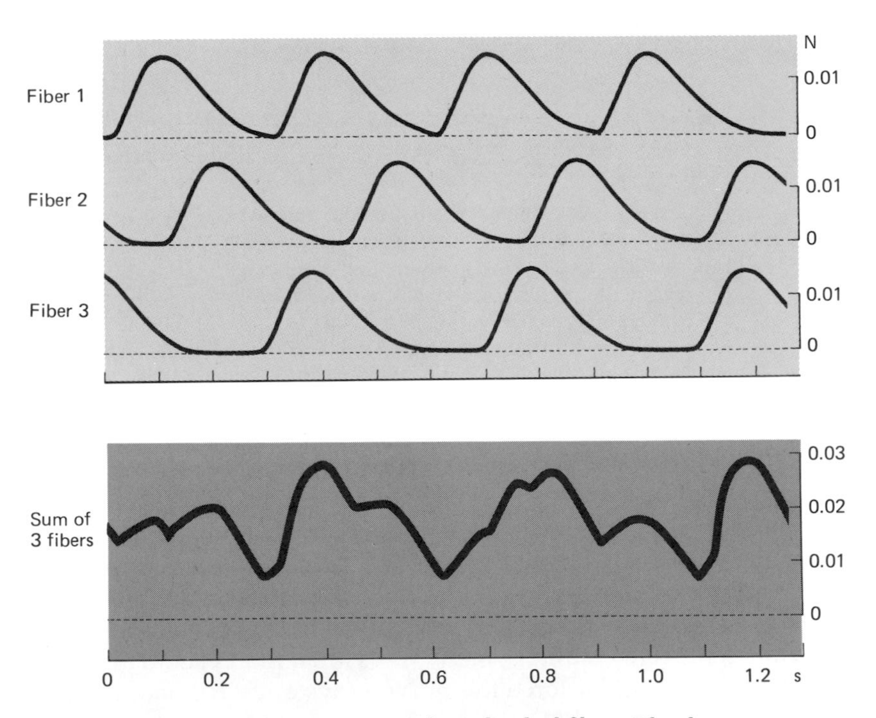

Fig. 5-12. Summation of force generation by individual fibers. The three *upper curves* diagram the sequence of uniform twitches in each of three parallel fibers of a muscle. The *bottom curve* shows the summed contractions of the three fibers; in this curve the force never falls to zero (*dashed line*), even though there was no summation of the twitches of each individual fiber.

bottom curve in Fig. 5-12 depicts the summated contractile force of the three individual fibers at each instant. The curve thus corresponds to the force measured at a tendon to which all three fibers are joined. The maximum force in this summation curve is more than twice that of the twitches, and at no time does the force drop to zero (dotted line). If the contractions of a larger number of fibers were summed, the combined curve would run more uniformly and attain a higher maximum force than the combined curve in Fig. 5-12.

Muscle Tonus. Summation of the twitches of many fibers, excited asynchronously at low frequencies up to 5/s, generates a total force that does not fluctuate very much, with an amplitude that must be approximately proportional to the average frequency of excitation. The "background" tension produced in this way by summation of the twitches of many fibers is called *tone,* or *tonus.* All the muscles in a living organism possess such tone. Even in a relaxed limb, the motor nerves are activated at low frequency. The resulting tone is detectable as a *resistance to passive bending of the limb.*

The tone of the muscles is chiefly involved in their *postural function.* Even when we sit relaxed, our limbs are not fully passive; instead they adopt a certain posture. This posture is determined by the relative degree of tone in the various groups of muscles. When some sort of disturbance is anticipated (example: someone is driving a car and a potential hazard situation develops on the road ahead), the tone increases—that is, the basic tension of all the muscles is increased and the posture adopted is more firmly held.

The tone of the muscles also plays an important role in the *thermoregulatory processes* of the organism. As described above, during each contraction, part of the converted energy appears in the form of heat, and in the case of tonic contractions, which perform no externally evident work, the energy associated with them is also finally converted into heat. Thus, by varying the muscle tone the body can greatly vary the amount of heat it produces, and this capability can be put to use in regulating body temperature at various ambient temperatures. When vigorous muscular activity occurs, a great deal of heat is released by the chemical reactions; the body also has special mechanisms, such as sweating, to eliminate the excess heat.

Generation of Maximum Muscular Force. If the frequency of excitatory impulses in the motor-nerve fibers is increased to more than 5/s, the force of the contraction increases through summation of the twitches in the individual fibers as well as through summation of the contractions of the individual parallel fibers. The *maximum muscular force* is developed when all the parallel fibers develop *tetanic ten-*

sion. This force can be attained in most muscles of warm-blooded animals at frequencies of about 50/s; the summation of the contractions of many fibers compensates for the slight fluctuations in force of the individual fibers, which still occur at this frequency (see Fig. 5-8D). Since the motoneurons and the motor end plates (see p. 73) can function at frequencies of several 100/s, the motor system is easily able to attain maximum contractile force.

Motor Unit; Electromyogram. The above statements concerning the parallel but independently contracting individual fibers need qualification. Not all the fibers are stimulated asynchronously and independently of each other. The number of motor nerve fibers that innervate the muscle is smaller than the number of muscle fibers. Within the muscle, the nerve fibers branch and innervate several muscle fibers. Thus, stimulation of a single nerve fiber in each case results in simultaneous excitation of a group of muscle fibers. The motor *nerve fiber* together with the *muscle fibers that it innervates* are referred to as a *motor unit.*

The motor units vary in size. In densely innervated muscles such as the outer muscles of the eye, units contain on the average seven muscle fibers. In muscles of the lower leg, they contain on the average 1,700 muscle fibers. As a result of the combining of muscle fibers in motor units, it is slightly more difficult to grade muscular contractions because all the fibers in the motor unit contract simultaneously. Therefore, in muscles that must produce very finely regulated amounts of force (for example, the eye muscles), the motor units contain only a few fibers.

The excitatory impulses of the motor units can be recorded in an *electromyogram* (EMG). The EMG is an extracellular recording of the potential of the muscle. The electrodes are either placed on the skin above the muscle or inserted into the muscle (extracellularly). Figure 5-13 illustrates an electromyogram of a human eyelid muscle. No changes in potential are visible in A, and the muscle is fully relaxed. In B, C, and D the lid is closed with increasing force. In the process, extracellularly recorded action potentials, or *impulses,* show up in the electromyogram. These impulses take the form of rapid rises in potential followed by a small drop in potential. The large impulses in Fig. 5-13B to D all stem from one motor unit. The first five such impulses are denoted by arrows in Fig. 5-13B. The frequency of excitation in this motor unit increases with increasing development of force. In B, 13 large impulses per second can be seen for the weak contraction; this rate increases to 31/s for the strongest contraction (Fig. 5-13D). At this frequency, this motor unit probably approaches tetanus.

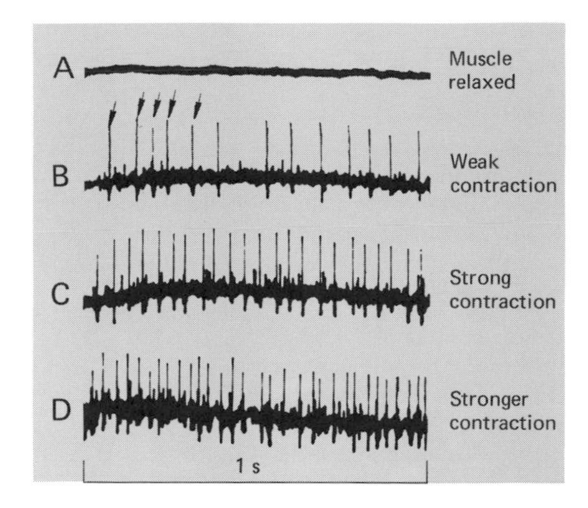

Fig. 5-13. Electromyogram of an eyelid muscle. During the recording of the EMGs, the muscle was fully relaxed in **A** and underwent increasing contraction in **B** and **C**. The large impulses in **B** to **D** (marked by *arrows* at the beginning of **B**) are all produced by one motor unit. The frequency of discharge in this motor unit increases from **A** to **D**. In addition, as contraction becomes stronger, progressively more small impulses are visible, indicating increased excitation in neighboring motor units as well. Modified from Bell, Davidson, and Scarborough (1968) *Textbook of Physiology and Biochemistry*, Livingston LTD, Edinburgh.

Figure 5-13 shows that with increasing contractile force not only does the frequency of excitation in a single motor unit increase, but also more motor units are activated. An increasing number and variety of small pulses appear alongside the "large pulses" in C and D. These small pulses are generated by nearby motor units that are further from the measuring electrodes and, therefore, generate only small changes in potential at the electrodes. The increasing number of small pulses also shows that a larger number of motor units are activated during strong contractions.

The EMG is a much-used neurophysiological tool for diagnosing muscular diseases. These diseases take the form of paralysis or reduced ability to generate force (myasthenia) or manifest themselves in uncontrolled excess development of force (myotonia). In many case histories, the reactions of the musculature reflect damage or disease of the motor nervous system. In other cases, it is the neuromuscular transmission that is affected. Recording the excitation pattern of the motor units in the electromyogram, apart from establishing exactly the type of motor disturbance involved, is, therefore, also a great aid in diagnosis. Diseases of the actual contractile system of the muscle fibers occur relatively rarely. In such cases, the trouble is due to degenerative changes in the muscle fibers (muscular dystrophies), which can usually be traced to hereditary enzyme defects or hormonal disturbances.

Q 5.13 The contractile force of a muscle can be controlled by
- a. varying the frequency of excitation of the individual motor units.
- b. varying the number of muscle fibers that belong to a motor unit.
- c. varying the number of activated motor units.
- d. increasing the ATP level in the muscle fiber.
- e. varying the number of activated motor units *only*.

Q 5.14 The electromyogram (EMG) records
- a. the amplitude of muscle contraction.
- b. the duration of muscle contraction.
- c. the excitation in the motor nerve fibers.
- d. the excitation in the motor units.
- e. variations in the number of activated motor units.

Q 5.15 A motor unit
- a. consists of all the muscles that together execute the same movement.
- b. consists of a motor nerve fiber with the muscle fibers that are innervated by it.
- c. always comprises at least 100 muscle fibers.
- d. is excited approximately simultaneously in all its elements.

6
MOTOR SYSTEMS

R.F. Schmidt

Only by using our skeletal muscles can we affect—or, indeed, interact on any level with—our environment. The musculature is involved in all actions, from the roughest manual labor to the communication of the most subtle thoughts and feelings, by speech or writing, expression or gesture. Moreover, no movement can be performed effectively unless a posture appropriate to the action is assumed, by suitable arrangement of the limbs and the body as a whole.

For these reasons, the control of posture and movement is one of the most important functions of the CNS. The structures chiefly responsible, called *motor centers,* are located in quite different parts of the nervous system. They are arranged on a *pyramidal plan,* extending from the phylogenetically oldest part, the spinal cord, to the newest, the cerebral cortex. Study of the motor functions of various parts of the brain has shown that the additions to the nervous system made necessary by progressive differentiation within the animal kingdom were brought about less by remodeling of existing structures than by superposition of supplementary, more proficient reflex and control systems. In this sense, the motor organization has a *hierarchic character.* In the following sections the operation of the centers lowest in the hierarchy will be discussed first, and then we shall examine the extent to which their capabilities are modified and extended by the higher centers.

In considering the motor functions of the nervous system, we must keep in mind that the CNS centers involved in posture and movement can perform appropriately only if they receive an uninterrupted stream of afferent (sensory) information. To emphasize the role of the sense organs in the control of posture and movement, one often uses the term *sensorimotor system* to connote the totality of afferent and efferent functions participating in muscular activity. The dependence of motor performance upon afferent input is particularly clear at the level of the spinal cord, where individual receptor types (such as muscle-spindle receptors) are linked to motoneurons in a relatively stereotyped manner, forming reflex arcs. In the first two sections of this chapter, which deal with the motor functions of the spinal cord, we shall, therefore, pay special attention to the sensory afferents.

6.1 Spinal Motor Systems I: Roles of Muscle Spindles and Tendon Organs

The structure of the muscle spindle and the central connections of the Ia afferents to the homonymous motoneurons have been described in Sec. 4.2. It was shown that the monosynaptic stretch reflex can be triggered either by stretch of the whole muscle or by intrafusal contraction. In Secs. 4.1 and 4.3 it was mentioned that the Ia fibers have both excitatory and inhibitory synaptic connections, the former with homonymous motoneurons and the latter with antagonistic motoneurons (see Fig. 4-4A). In this section, we shall proceed to consider in more detail the mechanical arrangement and pattern of discharge of the two most important receptor organs associated with the muscles, the muscle spindles and the Golgi tendon organs. Then the reflex connectivity of their afferent fibers, the Ia and Ib afferents, and its significance in motor function will be discussed.

Structure and Position of Muscle Spindle and Tendon Organ. The structure of the muscle spindle was illustrated in Fig. 4-7. Implicit in the discussion of this figure and of the activation of the receptor by muscle stretch or intrafusal contraction (Fig. 4-10) was the fact that the muscle spindles, as shown in Fig. 6-1, lie mechanically parallel to the extrafusal musculature.

A second important type of stretch receptor, in addition to the muscle spindle, is situated in the tendon. Each such receptor is associated with the tendon fascicles of about 10 extrafusal muscle fibers; it is surrounded by a capsule of connective tissue and sends out one or two thick myelinated nerve fibers. These stretch receptors are called *tendon organs* (or Golgi tendon organs), and their afferent nerve fibers are *Ib fibers* (conduction velocity about 75 m/s; see Table 2-2). As Fig. 6-1 shows, the tendon organs are arranged mechanically in series with the extrafusal musculature.

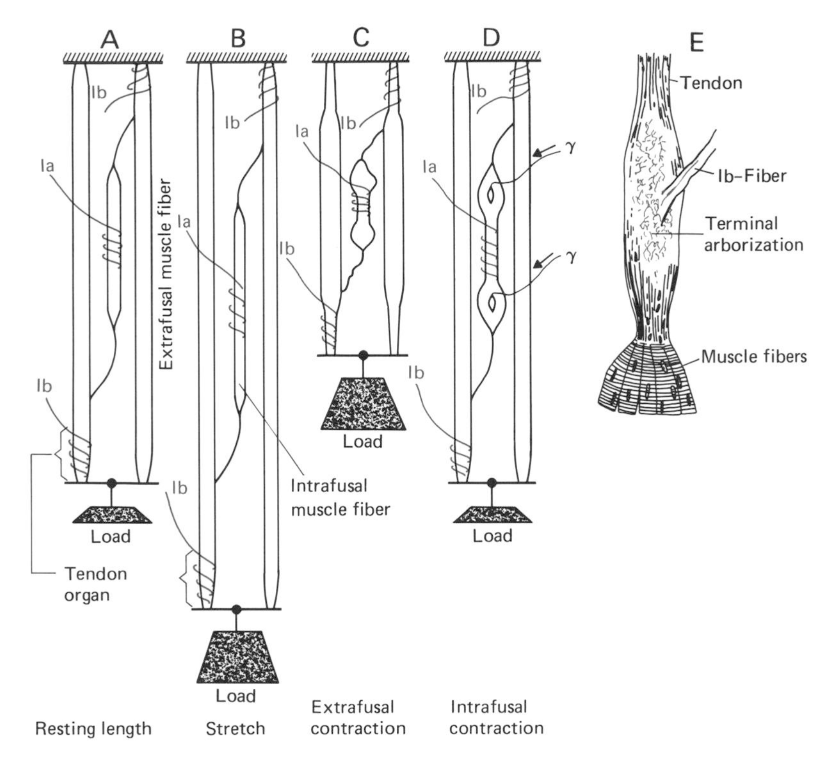

Fig. 6-1. Diagram of the mechanical arrangements of muscle spindles and Golgi tendon organs in a resting muscle (**A**), and their changes during passive stretch (**B**), isotonic contraction of the extrafusal muscle fibers (**C**), and contraction of the intrafusal muscle fibers alone (**D**). Only one intrafusal muscle fiber of a single muscle spindle is shown. **E** is a drawing of a Golgi tendon organ as seen in the light microscope by Ramon y Cajal.

Discharge Patterns of the Muscle Spindles and Tendon Organs. Because of the different arrangement of spindles and tendon organs in the muscle, the two types of receptor produce different patterns of nerve-impulse discharge when the muscle contracts. Comparison of Figs. 6-1 and 6-2 reveals this difference.

If a muscle is stretched to near its resting length (Figs. 6-1A and 6-2A, the primary muscle-spindle endings (with Ia fibers) discharge, whereas the tendon organs (with Ib fibers) are silent. This situation results from the fact that the thresholds of the tendon organs are somewhat higher than those of the muscle spindles; moreover, for a given muscle length (see B in Figs. 6-1 and 6-2) the discharge frequency of the tendon organ is always lower than that of the muscle spindle.

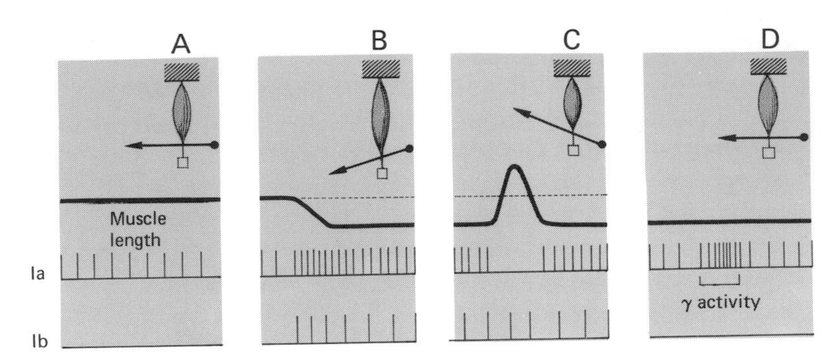

Fig. 6-2. Discharge patterns of muscle spindles (in Ia fibers) and tendon organs (in Ib fibers) under resting conditions (**A**), during stretch (**B**), during transient isotonic contraction of the extrafusal musculature (**C**), and during contraction of the intrafusal muscle fibers after activation via the γ motor fibers (**D**).

Further stretch (B in Figs. 6-1 and 6-2) causes an increase in discharge rate of the muscle spindles and initiates discharge of the tendon organs. While the stretching is in progress, the discharge rates of both fiber types are higher than when the muscle is held steady at its new length. This fact indicates that their discharge rates are determined not only by the length of the muscle but also depend upon the rate of change of length (that is, the first derivative of length with respect to time). The latter, dynamic component is considerably more pronounced in muscle spindles than in tendon organs.

Isotonic contraction of the extrafusal musculature (C in Figs. 6-1 and 6-2) releases the muscle spindles from stretch and thus causes an interruption of the Ia discharge. But the tendon organ is still stretched—in fact, its discharge rate actually increases temporarily during a contraction because acceleration of the load brings about a brief increase in stretch of the tendon organ. We may thus conclude that the *muscle spindles* measure primarily the *length of the muscle*, whereas the chief role of the *tendon organs* is to monitor *tension*. Accordingly, in *isometric conditions* (in which tension is increased without change in length), there is a marked increase in discharge rate of the tendon organs, while the muscle spindles continue to discharge at about the same rate as before.

Intrafusal contraction, resulting from activation of the γ motoneurons (D in Figs. 6-1 and 6-2), does not affect the discharge of the Ib fibers; as was pointed out in the discussion of Fig. 4-10, contraction of the relatively few intrafusal fibers alone produces no detectable change in muscle tension. However, the stretching of the receptor elements in the muscle spindle (Figs. 4-10C and 6-1D) increases the discharge rate of the Ia afferents (Fig. 6-2D). The implications of this mechanism with respect to motor activity will be brought out in the discussion below of the central connectivity of Ia afferents.

Stretch Reflex and Reciprocal Antagonist Inhibition. In Fig. 6-3A the example of flexion and extension of the elbow joint illustrates the central connections of the Ia muscle-spindle afferents coming from antagonistic muscles (the biceps and triceps, respectively). A preliminary description of this arrangement was given in Chapter 4 (see Figs. 4-4, 4-8, and the corresponding text). Now the processes associated with *passive change of joint position,* brought about by an external force, can be discussed in terms of the interaction among the neuronal elements. It will become apparent that all the changes in reflex activity originating in the muscle spindles of the opposed muscles collaborate to prevent or reverse the imposed change in position of the joint—that is, they tend automatically to *keep the muscle length constant.*

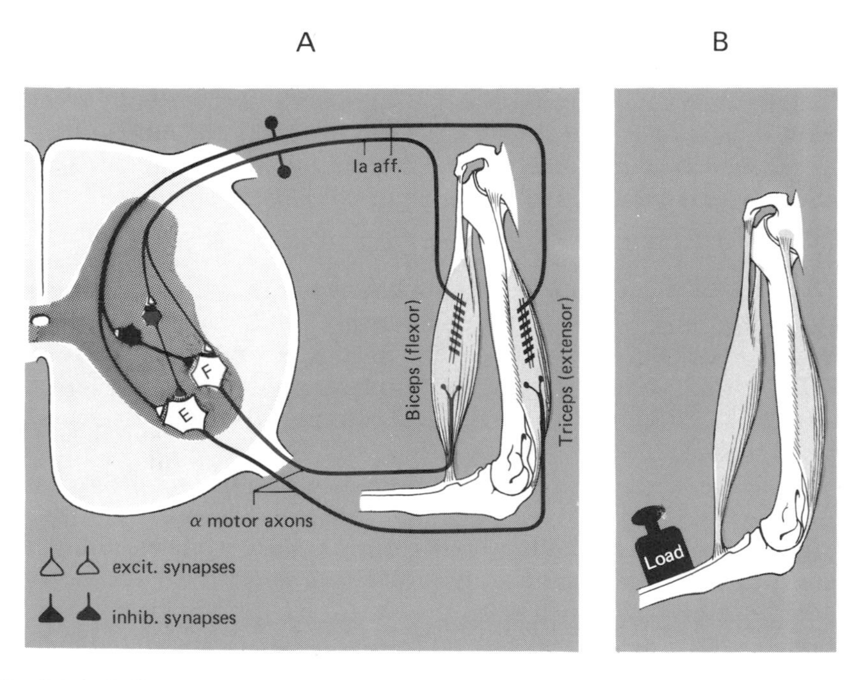

Fig. 6-3.A. Reflex paths underlying the stretch reflex and reciprocal antagonist inhibition. F, flexor motoneuron of the elbow joint; E, extensor motoneuron. The flexor (biceps) and extensor (triceps) of this joint and the polarity of the synapses are indicated in the drawing. The consequences of the passive extension of the joint by an external force (the load) shown in **B** are described in the text.

Let us assume that the original position of the elbow, shown in Fig. 6-3A, is altered by a load set onto the lower arm, as shown in Fig. 6-3B. The *stretching of the biceps* will cause an increase in the activity of its muscle-spindle receptors and thus more strongly (1) excite the biceps motoneurons and (2) inhibit the triceps motoneurons. Whereas the biceps is stretched by the load, the triceps is simultaneously released from tension. This *passive shortening of the triceps* reduces the activity of the triceps muscle spindles, so that (3) the homonymous excitation of the triceps motoneurons is diminished and (4) there is less reciprocal inhibition of the biceps motoneurons. (This latter effect—a partial or complete "removal" of inhibition—is frequently termed *disinhibition.*) An externally imposed extension of the elbow joint thus leads to *increased activation of the biceps motoneurons,* because the homonymous excitation increases and the reciprocal antagonist inhibition decreases. At the same time, the *activity of the triceps motoneurons is reduced,* because their homonymous excitation decreases and the reciprocal antagonist inhibition increases. Altogether, the changes of activity produced in the four mono- and disynaptic reflex arcs achieve a maintained increase in tension of the biceps and a maintained decrease in tension of the triceps; as a result, the imposed change in joint position is opposed—the net effect tends to keep the original muscle length constant. The four reflex arcs, then, together constitute a *length-control system* for the muscles involved. The application of control-systems technology to such systems is discussed in Chapter 7.

Functions of the γ-loop. It was shown in Sec. 4.2 that the discharge rate of the muscle-spindle afferents can be affected via the efferent innervation of the intrafusal muscle fibers (the γ motor axons; see Fig. 6-2D), with a consequent change in muscle length. We shall now take a closer look at this mechanism; but first it may be useful to read again the paragraphs on this subject in Sec. 4.2.

Curve *a* in Fig. 6-4 shows the relationship between muscle length (abscissa) and frequency of the Ia-afferent impulses from the muscle-spindle receptors (ordinate) when there is little γ-fiber activity, and thus little intrafusal contraction. The *discharge rate of the Ia fibers* is seen there to be *linearly related to muscle length.* As a result of increased γ-fiber activity, the muscle spindles that had been firing at a low frequency (for example, Point 1 in *a*) discharge at a greater rate (Point 2 in *b*), with no change in length of the muscle. The effect of this increased activity of the Ia fibers is an enhanced excitation of the homonymous motoneurons and more pronounced inhibition of the antagonistic motoneurons. At first, the discharge rate of the antagonis-

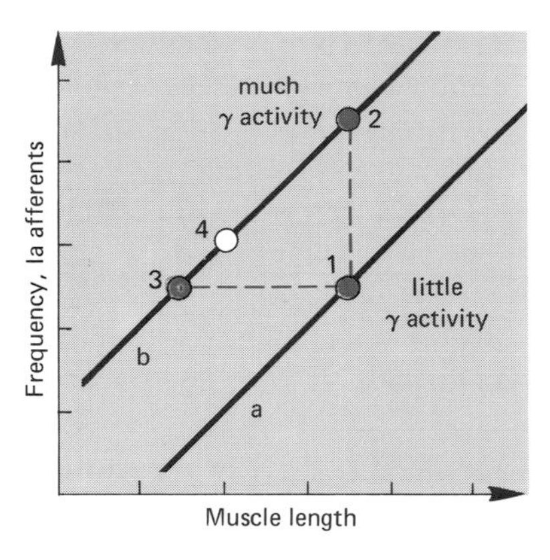

Fig. 6-4. The relationship between muscle length (abscissa) and frequency of the afferent impulses from the primary endings of the muscle spindles (ordinate). The discharge rate in the γ motor fibers is low in the case of the line labeled *a* and high in *b*, so that the tonus of the intrafusal muscle fibers differs in the two cases. The points labeled *1* to *4* are discussed in the text.

tic muscle-spindle afferents is unchanged, for in contrast to the situation of Fig. 6-3B, the lengths of the two muscles acting in opposition—the agonist and the antagonist—have not changed. The increased activity of the agonist Ia fibers thus leads to a contraction (rise of tonus) of the agonist with simultaneous relaxation (lowering of tonus) of the antagonist—that is, to a movement in the joint concerned. The movement will stop as soon as Point 3 in *b* is reached, when the (now shortened) muscle-spindle afferents again discharge impulses at a rate corresponding to that at Point 1 in *a*. Muscle length, then, can be altered by the γ *efferents without maintained change* in the discharge rate of the muscle-spindle receptors.

Since the changed position of the joint has stretched the antagonist, if Point 3 on the curve is to have precisely the same value on the ordinate as Point 1, the tension in the antagonistic intrafusal muscle fibers must also be changed somewhat. It must be reduced sufficiently to bring the slightly increased afferent output from the (stretched) antagonist back to the original level. If the intrafusal tension of the antagonist muscle spindle is not reduced, there results an enhanced reciprocal inhibition of the agonistic motoneurons, so that the muscle length returns not to Point 3 in Fig. 6-4, but, for example, only to Point 4. In this case, the agonistic Ia discharge rate, slightly increased in comparison to its value at Point 1, compensates for the increased discharge rate of the antagonistic Ia fibers.

All these considerations illustrate that contractions of the musculature can be brought about either by way of the γ loop or by direct activation of the α motoneurons. The advantage of direct *activation of the α motoneurons by supraspinal centers* is its short latency; a disadvantage is that it can cause a pronounced disturbance of the delicately balanced length control of the stretch reflex. That is, the affected muscle spindles may no longer be adequately stretched (subthreshold excitation) or may be stretched too much (saturation). By contrast, *activation of the γ loop* can cause a shortening of the muscle with only a transient change in the discharge rate of the muscle-spindle afferents.

More recent studies, on man and other animals, have shown that changes in muscle length are *not induced by an exclusive activation of the γ loop*. It is true that an increase in discharge rate of the spindles can be observed when muscles contract (which unambiguously indicates contraction of the intrafusal fibers in the shortening muscle), but the increase does not precede the movement, as would be observed if the movement were initiated by γ activity, but rather appears simultaneously with it. That is, the α and γ motoneurons are activated at about the same time; for this reason, the term *α–γ coupling* (or *coactivation*) is used. The action of the α motoneurons is assisted by that of the γ-motoneurons. The two collaborate rather as in the mechanism of power-assisted braking in an automobile. The *role of the γ loop* can thus be described as a *servoassistance of movement* (see Chapter 7).

Segmental Connectivity of the Ib Fibers. Role of the Golgi Tendon Organs. In a functional sense, the segmental connections of the Ib fibers are the mirror image of those of Ia fibers (compare Figs. 6-3A and 6-5). The tendon organs make *inhibitory connections* with their *homonymous* motoneurons and *excitatory connections* with the *antagonistic* motoneurons. None of these connections, however, are monosynaptic; all involve at least two synapses. Because the tendon organs are *activated by muscle tension,* a marked increase in this tension, whether it results from stretch, contraction, or a combination of the two, brings about inhibition of the homonymous motoneurons via the Ib fibers. Thus, the Golgi organs *protect against overloading,* acting to prevent too rapid rises in tension, which might tear the muscle or tendon. In experiments on animals, it has been found that increasing stretch of a muscle produces steadily increasing tension in the muscle (via the stretch-reflex arc) until a degree of stretch is reached at which muscle tonus suddenly falls off. (The phenomenon, the "jackknife reflex," rather resembles the sequence of forces applied as one closes a pocket knife.) The sudden release of tension is

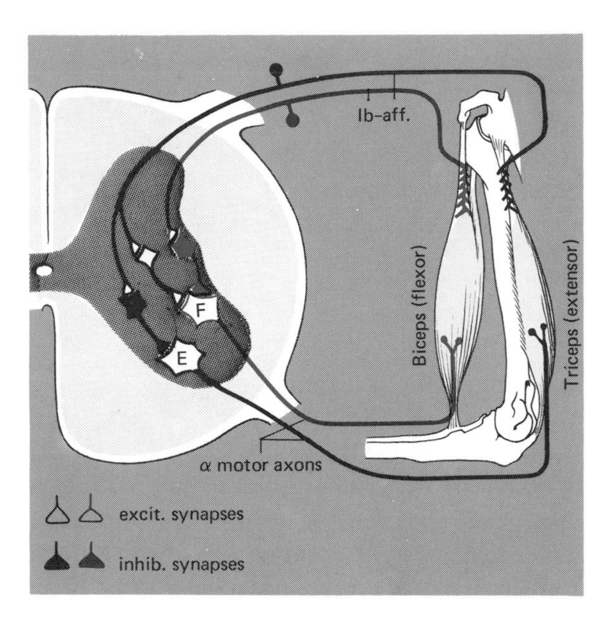

Fig. 6-5. Intrasegmental connections of the Ib fibers from the tendon organs in the muscle, represented as in Fig. 6-3. The excitatory connection between the flexor Ib fiber and the extensor motoneuron (E) has been omitted, because this reflex path is not present for all joints.

ascribed to the inhibitory action of the homonymous tendon organs. This observation has suggested that the Ib reflex arc is primarily a *protective reflex*.

But such protection from overstretching is only one aspect of tendon-organ function. Whereas increase in muscle tension produces inhibition of the homonymous motoneurons via the tendon organs, decrease in tension leads to a decrease of impulse frequency in the Ib fibers and thus to disinhibition (removal of inhibition) of the homonymous motoneurons, so that the tension in the muscle tends to increase again. In other words, the *tendon-organ reflex arc* is connected in such a way that it can tend toward *keeping muscle tension constant*. In every muscle, then, there are two feedback systems—a length-control system with the muscle spindles as sensors, and a tension-control system with the tendon organs as sensors. (In Chapter 7, the length-control system is discussed more explicitly in the light of control theory.)

It is not immediately obvious, from the viewpoint of systems theory, why both a tension-control and a length-control system should be necessary (see Chapter 7). In a simplified length-control system, the force developed by the muscle might simply be proportional to the efferent impulses in the α motor axons, and a tension-control system would seem superfluous. But we have learned in Chapter 5 that muscular force depends on other factors as well—the previous degree of stretch, the speed of contraction, and the amount of work the muscle has already done. In a sense, then, departure of muscular tension from the desired value, as a result of such factors, is monitored by the tendon organs and corrected by way of the tension-control system.

Q 6.1 Which of the following statements is/are correct?
a. The muscle spindles lie parallel to the intrafusal musculature.
b. The tendon organs lie in series with the extrafusal musculature.
c. The tendon organs are innervated by Ia afferents.
d. The Ib afferents make disynaptic excitatory connections with the homonymous motoneurons.
e. The efferent innervation of the tendon organs comprises γ fibers.

Q 6.2 Which of the following inputs to a motoneuron are excitatory?
a. Afferent activity of homonymous muscle spindles.
b. Afferent activity of homonymous tendon organs.
c. Afferent activity of antagonistic muscle spindles.
d. Afferent activity of antagonistic tendon organs.

Q 6.3 Increased activity of the γ efferents to a flexor muscle
a. leaves the position of the joint unchanged.
b. causes extension of the joint via the tension-control system.
c. increases the tonus of both flexor and extensor muscles with no change in joint position.
d. causes a reflex bending of the joint.
e. reduces the Ia activity of the antagonist extensor.

Q 6.4 The sudden relaxation of muscle tonus observed after extreme stretch is caused by
a. strong excitation of the homonymous muscle spindles.
b. strong excitation of the antagonist muscle spindles.
c. complete removal of stimulation of the antagonist muscle spindles.
d. strong excitation of the homonymous tendon organs.
e. complete removal of stimulation of the heteronymous tendon organs.

Q 6.5 Which type of receptor is
a. the sensor in the length-control system of the muscle.
b. the sensor in the tension-control system of the muscle?

6.2 Spinal Motor Systems II: Polysynaptic Motor Reflexes; the Flexor Reflex

In the preceding section we examined the role of the receptors located within the muscle itself, the muscle spindles, and tendon organs. Many of the other receptors in the body—for example, those in the skin—can also trigger motor reflexes (examples were given in Sec.

4.3). Experiments on spinal animals have shown that the reflex arcs of many of these reflexes are restricted to the spinal cord. A common characteristic of them all is the presence of more than one interneuron in the reflex pathway; that is, they are polysynaptic. In this section we shall become familiar with the most important of these polysynaptic, spinal, motor reflexes and their properties. The most prominent examples is the *flexor reflex;* this will, therefore, be the starting point and central subject of the discussion. But we shall see that other reflex arcs are always activated together with the flexor reflex, so that the disturbances of the body's equilibrium caused by the flexor reflex are limited and compensated. In the last part of this section, we shall discuss what can be achieved by *reflexes in the isolated human spinal cord.* This question is of great practical significance, since accidental severing of the spinal cord (particularly in automobile crashes) is becoming increasingly common. The clinical syndrome is referred to as *paraplegia.*

Flexor Reflex and Crossed Extensor Reflex. If one hind paw of a spinal animal is painfully stimulated (by pinching, by a strong electrical stimulus, or with heat), the stimulated limb is pulled away; that is, there is flexion of the hock, knee, and hip joint. This phenomenon is called the flexor reflex. Painful stimulation of the fore paw also elicits withdrawal of the stimulated limb—another example of a flexor reflex. In this case wrist, elbow, and shoulder joints are flexed. The receptors for this reflex are located in the skin of the limbs, and the effectors are the flexor muscles. That is, we are dealing with an *extrinsic reflex.* The flexor reflex obviously serves to remove the limb from the site of the painful, and thus potentially damaging, stimulus. It is, therefore, also a *protective reflex.* Variation of the pressure applied to a paw shows that both the delay and the magnitude of the reflex action are very much dependent on stimulus intensity. As stimulus intensity increases, the reaction time becomes shorter, and the limb is withdrawn more quickly. This reflects an underlying process of summation—a typical property of polysynaptic reflexes, as we have seen in Sec. 4.3. In summary, we can conclude from such observation that the flexor reflex is an extrinsic reflex (because of the anatomic position of the receptors and effectors) comprising a spinal polysynaptic reflex arc (because it occurs in a spinal animal and has variable delay and amplitude). From a functional point of view, it is a protective reflex.

By feeling the limb musculature of the flexed leg during a flexor reflex, one can determine that the *extensor musculature relaxes during flexion.* This implies that the extensor motoneurons of the flexed limb are inhibited during this time. Moreover, it can be observed that the flexion of a hind or fore limb is always accompanied by the extension of the limb on the other side of the body. That is to say, painful stimulation of a limb results in an ipsilateral flexor reflex and a contralateral extensor reflex. The contralateral extensor reflex is also called the *crossed extensor reflex,* since the afferent activity in the nociceptive fibers crosses to the opposite side of the spinal cord in order to induce an extensor reflex there. The flexor muscles of the contralateral limb can be felt to relax during the crossed extensor reflex, which indicates that while the contralateral extensor motoneurons are excited, the contralateral flexor motoneurons are inhibited.

At a segmental level, then, painful stimulation of a limb evidently activates a total of four motor reflex arcs. Such stimuli produce (1) excitation of all the ipsilateral flexor motoneurons (the flexor reflex, bending of all the joints), (2) inhibition of the ipsilateral extensor motoneurons, (3) excitation of the contralateral extensor motoneurons (the crossed extensor reflex), and (4) inhibition of the contralateral flexor motoneurons. Electrophysiologic experiments on these coupled reflexes have confirmed the conclusions drawn previously about the underlying neuronal reflex arcs. Figure 6-6 is a diagram of the polysynaptic reflex connections of an afferent from a cutaneous nociceptor (pain receptor) on the segmental level. By way of several interneurons (only two are shown in each case), the ipsilateral flexor motoneurons are excited and the ipsilateral extensor motoneurons inhibited. In addition, the input from the ipsilateral nociceptors is transmitted to the opposite side by interneurons with axons in the *anterior commissure.* There, again by way of polysynaptic reflex arcs, the extensor motoneurons are excited and the flexor motoneurons inhibited.

Spinal animals are used to demonstrate these reflexes in the laboratory so that the simple reflex is not obscured by signals from higher levels of the CNS. A similar situation exists in newborn animals, since during the days just after birth the higher levels of the brain are not fully developed. Thus, flexor reflexes can be easily studied in a very young dog, cat, or human infant. Although less practicable for study purposes, the strong flexor reflexes of adults in response to very painful stimuli (contact of the hand with a hot object or of the bare foot with a sharp stone) are also graphic examples.

Intersegmental Reflex Arcs. Some of the axons of neurons with somata in the gray matter of the spinal cord pass into the ventral roots as efferents (motor axons and autonomic efferents), while others project to higher levels of the nervous system, forming ascending (somatosensory) pathways (see *Fundamentals of Sensory Physiology* for a detailed account). The great majority of axons leaving the gray matter, however, end within the spinal cord. These neurons, with axons lying

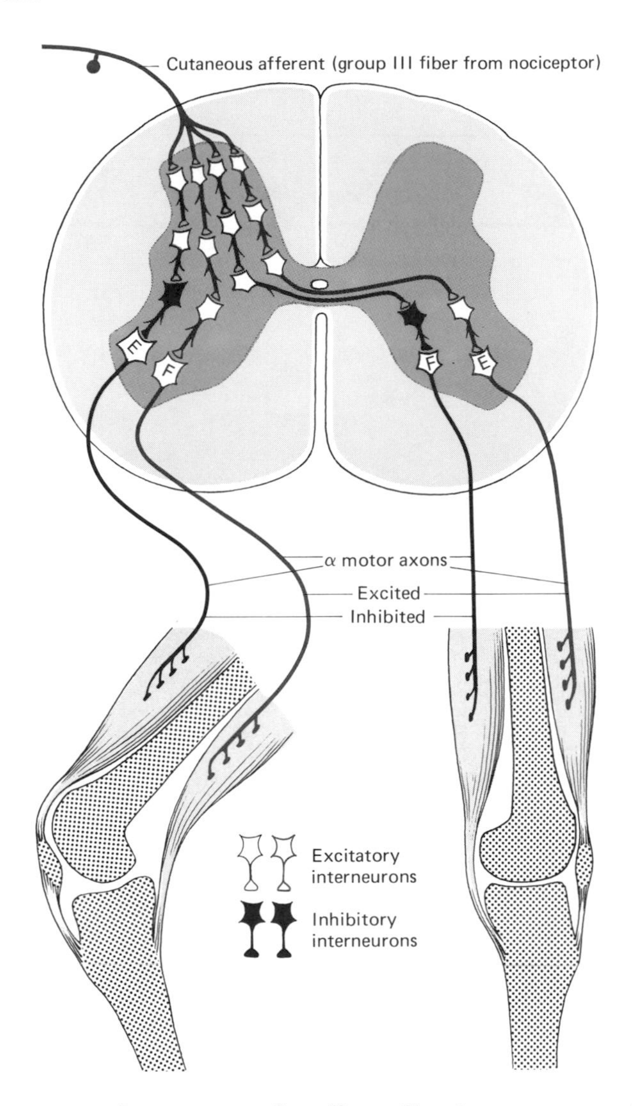

Fig. 6-6. Intrasegmental connections of an afferent fiber from a nociceptor (pain receptor) in the skin of the foot. The Group-III afferent fiber and the reflex paths underlying the ipsilateral flexor and the contralateral extensor reflexes are shown in red. E, extensor motoneurons; F, flexor motoneurons.

entirely within the spinal cord, are called *propriospinal neurons.* Bundles of ascending and descending axons with both origin and termination in the spinal cord are called *propriospinal tracts.*

Most neurons in the gray matter of the spinal cord, then, are propriospinal neurons. This in itself indicates the great significance of *connections between the different spinal-cord segments.* Figure 6-7

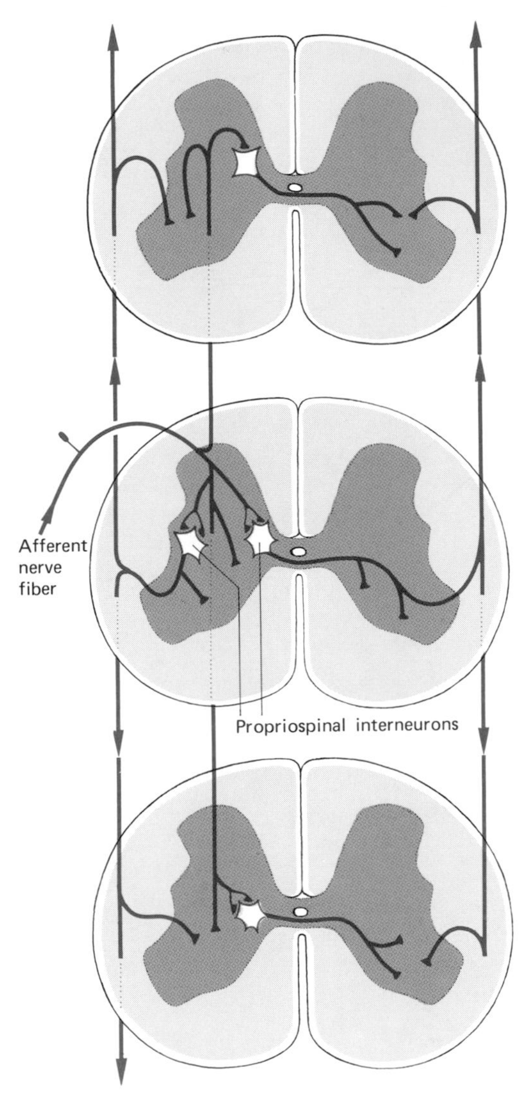

Fig. 6-7. Intersegmental connections of an afferent nerve fiber. After entering the spinal cord, the fiber sends out three kinds of collaterals: segmental, ascending, and descending. Each ends on an interneuron, and the axons of these interneurons either terminate within the spinal cord (propriospinal neurons) or proceed to supraspinal structures.

shows that an afferent fiber, in addition to its segmental (Figs. 6-3, 6-5, and 6-6) and ascending connections, as a rule is connected by several paths to neighboring and more distant segments. One type of connections is direct; that is, after entering the spinal cord, an afferent fiber divides into several collaterals, some of which pass directly into adjacent segments. In addition, collaterals of the afferent fibers may form excitatory synapses with propriospinal interneurons in the segment of entry; their axons then complete the reflex circuit, synapsing either ipsilaterally or, after crossing in the anterior commissure, contralaterally.

Details of the propriospinal pathways will not be treated here. It should be emphasized, though, that in certain pathways the propriospinal axons are very short (a few millimeters or less), so that their activity is confirmed to the immediate vicinity; others extend over many segments, as indicated in Fig. 6-7. The *role of the propriospinal tracts* is to link the different segments of the spinal cord. That is, they participate in *intersegmental reflex arcs.* For example, painful stimulation of one limb of a spinal animal produces not only a crossed extensor reflex, but also an excitation of the two remaining limbs. When the noxious stimulus is maintained this extension may give way to a rhythmic extension and flexion of all three nonstimulated limbs—to a pattern of movements related to running.

Reflexes That Persist in the Isolated Spinal Cord. As already implied, a spinal animal may exhibit standing and walking reflexes in response to stimuli other than pain, such as pressure on the soles of the feet (see Sec. 6.1). All such observations emphasize that the connectivity of spinal-cord neurons is organized so as to generate complex motor actions when an appropriate signal arrives from the periphery or from higher centers of the nervous system. When we speak of the *integrative function of the spinal cord,* it is this ability that is meant. We can also regard the spinal reflexes as a *set of elementary posture and movement "subroutines,"* which the organism can use as required; thus, the higher levels of the CNS need not be concerned with the details of executing these programs.

But we are far from a complete understanding of the functional significance of all known spinal phenomena. One example is *Renshaw inhibition* (illustrated in Fig. 4-4B), in which collaterals branching from motor axons before they leave the spinal cord synapse with an inhibitory interneuron (the *Renshaw cell*), which acts on the motoneuron itself. This is a negative-feedback circuit. Such circuits are commonly employed, both in the CNS and in technology; by its nature, we may infer that this system of negative feedback acts to prevent instability in the activity of the motor system, but its precise role is not yet clear.

In the more highly developed vertebrates, particularly the mammals, the higher levels of the CNS progressively assume control of the functions of the spinal cord—a process called *encephalization* or *cerebral dominance.* In such animals, the isolated spinal cord retains only a very modest ability to direct and regulate actions. Complete transection of the human spinal cord brings about an immediate and permanent paralysis of all voluntary movement in muscles supplied by the caudally located segments of the cord (*paraplegia*). Conscious sensation of the affected parts of the body is also irretrievably lost. And at first, all the motor and autonomic reflexes are extinguished (*areflexia*).

But during the following weeks and months, the *motor reflexes* recover. If the patient is properly cared for, in six months to a year there appear certain basic patterns of the recovery process on which prognoses may be based. The *autonomic reflexes* also reappear, to varying degrees, over periods of weeks and months. These processes will not be discussed further; the chief point for our purposes is that with well-planned intensive care and the proper schedule of using and strengthening the residual functions, such patients can be returned to a tolerable and useful existence.

The period of depression of all reflexes following spinal-cord transection is called *spinal shock*. In animal experiments it can be shown that a simulated transection, obtained by local cooling or anesthesia, also produces spinal shock. Moreover, when the spinal cord has been cut at one level and the reflexes have returned, a second transection caudal to the first does not result in spinal shock. A critical factor governing its appearance, then, is interruption of connections with the supraspinal CNS. As far as the *causes of spinal shock* and the events leading to return of the reflexes are concerned, our knowledge is very incomplete and unsatisfactory. When the descending pathways are cut, excitatory inputs to α and γ motoneurons and other spinal neurons are eliminated. In addition, there may be disinhibition of inhibitory spinal interneurons. The two together bring about a severe suppression of reflex activity, which appears as the clinical syndrome of areflexia. Unfortunately it still is unclear what accounts for the reestablishment of certain spinal-cord functions, and why the recovery period for humans lasts many months.

Q 6.6 Which of the following terms describe the flexor reflex? (Choose three.)
- a. Intrinsic reflex.
- b. Extrinsic reflex.
- c. Monosynaptic reflex.
- d. Disynaptic reflex.
- e. Polysynaptic reflex.
- f. Nutritional reflex.
- g. Protective reflex.
- h. Locomotor reflex.

Q 6.7 Which spinal afferents can activate the flexor reflex?
- a. Ia fibers of the primary muscle-spindle receptors.
- b. Group-III fibers from cutaneous nociceptors.

c. Ib fibers of the Golgi tendon organs.
d. Any spinal afferent can activate the flexor reflex under intense suprathreshold stimulation.

Q 6.8 Painful stimulation of a limb activates the flexor reflex arc. An additional result is
a. inhibition of the ipsilateral flexor motoneurons.
b. inhibition of the ipsilateral extensor motoneurons.
c. excitation of the contralateral flexor motoneurons.
d. excitation of the contralateral extensor motoneurons.
e. inhibition of the contralateral flexor motoneurons.

Q 6.9 Which of the following statements about propriospinal tracts is correct:
a. All propriospinal tracts leave the spinal cord via the ventral roots.
b. All propriospinal tracts enter the spinal cord via the dorsal roots.
c. Bundles of ascending and descending axons which originate and terminate in the spinal cord are called propriospinal tracts.
d. The neurons belonging to the propriospinal tracts are located exclusively in the dorsal horn of the cord.
e. Propriospinal tracts are only a small fraction of all the ascending and descending tracts in the white matter.

Q 6.10 Spinal-cord transection produces spinal shock in most vertebrates. During spinal shock
a. all motor and autonomic reflexes disappear.
b. the motor reflexes disappear, the autonomic are enhanced.
c. the extensor reflexes disappear, the flexor reflexes are enhanced.
d. all reflexes are unchanged.

6.3 Functional Anatomy of Supraspinal Motor Centers

The CNS is customarily subdivided into regions on the basis of phylogenetic age. Within the various regions, aggregations of anatomically and functionally related neurons are distinguished and set off from one another as separate *nuclei* or *ganglia*. *"Tract"* is the term used for a bundle of nerve fibers (axons) that joins different regions of the brain. Because of the sheaths of the myelinated fibers of which they are composed, the tracts appear white in unstained histologic sections, whereas the nuclear areas are gray. In the spinal cord the gray matter (that is, the zones containing the somata of the neurons) is surrounded by white matter (Fig. 1-9). In the cerebrum the situation is reversed. Here the cortex appears gray, because the somata of the cortical cells lie in the surface layers; the tissue below is white because it consists of axons running to other regions of the CNS (cf. Fig. 9-1, p. 272).

One way to study the *courses of the individual tracts* in the brain is by *transection experiments;* these rely on the fact that axons degenerate within a few days after they are separated from their somata. For example, degeneration of a nerve bundle below (caudal to) the point of transection means that the cell bodies of these axons are above (cranial to) the cut. The degenerating axons, then, belong to an efferent pathway, conducting from the center to the periphery. This technique has been used to reveal very many—but far from all—longitudinal and cross connections in the CNS. *Electrophysiologic stimulation and recording techniques* are currently employed to supplement and extend the histologic methods.

This section provides a schematic, greatly simplified *description of the most important motor nuclei* and their connections; no attempt will be made to present these in their phylogenetic context, and many of the smaller nuclei, especially in the brainstem, will be grouped into functionally related units (centers). But it must be emphasized that the anatomy of the supraspinal structures of the CNS is a difficult subject, with respect both to the macroscopic organization and to the fine structure of the individual components. A number of textbooks are available to the student wishing to explore this area further.

Supraspinal Motor Centers: Nomenclature, Position in the CNS. Above the spinal cord (at a supramedullary or supraspinal level), there are important motor centers, the functions of which we must know. In the block diagram of Fig. 6-8, four supraspinal motor centers are drawn in red; these are labeled *brainstem, motor cortex, basal ganglia,* and *cerebellum.* The arrows connecting the various centers show the main direction of information flow underlying muscular movement. On the right in the picture is an indication of the role each of these centers plays in producing such a movement.

Each center is a site of processing and redirection of the incoming information. A conspicuous aspect of the diagram is the key position of the *motor cortex.* This is connected to the motor centers in the spinal cord, both indirectly by way of the brainstem and directly, with no interposed neurons, via the corticospinal tract. In addition, collaterals leave the corticospinal tract, some passing to the brainstem and others acting on other high-level motor centers, the cerebellum, and the basal ganglia.

Figure 6-9 shows, in side view, the approximate *position of the motor centers* (red shading) in brain and spinal cord. All the motor centers are paired; that is, each is represented on both the right and the left side of the brain, as the schematic cross section of the spinal cord in the lower part of Fig. 6-9 shows for the motoneurons. The *motor centers of the brainstem* include a number of smaller nuclei located in quite different parts of the brainstem (see Fig. 6-13). The red shading, then, should be regarded as only a rough indication. The

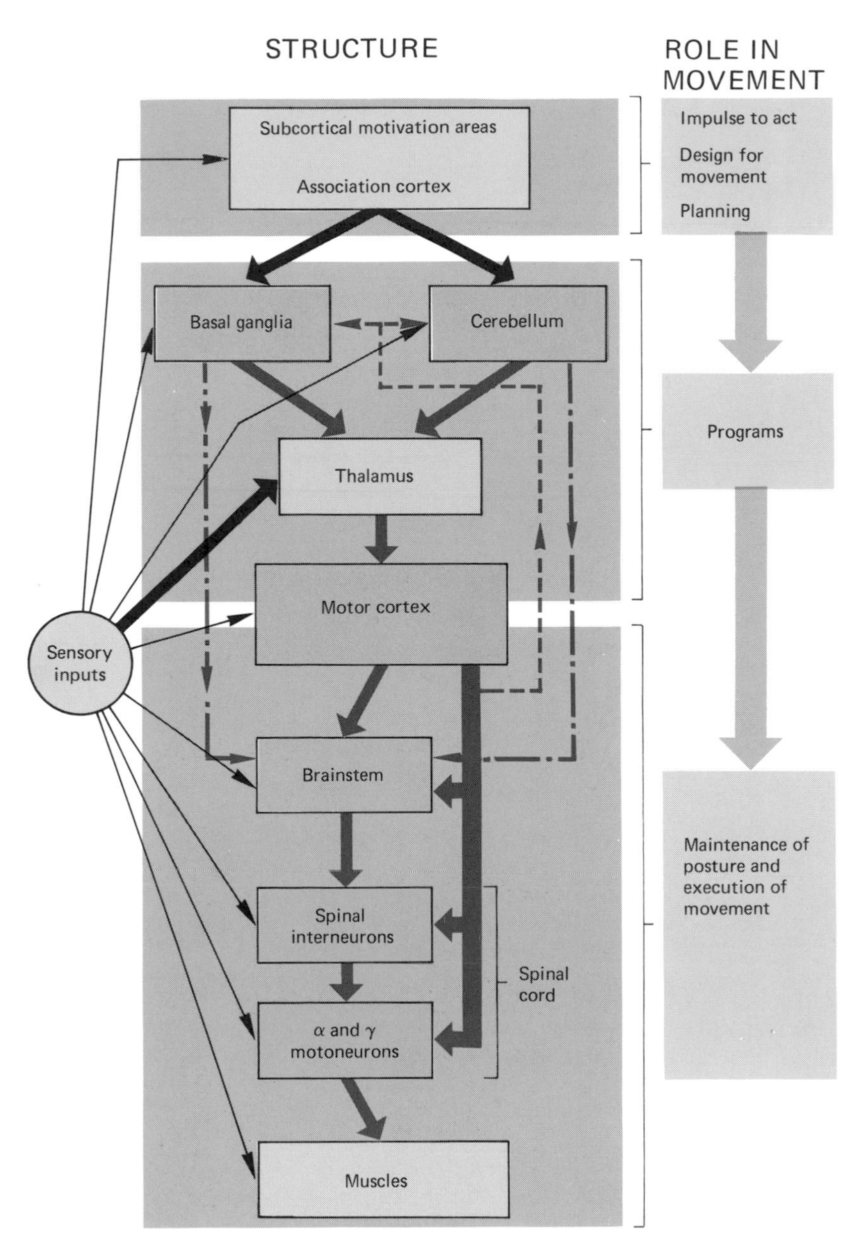

Fig. 6-8. Block diagram of the spinal and supraspinal motor centers and their most important connections. For the sake of simplicity, all sensory inputs are summarized at the *far left*. The *right-hand column* indicates the chief role played by the structures in the middle of the diagram during the performance of movements. The parallel position of the basal ganglia and the cerebellum is elaborated in Figs. 6-16 and 6-17. Note that the motor cortex is assigned to the transition between program and execution.

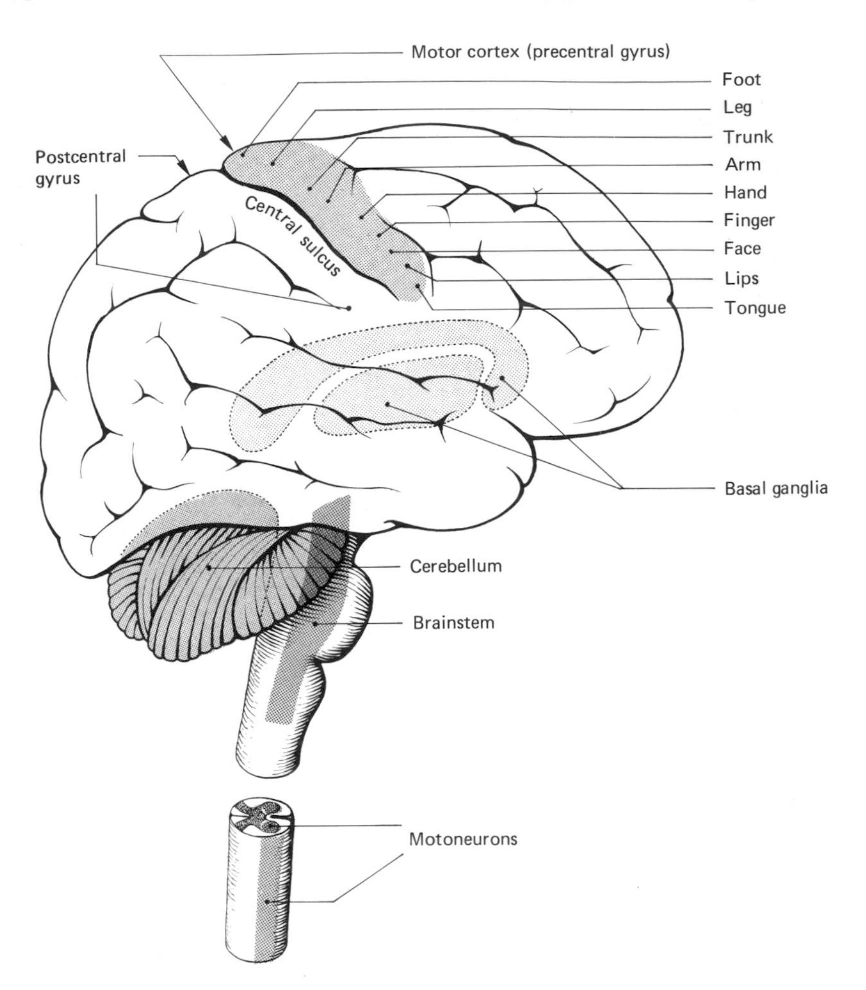

Fig. 6-9. Anatomic orientation of spinal and supraspinal motor centers (*red shading*). The drawing also shows the somatotopic organization of the precentral gyrus. The regions of the body involved in fine movements (skin, fingers, lips, tongue) take up large areas of the gyrus. All these centers are represented on both the right and the left side of the brain.

basal nuclei or ganglia, on the other hand, are relatively large, clearly demarcated structures. The most important of these are called the *striatum* (putamen and caudatum) and the *pallidum.* The basal nuclei are immediately adjacent to the *thalamus,* which is the most important sensory nucleus of the brain and is also associated with the motor system via some of its nuclei (see Fig. 6-8). All of these nuclei are covered by the cortex, so that they cannot be seen from outside and can be reached surgically only by penetrating the cortex. The *motor cortex,* by contrast, lies mostly on the surface of the brain. The most important, but not the only, cortical motor area is the gyrus just anterior to the central sulcus—the precentral gyrus. The corticospinal tract and the cortical motor efferents to the brainstem originate here and in surrounding areas.

The *cerebellum* is quite distinct from the rest of the brain, and is connected to it by thick strands of afferent and efferent fibers (arrows in Fig. 6-8). Here the incoming afference first affects the cerebellar cortex, which in turn projects to the cerebellar nuclei. The efferents from the cerebellar nuclei act in part on the motor cortex, via the thalamus, and in part directly on the motor centers in the brainstem. The cerebellar cortex differs markedly from that of the cerebrum. Its substructure is considerably simpler, consisting of three rather than six layers, and its pattern of surface folding is also distinctive; even an unpracticed eye can at a glance identify the cerebellum in a dissected brain.

The Corticospinal Tract. Figure 6-10 shows the course of the corticospinal tract in greater detail. This pathway proceeds without interruption from the motor cortex—that is, from areas in and around the precentral gyrus—into the spinal cord. At the level of the brainstem, the corticospinal fibers run in large paired bundles near the ventral surface. These structures, because of their shape, are called the pyramids—hence the alternative name for the corticospinal tract, the *pyramidal tract.* The axons of this tract, some of them over a meter long in humans, first pass between the thalamus and the basal nuclei in the brainstem. This region is called the *internal capsule* because of the way the pyramidal tract and other tracts encapsulate the thalamus. This region is of great clinical significance, because hemorrhage and occlusion of the blood vessels (for example, as a result of arteriosclerosis) here frequently cause block of conduction in motor pathways, with potentially lethal effects (the phenomenon is referred to as a *stroke*). When this happens, not only the corticospinal tract but *other motor tracts* from cortex to brainstem *are always affected* as well.

After emerging from the internal capsule, the pyramidal tract enters the *brainstem.* The majority of the fibers cross over to the opposite sides at this level, and then proceed caudally in the postero-lateral quadrant of the spinal cord. The other, smaller fraction follows an ipsilateral path, running caudally in the antero-medial parts of the cord. This fraction of the pyramidal tract extends as a rule only as far as the cervical (neck) and thoracic (chest) levels of the cord, and does not reach the lumbar region. Of the million or so fibers in each corticospi-

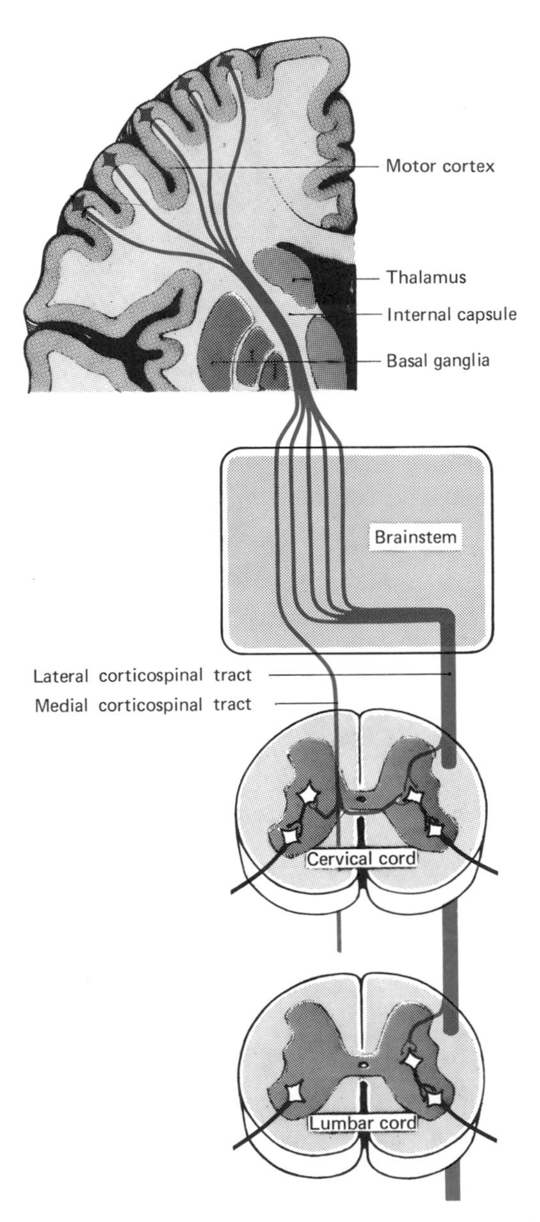

Fig. 6-10. Schematic diagram of the course of the corticospinal tract (*red*) from the motor cortex to the spinal cord. The collaterals to the basal ganglia, the cerebellum, and the motor centers of the brainstem have been omitted for simplicity (see Fig. 6-8). For further description see text.

nal tract (only one tract is shown in Fig. 6-10), 75 to 90% cross in the lower part of the brainstem; this region of crossing forms a prominent anatomic structure called the *decussation of the pyramids*. Interruption of a pyramidal tract and the other motor efferent pathways (see Fig. 6-11) in the internal capsule, therefore, results primarily in clinical *symptoms on the side contralateral* to the injury (see Fig. 9-12).

The axons of the corticospinal tract end in the *spinal cord*. In the segments in which they end, some of the uncrossed axons cross to the opposite side, so that the percentage of crossed axons is increased. Only a small proportion of the pyramidal-tract axons end directly on motoneurons; more commonly, they act on the spinal motor nuclei by way of segmental interneurons (see Fig. 6-8). The axons are so connected that those from a particular area of the motor cortex affect peripheral muscles of a particular part of the body—that is, the *motor cortex is organized somatotopically*. The somatotopic organization of the precentral gyrus is illustrated in Fig. 6-9; the neurons controlling the foot motoneurons are furthest medial and those affecting the face, the lips, and the tongue, furthest lateral. A striking feature of this mapping is the relatively large fraction of the precentral gyrus occupied by forelimb and face areas. The functional implications of this finding are discussed in Sec. 6.5.

In Summary: The corticospinal tracts originate in cells in the motor cortex. Most of the corticospinal axons cross to the opposite side in the brainstem, proceeding caudally in the lateral corticospinal tract of the spinal cord. At the spinal segmental level, the axons of the pyramidal tract end predominantly on interneurons. Local groups of muscles at the periphery are mapped onto local areas of the motor cortex; that is, the motor cortex is organized somatotopically.

Cortical Motor Efferents to the Brainstem. The cortical sites of origin of motor efferents to the brainstem are in the same motor areas that give rise to the corticospinal tract. These tracts differ from the pyramidal tract in that they do not enter the spinal cord without synaptic relay, and their postsynaptic fibers (for example, the reticulospinal and rubrospinal tracts) do not enter the pyramids. For this reason they, and all the other tracts descending into the brainstem and there synapsing to continue into the spinal cord, are frequently termed collectively the *extrapyramidal system*. The distinction is a purely anatomic one; the question of whether it is reflected in function (as is frequently claimed) will be examined more closely in Sec. 6.5.

Figure 6-11 illustrates the four most important extrapyramidal *connections between motor cortex and brainstem:* (1) direct from cortex to brainstem (via the internal capsule), (2 and 3) with one synapse,

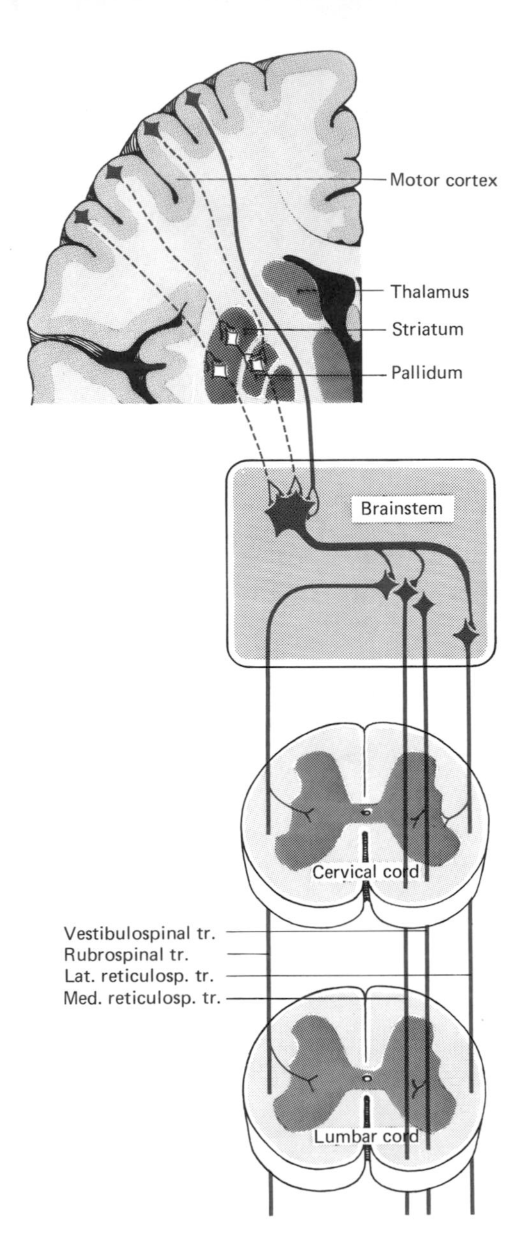

Fig. 6-11. Schematic diagram of the courses of the most important extrapyramidal tracts (*red*) from the supraspinal motor centers into the spinal cord. The neuron with a thick axon in the brainstem symbolizes the crossing of most of the extrapyramidal motor fibers to the opposite side at that level, and does not imply convergence. The pathways from motor cortex to basal nuclei are partly collaterals of the corticospinal tract and partly separate efferents. The details of connectivity among the brainstem structures involved in motor activity are extremely complicated; the representation here is greatly simplified. With regard to the basal nuclei, it should be emphasized that their connections to the motor cortex by way of the thalamus (see Figs. 6-8 and 6-17) are of greater functional importance than the descending tracts shown here.

either in the striatum or in the pallidum, and (4) with two synapses, the first in the striatum and the other in the pallidum (but not the reverse). The axons terminate in the brainstem without crossing to the contralateral side. Crossing occurs, as a rule, after these axons have synapsed with the brainstem neurons (Fig. 6-11). From the brainstem, a number of extrapyramidal motor tracts emerge. In Fig. 6-11 the four most important of these are shown. Their names are derived from their points of origin in the brainstem—the reticular formation, vestibular nuclei, and red nucleus (or nucleus ruber)—and their position in the spinal cord (medial, lateral). These tracts are the lateral reticulospinal, medial reticulospinal, vestibulospinal, and rubrospinal.

It is possible only within certain limits to say which extrapyramidal tracts leaving the motor cortex enter which motor centers of the brainstem, because of the complex intermingling of the motor pathways from cortex, cerebellum, and basal ganglia as they enter the brainstem (see Fig. 6-8). In essence, most of the fibers form *corticorubral connections,* the fibers with which they synapse in the red nucleus continuing as the rubrospinal tract, or form *corticoreticular connections* that eventually give rise to the medial and lateral reticulospinal tracts (Fig. 6-11). The *red nucleus* also receives major inputs from the cerebellum, while the cells of origin of the vestibulospinal tract are affected chiefly by the organs of equilibrium.

The positions of the most important motor centers of the nervous system, and the main pathways that connect them, have now been described. The microscopic structure of these centers—that is, the arrangement and connections between the neurons in each of the nuclei mentioned—will not be treated in depth here, since too little is known as yet about the relationships between function and substructure of the supraspinal regions of the brain. The part of the brain best studied in this respect is the cerebellar cortex, the relatively simple histologic organization of which (see below) makes it possible to consider its physiologic correlates. By contrast, the motor cortex—and, indeed, the cortex of the whole cerebrum—is considerably more complex in structure and, thus, far more difficult to analyze. Only a brief survey of its structural properties will be given.

Structure of the Motor Cortex. In the motor cortex, as throughout the cerebral cortex (see Sec. 9.1), layers composed chiefly of somata alternate with layers containing mainly axons, so that the freshly cut cortex presents a striped appearance. The precentral gyrus is distinguished primarily by its thickness, 3.5 to 4.5 mm, and by the *large pyramidal cells* (*Betz cells,* 50 to 100 μm in diameter) in the fifth layer (counting from the surface down). These and other smaller pyramidal cells in the third layer are the cell bodies of the corticospinal-tract neurons;

their axons run down toward the internal capsule, and most of their dendrites spread up toward the surface of the cortex. The pyramidal cells were named for their shape long before it was known that some of them give rise to the pyramidal tract; the correspondence in terminology is quite accidental, and in fact pyramidal cells also exist in other areas of the brain. The large pyramidal cells have the most rapidly conducting axons in the corticospinal tract (conduction velocity 60 to 90 m/s), but they contribute only about 3% (30,000 fibers, as compared with a total of 10^6 on each side of the brain) of the pyramidal-cell axons; all the others conduct considerably more slowly. Neurons such as the pyramidal cells, the axons of which carry integrated information from the cerebral cortex into the periphery, are called *projection neurons*. They are far fewer in number than the other cortical neurons, with axons restricted to the local cortex or passing to other ipsi- or contralateral areas of cortex. The latter serve in intracortical information processing, and are called *association neurons* (see also Fig. 9-3, p. 274). On the whole, the association neurons tend to lie in the superficial cortical layers, and the projection neurons in the deeper layers.

Structure of the Cerebellar Cortex. In contrast to the cortex of the cerebrum, that of the cerebellum consists of just three clearly defined layers; these look essentially the same in all parts of the cerebellar cortex. The superficial layer, labeled *molecular layer* in Fig. 6-12, is separated from the lowest, the *granular layer*, by an array of conspicuous cells, the *Purkinje-cell layer*. The top and bottom layers were so named because of their finely dotted and granular (respectively) appearance in fresh cross sections of the cortex. The *Purkinje cells* lying between these two layers have large somata and a highly branched dendritic tree extending far into the molecular layer. One such Purkinje cell was illustrated in Fig. 1-3C. Apart from the Purkinje cells, there are two other main cell types in the cerebellar cortex, the *granule cells* in the granular layer and the *basket cells* in the molecular layer. In addition to these three chief types of cell, there are three other types that will not be described here.

Two kinds of axons (*fibers*) enter the cerebellar cortex. One of these is called the *climbing fiber* (Fig. 6-12A). The climbing fibers pass through the granular layer and end in the molecular layer, at the dendrites of the Purkinje cells. The branches of these fibers "climb" up the limbs of the dendritic tree, twining like ivy around its twigs. The other incoming fiber type is called the *mossy fiber* (Fig. 6-12B). These end in the granular layer, with terminals at the granule cells. The axons of the granule cells pass between the Purkinje cells into the

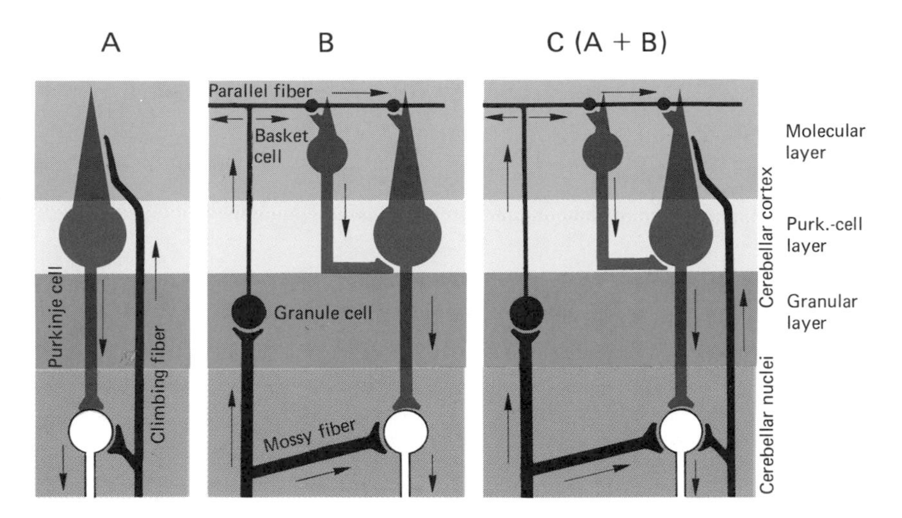

Fig. 6-12. The most important neuronal circuits of the cerebellum. **A**, Synaptic connections of the climbing fibers; **B**, synaptic connections of the mossy fibers; **C**, the afferent inputs to a Purkinje cell from mossy and climbing fibers. An extensive discussion of the anatomic connections is given in this section, and details of functional relationships are in Sec. 6.5.

molecular layer, where they divide into two collaterals to form a T shape. These axonal collaterals are called *parallel fibers* (in cross section they look like small dots—hence the name "molecular" layer). Each parallel fiber is about 2 to 3 mm in length; over this distance it makes synapses with some dozens to hundreds of dendrites of two types of cells—the basket cells and the Purkinje cells. The *basket cells* in turn send their axons to the soma of the Purkinje cells. The mossy fibers, then, do not reach the Purkinje cells directly (like the climbing fibers), but rather via one or two interneurons, the granule cells, and the basket cells.

The only *axons leaving the cerebellar cortex are those of the Purkinje cells;* they run to the neurons of the cerebellar nuclei. These cerebellar nuclei also receive collaterals of the climbing fibers (Fig. 6-12A) and the mossy fibers (Fig. 6-12B). We can say, then, that the cerebellar cortex has *two inputs,* the mossy fibers and the climbing fibers, and *one output,* the axons of the Purkinje cells. All the information processing in the cerebellar cortex occurs in neuronal networks like that shown in Fig. 6-12C (a composite of parts A and B of that figure). Altogether the human cerebellar cortex contains about 15 million Purkinje cells. Each of these receives synaptic input from only one climbing fiber, but from many thousand parallel fibers and a few dozen basket cells. The function of the cerebellar cortex must involve the spatial connectivity of these pathways, the excitatory or inhibitory nature of the synapses, and the temporal sequence of synaptic and action potentials. Some of our knowledge of these parameters will be described in Sec. 6.5.

Q 6.11 Which of the following structures has/have a predominantly motor function?
 a. Thalamus.
 b. Postcentral gyrus.
 c. Central sulcus.
 d. Pallidum.
 e. Dorsal horn of the spinal cord.

Q 6.12 Which (one or more) of the following is correct? The corticospinal tract
 a. has a relay point only in the brainstem.
 b. crosses to the contralateral side with 75 to 90% of its fibers.
 c. originates primarily in the postcentral gyrus.
 d. ends primarily on medullar (spinal) interneurons.
 e. is also called the pyramidal tract.

Q 6.13 Which of the following statements is/are false?
 a. An interruption of the motor pathways in the internal capsule causes motor disturbances (paralysis) on the side of the body opposite the injury.
 b. Extrapyramidal efferent axons from the motor cortex end in the brainstem, and do not reach the spinal cord.
 c. The precentral gyrus is organized somatotopically; that is, adjacent areas supply spatially related peripheral muscles or muscle groups.
 d. All the extrapyramidal tracts leaving the brainstem are uncrossed.

Q 6.14 The *dendrites* of the cerebellar Purkinje cells receive synaptic input from
 a. the parallel fibers of the granule cells.
 b. the mossy fibers.
 c. the climbing fibers.
 d. the axons of the basket cells.

Q 6.15 The following axons enter the cerebellar cortex as afferent inputs:
 a. mossy fibers.
 b. parallel fibers.
 c. Purkinje-cell axons.
 d. climbing fibers.

6.4 Reflex Control of the Posture of the Body in Space

This section describes the functions of the *motor centers of the brainstem*. These can be studied experimentally by cutting the connections between the brainstem and the higher motor centers in the basal ganglia and the cortex; in some cases the communication with the cerebellum is interrupted as well. In addition to such complete isolation of the lower centers in the brainstem, experiments involving more localized stimulation and ablation of brain areas have been informative. It has been found, in fact, that these centers are chiefly responsible for *reflex control of the posture and spatial orientation of the body.* To accomplish this they evaluate the afferent signals of many receptors throughout the body. The most important of these are the receptors in the organs of equilibrium (the vestibular organs of the inner ear, one set on each side of the head) and the stretch and joint receptors of the neck musculature. Inputs from these sources enable the brainstem motor centers to provide a continuous regulatory output, so that the normal body posture is adopted and maintained entirely without voluntary control.

Subregions of the Brainstem and Their Afferents. The *brainstem* in the physiologic sense is considered to be the portion of the CNS shaded red and labeled 1 to 3 in the longitudinal (sagittal) section of Fig. 6-13. At its caudal end, the brainstem merges with the spinal cord, and rostrally (cranially) it adjoins the diencephalon. (The chief components of the diencephalon are the sensory nuclei of the thalamus and the hypothalamus, along with other centers of the autonomic nervous system). The three divisions of the brainstem indicated in the figure can be distinguished histologically, phylogenetically, and to some extent functionally. They are, from caudal to cranial, (1) the *medulla oblongata,* (2) the *pons,* and (3) the *mesencephalon* (midbrain). Cranial to the brainstem also lie the motor centers of the basal ganglia and the motor cortex (see Figs. 6-8 and 6-9); these have inputs to the brainstem via the extrapyramidal tracts and collaterals of the corticospinal tract (see Fig. 6-11). Still other important inputs to the brainstem motor centers are shown in Fig. 6-14. These come from the peripheral receptors (those in skin, muscles, and joints), the cerebellum, and the equilibrium organs.

The *organ of equilibrium* (the vestibule plus semicircular canals) on each side lies immediately adjacent to the cochlea in the inner ear. Vestibule, canals, and cochlea are linked to the brain by a common nerve, the stato-acoustic nerve (eighth cranial nerve), and the fluid-filled spaces within them are all in communication. Because of their very complex shape, the three structures together are called the *labyrinth.* The labyrinth is entirely embedded in the temporal bone and is thus very difficult to reach for clinical or experimental surgery. The parts concerned with equilibrium provide information about the *position of the head in space* (which we know quite accurately even with our eyes closed and with no other external cues) and about *angular acceleration* (during rotation) and *linear acceleration* (for example, in

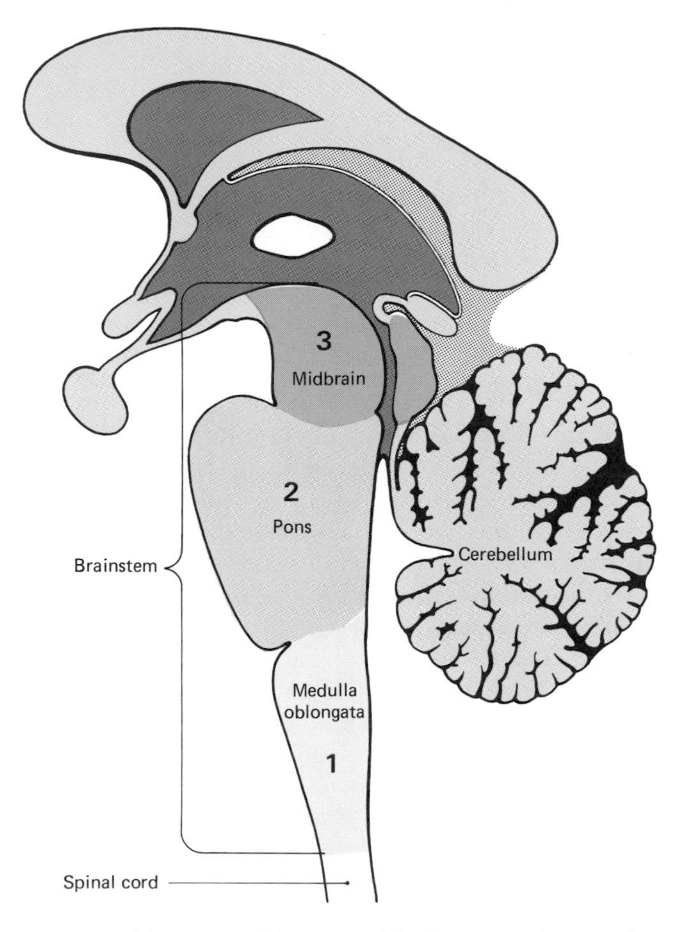

Fig. 6-13. Positions of the main subdivisions of the brainstem. See text for explanation.

a train or elevator). Both positive and negative accelerations are signaled. Strong or unfamiliar accelerations can have aftereffects for many seconds, reflecting the inertia of the labyrinth mechanism. For example, if a person is rotated for some time and then suddenly stopped, the semicircular canals report "angular deceleration" for some time thereafter. If the eyes are open, the visual and labyrinthine information conflict, producing dizziness and loss of motor coordination. There are also eye-movement reflexes controlled by the labyrinths, which can be demonstrated in similar experiments (for more details see *Fundamentals of Sensory Physiology*).

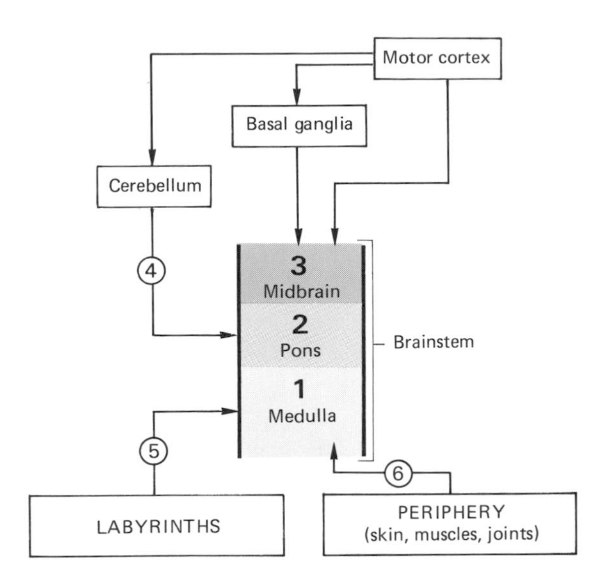

Fig. 6-14. Inputs to the motor centers of the brainstem. The numbers *1* to *3* indicate parts of the brain identified by the same numbers in Fig. 6-13.

Transection of the Brainstem. Inputs to the brainstem from more cranially located motor centers (the basal ganglia and the cerebral cortex) can be eliminated by making a cut at the cranial boundary of the brainstem. To ensure that the cut is effective, the usual practice is to remove all the brain tissue rostral to it. An animal so treated is referred to as a *midbrain animal* (Fig. 6-15), because the midbrain is the highest level of the CNS still intact. If the cut is made somewhat further caudally, about at the boundary between midbrain and pons, a *decerebrate animal* is the result (Fig. 6-15B). In such an animal only the medulla oblongata and the pons are in communication with the body, via the spinal cord. The same afferent inputs are available to both the midbrain and the decerebrate animals, and the connection to the cerebellum is preserved in both cases.

Motor Abilities of the Decerebrate Animal; Postural Reflexes. After acute transection of the spinal cord (spinalization, paraplegia), the peripheral musculature is either completely flaccid or the tonus of the flexors predominates. Neither the paraplegic human nor the spinal animal is able to stand (see Sec. 6.2). In the decerebrate animal, however, we find a marked *increase in tonus of the entire extensor musculature*. All four limbs are stretched out straight. Head and tail are bent

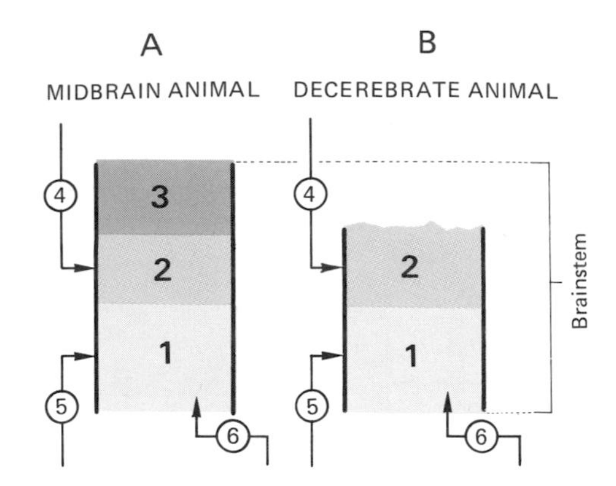

Fig. 6-15. Diagram showing the parts of the brainstem left in connection with the body in a midbrain animal (**A**) and a decerebrate animal (**B**), numbered *1* to *3* as in Figs. 6-13 and 6-14. The central and peripheral inputs associated with motor activity are indicated by the numbers *4* to *6*, corresponding to the labeling of Fig. 6-14 (cerebellum, labyrinths, and periphery of the body, respectively).

backward. This condition is called *decerebrate rigidity.* If a decerebrate animal is set on its feet, it remains upright, for the high tonus of the extensor muscles does not permit the joints to bend. The unnatural, stiff posture of the animal looks like a caricature of normal standing. From the facts that the decerebrate animal stands upright and the spinal animal does not, we may conclude that the medulla oblongata and pons contain motor centers that regulate muscle tonus in the limbs in such a way that they can bear the weight of the body. The much increased extensor tonus in the decerebrate animal (decerebrate rigidity) indicates that these centers have been disinhibited by the removal of the highest levels of the brain.

The *distribution of tonus in the musculature* of a decerebrate animal can be changed by manipulating the position of the head. Since moving the head changes both its position in space and its position relative to the body, these muscle-tonus changes can be produced by signals from the equilibrium organs and/or the neck musculature. To study the effects of either source of information on muscle tonus, the other must be eliminated. For example, if both labyrinths are removed, the absolute position of the head in space is no longer signaled, but the receptors in the cervical musculature and the joints between the cervical vertebrae continue to monitor every change in position of the head relative to the body. In fact, it is found that these signals to the motor centers of the brainstem lead to appropriate corrections in tonus distribution of the muscles of the body. We shall now describe examples of such *neck reflexes;* since they produce a change in tonus distribution, they are also called *tonic neck reflexes.*

In a decerebrate animal with labyrinths removed, if the head is bent upward (red arrow in Fig. 6-16A), the tonus of the limb musculature changes as shown in the figure: the extensor tonus in the hind limb is reduced, and that in the fore limb increases. When the head is bent downward (red arrow in Fig. 6-16B), the tonus distribution changes in the reverse manner: the extensor tonus in the fore limbs is reduced and that in the hind limbs increased. As a third example, if the head is turned to the side, so that the balance of the body is disturbed, this is compensated by a corresponding increase in tonus of the extremities. Turning the head to the right (and thus shifting the body weight toward the right side) results in an increase in extensor tonus in the two right limbs. In all three cases the new body posture is maintained as long as the head remains in the altered position. These tonic neck

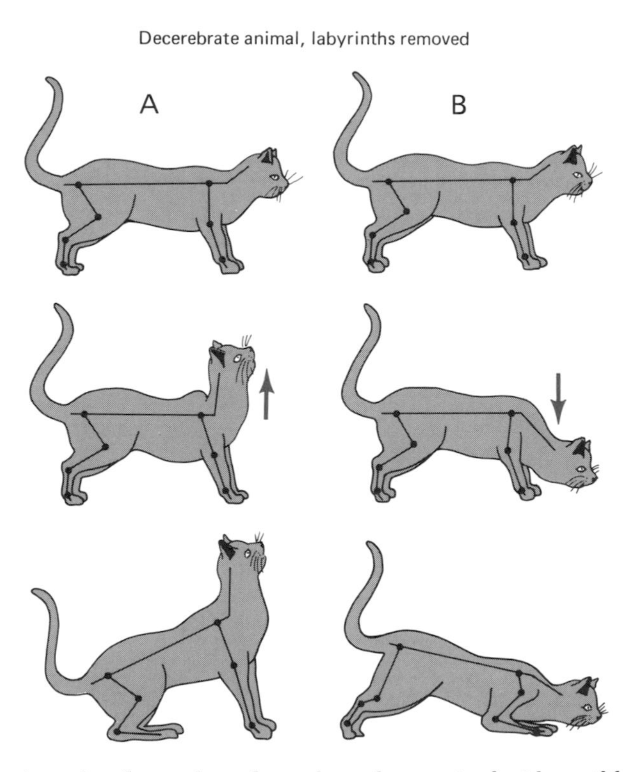

Fig. 6-16. Tonic neck reflexes elicited in a decerebrate animal with equilibrium organs (labyrinths) removed. Imposed upward bending of the head (*red arrow* in **A**) produces a reduction of extensor tonus in the posterior limbs and an increase of extensor tonus in the fore limbs. Downward bending of the head (*red arrow* in **B**) has the opposite effect.

reflexes are, thus, termed *postural reflexes*, or sometimes "standing reflexes," since they are observed in a quietly standing animal.

Postural reflexes can also be elicited by the labyrinths, in which case they are called *tonic labyrinthine reflexes*. These will not be discussed in detail here, but a general property of such reflexes worth noting is that the extensor muscles of all four limbs always respond in the same sense.

An interesting special case in the postural-reflex category is found in the *compensatory eye movements*. These movements of the eyeball ensure that the visual field is kept approximately constant during head movements; that is, the image on the retina is fixed. In humans and animals with frontally situated eyes, this is achieved primarily by visual signals from the overlapping field of view. In animals with eyes at the sides of the head, so that there is little or no overlap between the two fields of view, the distribution of tension in the eye musculature is controlled primarily by collaboration between labyrinth and neck reflexes. For example, if the head of a rabbit is turned so that the right half of the face moves downward, the right eye turns upward and the left eye (on the upper side of the head) turns down. These displacements to some extent enable the eyes to retain their orientation with respect to the horizon rather than following the angle of the head.

In Summary: The essential points to remember are as follows. The motor centers in pons and medulla oblongata are able not only to keep the tonus of the limb musculature so high (*decerebrate rigidity*) that the body stands upright, but also to *change appropriately* this high extensor tonus on receipt of signals from the labyrinths (about the position of the head in space) and from the receptors in the neck muscles and joints (about the position of the head with respect to the body). These *postural reflexes* are termed *tonic neck reflexes* and *tonic labyrinthine reflexes* according to the receptors involved. In an intact animal, these and related reflexes provide a *library of elementary postural subroutines* that the body can employ as needed.

Motor Ability of the Midbrain Animal; Righting Reflexes. A decerebrate animal remains standing when set on its feet. But it falls over if it is pushed, and after falling it does not get up again. Moreover, when it is upright, its tonus distribution is abnormal; there is a marked predominance of extensor tonus, whereas in ordinary standing the flexor and extensor muscles are about equally activated to hold a joint in position. But if not only the medulla oblongata and the pons, but the midbrain as well, are left in communication with the spinal cord (a midbrain animal), the motor performance of the animal is considerably improved and extended. The two most notable improvements over the decerebrate case are (1) that the *midbrain animal lacks decerebrate rigidity* (that is, there is no predominance of tonus in the extensor muscles), and (2) that the *midbrain animal can set itself in position*. Since the inputs to the brainstem relevant to motor function, as Fig. 6-15 shows, are no different in the midbrain and decerebrate animals, the improvement in motor performance of the midbrain animal must be brought about chiefly by the motor centers of the midbrain.

The absence of decerebrate rigidity in the midbrain animal with its body in a stable standing posture implies that inclusion of the midbrain nuclei has produced a tonus distribution that is more physiologic (that is, more normal) than in the decerebrate case. More important, though, is the ability of the midbrain animal to right itself; from any abnormal starting position, it returns to the basic posture reflexly and accurately. Reflexes that bring the animal into its normal posture are called *righting reflexes*. Study of these reflexes has revealed that they occur in a particular sequence—that is, there is a chain of righting reflexes leading to the final posture. In the first stage, signals from the vestibular organs bring the head into the normal position. These reflexes are called *labyrinthine righting reflexes*. This movement of the head changes its position with respect to the rest of the body, an alteration signaled by the receptors in the neck musculature. Messages from these receptors then result in movements that bring the trunk into its normal alignment. By analogy with the labyrinthine righting reflexes, these are called *neck-muscle righting reflexes*.

Role of the Righting Reflexes. Righting reflexes, then, return the body to its normal posture after it has been displaced from it in some way. That is, these reflexes serve to maintain normal body posture and balance without voluntary control. Apart from the labyrinthine and neck-muscle righting reflexes, a number of other reflexes affect head and body position—for example, those initiated by receptors on the surface of the body. If one includes the *visual righting reflexes*, which are eliminated in a midbrain animal but can be shown to exist under other experimental conditions, it becomes clear that the mechanism for bringing the body into normal posture has a very high "safety factor" indeed. In combination with the postural reflexes, the righting reflexes guarantee suitable posture at each instant. An important element of these responses is that the head, in which the eye, ear, and olfactory organ are situated, plays the leading role. As a result, even stimuli at some distance can act to bring the body into the appropriate posture, which will often be a defensive one.

Static and Stato-Kinetic Reflexes. The reflexes described so far are often lumped as *static reflexes,* because they control body position and balance during quiet lying, standing, and sitting in a variety of positions. The midbrain animal can also be shown to have a number of other reflexes that are elicited by movements and result in movement. These are in the category of *stato-kinetic reflexes.* Many of these are initiated by the labyrinths. The best known are the head and eye turning responses. For example, if an animal is rotated clockwise the head turns counterclockwise relative to the body, and so on. These responses are compensatory—eyes and head are moved so that the visual images remain as constant as possible during the movement of the body. Without movement they are kept in position by static reflexes (compensatory eye positioning, see above). Other important stato-kinetic reflexes keep the body in balance and in the correct positon during jumping and walking. For example, such reflexes ensure that a cat lands with its body in the proper position, regardless of its position when it began to fall.

In Summary: The *midbrain animal* differs little from an intact animal with respect to *postural, positioning, walking, and jumping responses.* But it lacks spontaneous movement; it behaves like an automaton so that an external stimulus is necessary to set it in motion. The experiments on decerebrate and midbrain animals show clearly that the mechanisms underlying the various postures of animals and man—features of behavior that we meet in ordinary life and in the works of painters and sculptors—actually lie in the intricate operation of the brainstem motor centers. There the many linked positioning and postural reflexes are integrated, coordinating the musculature of the body in purposeful behavior.

Q 6.16 Which parts of the brainstem of a decerebrate animal are still connected to the spinal cord and functional?
- a. Medulla oblongata.
- b. Pons and midbrain.
- c. Medulla and pons.
- d. Medulla and midbrain.
- e. Medulla, pons, and midbrain.

Q 6.17 Which of the possible answers given in Q 6.16 includes all the functional brainstem regions of the midbrain animal?

Q 6.18 Which of the following do *not* occur in a decerebrate animal?
- a. Decerebrate rigidity.
- b. Righting reflexes.
- c. Predominance of extensor tonus.
- d. Postural reflexes.
- e. Predominance of flexor tonus.

Q 6.19 The motor centers of the midbrain animal receive afferent inputs no different from those of the decerebrate animal. Which two inputs in the following list are of particular importance to postural and righting reflexes?
a. Inputs from the cerebellum.
b. Inputs from the vestibule and semicircular canals.
c. Inputs from the receptors on the body surface.
d. Inputs from the muscle and joint receptors of the body.
e. Inputs from the muscle and joint receptors of the neck.

Q 6.20 Which of the following postural and righting reflexes involve afferents from the labyrinths, and which from the neck muscles?
a. Increase of extensor tonus of the fore limb when the head is raised.
b. Decrease in extensor tonus of the left limbs when the head is turned to the right.
c. Raising the body to the normal position.
d. Raising the head to the normal position.

6.5 Functions of the Basal Ganglia, Cerebellum, and Motor Cortex

In our discussion of the physiology of motor systems, we now turn to the motor areas of the cerebrum, the basal ganglia, and the cerebellum; the functional anatomy of these structures has been described in Sec. 6.3 of this chapter. Here, as everywhere in the CNS, two questions are in the foreground: (1) What do these centers do? (2) How do they do it? We shall not be able to answer either question satisfactorily, partially because our knowledge of the "What" (and still more, of the "How") is preliminary; moreover, in this book we are concerned only with the essential, thoroughly documented facts of neurophysiology. The higher motor functions represent an area of research in which—understandably—established knowledge is mixed with an extraordinary amount of hypothesis and speculation. For this reason we shall have to treat the functions of the motor areas of the cerebrum and the basal ganglia without attempting to explain how they operate. Regarding the cerebellum, however, a number of the basic interactions are understood, at least in outline.

The Role of the Basal Ganglia. From the survey of the motor centers given in Fig. 6-8 it was evident that the *basal ganglia* are an important *subcortical link between the motor cortex and the rest of the cerebral cortex*. The associative cortex sends efferents to, and the motor cortex receives afferents from, the basal ganglia. Figure 6-8 also indicates sensory inputs to the basal ganglia, and efferents from the basal ganglia to the motor centers of the brainstem.

The *functions of the the basal ganglia* involve participation in the conversion of the plans for movement arising in the associative cortex into *programs for movement*. That is, they elaborate spatio-temporal patterns of nervous impulses, which are further processed by the executive motor centers (see Fig. 6-8, right column). Figure 6-17 shows, in more detail than Fig. 6-8, at what point the basal ganglia are involved in the information flow from the associative cortex to the motor cortex and basal ganglia (see also Fig. 6-11). The *striatum* receives the

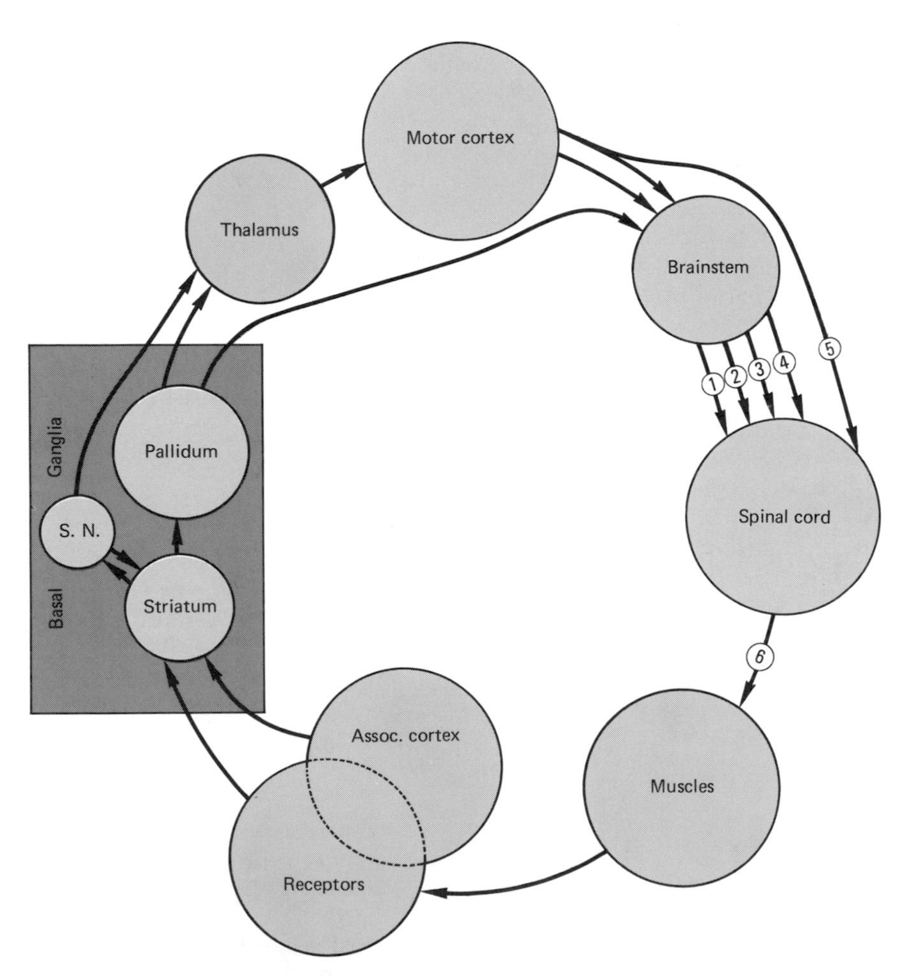

Fig. 6-17. Diagram of the most important afferent, efferent, and intrinsic connections of the basal ganglia, showing their involvement in the motor system. S.N., substantia nigra; motor centers shown in red. *Arrows 1 to 4* are the pathways from brainstem into spinal cord shown in Fig. 6-11; *Arrow 5* is the corticospinal tract (Fig. 6-10), and *Arrow 6* symbolizes α and γ motor axons. For further discussion see text.

majority of all afferents to the basal ganglia, whereas the most important *efferents leave via the pallidum* and proceed, as described above, chiefly by way of the thalamus to the motor cortex and in lesser numbers to the motor centers of the brainstem. The clinical (see also below) and experimental evidence available thus far indicates that the basal ganglia may be particularly significant with respect to the *initiation and execution of slow movements*.

The Role of the Cerebellum. As Fig. 6-8 illustrates, the projections of the cerebellum are quite analogous to those of the basal ganglia; its

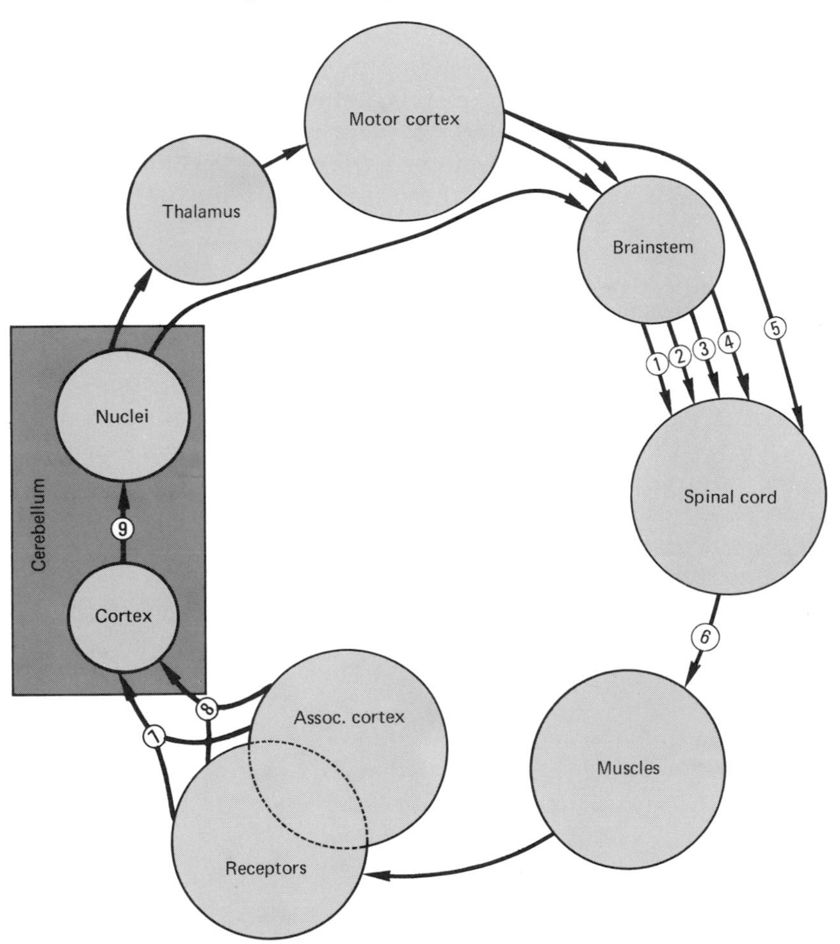

Fig. 6-18. Diagram of the afferent, efferent, and intrinsic connections of the cerebellum and their involvement in the motor system. Motor centers shown in red. *Arrows 1* to 6 are numbered as in Fig. 6-17; *Arrow 7* symbolizes mossy fibers; *Arrow 8*, climbing fibers; and *Arrow 9*, Purkinje-cell axons. For further discussion see text.

efferents pass primarily via the thalamus to the motor cortex or directly to the motor centers of the brainstem. In the hierarchy of brain organization, then, the cerebellum and basal ganglia are centers of equal rank, involved in the programming of cortically initiated movement patterns. This parallelism is emphasized by Fig. 6-18, and by its remarkable similarity to Fig. 6-17.

Whereas the basal ganglia seem to be responsible primarily for the execution of slow steady movements, the cerebellum appears to be chiefly concerned in (1) *programming rapid movements,* (2) *correcting the course of such movements,* and (3) *correlation of posture and movement.* Thus, cerebellum and basal ganglia apparently serve different but related functions, though only certain aspects of these are known thus far. The division of labor between cerebellum and basal ganglia will become particularly evident when comparison is made between the impairments of motor performance resulting from disease of and damage to the two structures, as described in the following paragraphs.

Pathology of the Basal Ganglia. The best-known complex of symptoms resulting from disturbance of basal-ganglion function is the *Parkinson syndrome.* These patients are conspicuous by the rigidity of their expressions, the poverty or absence of communicative gesture, the hesitant gait with small steps, and the trembling of the hands. That is, the most striking symptoms are (1) *akinesis,* the absence or impairment of slow movements, (2) *rigor,* an increased muscle tonus which is present regardless of joint position or movement (in which respect it differs from decerebrate rigidity), and (3) *passive tremor,* which occurs when the patient is at rest, is suppressed during movements, and begins again when the movement is completed. Other forms of basal-ganglia disease are also accompanied by inappropriate movements in the sense of overshooting, rigor, and tremor.

A probable *cause of Parkinson syndrome* is deterioration of the tract passing from the substantia nigra to the striatum (Fig. 6-17), the transmitter for which in the striatum is dopamine. Symptoms of Parkinsonism, in particular akinesis, can therefore be successfully treated by administration of L-dopa, the precursor of dopamine. (Dopamine itself is ineffective, since it cannot pass the blood-brain-barrier.)

Pathology of the Cerebellum. When all connections with the cerebellum are severed, three symptoms predominate: (1) *asynergy,* defined as the inability to achieve a properly timed and balanced activation of the muscles during a movement. The individual elements in a move-

Table 6-1. Comparison of the Cardinal Symptoms of Disease of the Cerebellum and of the Basal Ganglia.

Cerebellar Symptoms	Parkinson's Syndrome
Asynergy	Akinesis
Hypotonus	Rigor
Intention tremor	Tremor at rest

ment program are not carried out together, but rather in succession ("decomposition" of movement); the movements are carried too far or fall short, and the error is then overcompensated (dysmetria), the gait becomes uncertain, with the feet placed far apart and the steps overshooting (cerebellar ataxia), it is no longer possible to make movements in rapid succession (adiadochokinesis), and there are corresponding disturbances of speech; (2) *intention tremor,* trembling that appears not at rest but during movements, and (3) *hypotonus,* a condition in which the muscle tonus is too low (in many cases the muscles are weak and tire easily).

The tabular presentation of the chief symptoms of disease of cerebellum and basal ganglia (Table 6-1) brings out once again the *parallelism between the deficits* in the two cases. On each line of the table, the symptoms tend to differ in their sign (particularly in the case of hypotonus versus rigor). This suggests that in both cases the fundamental problem is a *disturbance of a program.*

The Role of the Motor Cortex. By electrical stimulation of the precentral gyrus and the adjacent motor areas in man (during therapeutically required operations) and in animals, contractions of single muscles and movements at joints can be elicited, but complex sequences of movement directed toward some goal are never produced. This finding has been made without exception in children and adults; it is as true of a skilled pianist as of a manual laborer. From these and other observations it can be concluded that the motor cortex is not responsible for the design of innate or acquired movement patterns. Rather, it is (as Fig. 6-8 shows) the *last supraspinal station* for conversion of the designs for movement arising in the associative cortex into *programs for movement.* At the same time, as can also be seen in Fig. 6-8, it is the beginning of the chain of structures responsible for the *execution of movement.* Remember, though, that many complex sequences of movement involve only subcortical structures—for example, recall the acid-wiping reflex of the decerebrate frog described in Sec. 4.3. Even mammals from which the entire cerebral cortex has been removed (decorticate animals) display motor activity—in some cases in an abnormally lively way—but it is undirected; often such movement is continued until the animal becomes exhausted.

Pathology of the Motor Cortex and its Efferents. By far the most important impairments from a clinical point of view are partial or complete interruptions of the cortical efferents in the region of the internal capsule—where the tracts pass between the basal nuclei and the thalamus (Figs. 6-10 and 6-11). Sudden hemorrhages or blockages of blood vessels (thromboses) in this region (they tend to occur here because of the sharp bends in the vessels) produce the syndrome of *stroke* (apoplexy). This leads to an initial stage of shock with flaccid paralysis of the contralateral side of the body (hemiplegia). When the state of shock passes off, the paralysis usually becomes spastic; that is, the paralyzed muscles have high tonus, expressed (in contrast to rigor) particularly as an increasing resistance to stretch. Patients with severe cases of this spastic hemiplegia present a picture reminiscent of experimentally produced decerebrate rigidity.

The *paralysis* in spastic hemiplegia was once attributed to the interruption of the corticospinal tract, while the absence of the other cortical efferents was considered responsible for the *spasticity*. That is, the paralysis was regarded as a *pyramidal-tract symptom*, and the spasticity the result of damage to the *extrapyramidal motor system*. This neat subdivision is no longer acceptable in the light of our present knowledge. The discrepancy has perhaps already become apparent from the preceding description of the motor system (see in particular Figs. 6-8, 6-17, and 6-18 and the associated parts of the text.) An additional consideration is that transections confined to the corticospinal tract in monkeys (in the pyramids or cerebral peduncle) and clinical observations of humans with corresponding brain damage have shown that after the recovery phase the only deficit is a certain limitation on manual dexterity; there is little or no increase in muscle tonus or in the magnitude of the stretch reflexes. Thus, the pronounced hemiplegia after a stroke is always ascribable to the simultaneous destruction of various descending pathways from the motor cortex.

The Impulse to Act and the Design of the Motor Pattern. Practically nothing is known of the generation of the patterns of neuronal discharge that convert the initial conception of an action into a program for the required movements (see Fig. 6-8), regardless of whether the action is elicited by innate releasing mechanisms or is an expression of free will. But when thoughts lead to actions, the neurophysiologist is forced to accept that *thinking can change the neuronal activity of the brain*—the efferent flow of signals from the motor centers in fact results in the desired movement of the muscles. Such conversion of thinking and intent into cortical impulse patterns remains, for the time being, far beyond the limits of our understanding.

Nevertheless, the first steps on the road to discovery of the neurophysiologic processes that precede a movement have already been taken. For example, if the subject of an experiment is instructed to make a movement in response to the second of two successive stimuli (light signals, clicks, and so on), a slow negative wave can be recorded from the cerebral cortex just prior to the movement. This has been called an *expectancy potential*, for it is associated with the conscious expectation of the second stimulus which the first stimulus produces.

If, on the other hand, it is left to the subject to make a voluntary movement (for example, bending a finger) at irregular intervals independently of any sensory stimuli, a slowly increasing surface-negative brain potential develops, beginning about 800 ms before each movement. This has been called a *readiness potential;* it can be recorded over the entire convexity of the skull (Fig. 6-19). The readiness potential is thought to be associated with the processes that precede the sending out of a movement program from the motor cortex (see Figs. 6-8, 6-17, and 6-18). This event is followed by more rapid and spatially restricted potentials that are especially pronounced over the contralateral sensorimotor areas; they reflect the activity of these areas before and after the onset of the relevant movements (see the legend of Fig. 6-19).

The *readiness potential* can thus be regarded as the *neuronal correlate of a design for voluntary movement.* The spatial extent and the slow development of this potential is astonishing. Evidently, this stage of development of a voluntary action *requires the collaboration of large parts of the cerebral cortex,* which take considerable time to play their part in preparation of the program.

Neural Correlates of Central Processing: the Cerebellar Cortex as an Example. It was stated in the introduction to this section that practically nothing is known about the way in which the higher motor centers perform the tasks we have just discussed. The most progress has been made in analysis of the neuronal networks of the cerebellar cortex and the associated nuclei of the cerebellum. Some current knowledge of these networks will, therefore, be outlined briefly in the following paragraphs.

The diagram of inputs and outputs of the cerebellar cortex in Fig. 6-12 shows two inputs, the *mossy fibers* and the *climbing fibers,* and one output, the *Purkinje-cell axons* (these were discussed in Sec. 6.3). The cerebellar nuclei connect the cerebellar cortical outputs with the other motor centers. Information processing in the cortex takes place

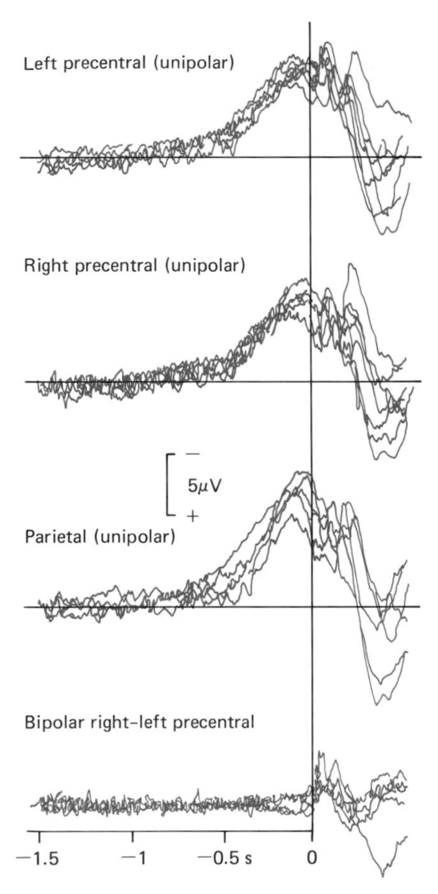

Fig. 6-19. Electric activity of the brain, recorded from the human scalp prior to voluntary rapid flexion of the right index finger; means of 8,000 trials with the same subject. The *upper three rows* are unipolar recordings from the indicated positions (see Fig. 9-1) with the two ears serving as the indifferent electrodes. The *bottom curve* shows bipolar recordings of the potential between the left and the right precentral hand regions of the motor cortex (see Fig. 6-9). The readiness potential (in the *three upper curves*) begins about 0.8 s before the onset of movement (which is at time zero). About 90 ms before the onset of movement, this gives way to a "premotor positivity" (the traces are deflected downward). In the *lowest curve,* a "motor potential" appears simultaneously with the movement; this can be recorded only over the left precentral hand region (with movement of the right index finger). The potentials occurring after the beginning of movement (to the *right* of the zero point in the upper three curves) will not be discussed here. (Recordings from Kornhuber and co-workers.)

in neuronal circuits, one of which is drawn in Fig. 6-12C. The individual components are shown in Figs. 6-12A and B, and the direction of flow of excitation is indicated by arrows. The polarity of the synapses is given by the color of the individual components: black axons form excitatory synapses and red axons, inhibitory synapses.

The *climbing fibers*, then, make excitatory synapses with the dendrites of the Purkinje cells. The effects of the *mossy fibers* are more complex; they excite the granule cells, the *parallel fibers* of which in turn have an excitatory action on the basket cells and the Purkinje cells. An interesting complication is introduced by the fact that the basket cells inhibit the Purkinje cells. The parallel fibers (and hence the mossy fibers) thus have a dual effect on the Purkinje cells—excitatory on the dendrites, inhibitory on the soma via the basket cells. This inhibition of the Purkinje cells by way of the basket cells is a typical example of *forward inhibition;* in contrast to feedback inhibition, this inhibition depends upon the inputs to the cells and not upon their outputs (see Fig. 4-4B). The inhibitory synapse of the basket cell is probably especially effective because of its position at the axon hillock of the Purkinje cell. The Purkinje-cell axons, finally, make *inhibitory synapses on the cells of the cerebellar nuclei.* Since the Purkinje cells have a resting discharge that *produces a tonic inhibition of the cerebellar nuclei,* an increase in the discharge rate of the Purkinje cells causes a more profound inhibition of the cerebellar nuclei, whereas a decrease results in disinhibition.

Afferent activity in the climbing fibers enhances the tonic activity in the cerebellar cortex. *Afferent activity in the mossy fibers,* on the other hand, has a dual effect, in part excitatory via the parallel fibers and in part inhibitory via the basket cells. It is not yet entirely clear which sensory modalities, or which parameters of the sensory stimuli, are transmitted by the mossy fibers and which by the climbing fibers. Nor is enough yet known about the connectivity of the parallel fibers, although it has been shown that mossy-fiber activity usually leads to excitation of circumscribed groups of Purkinje cells, while neighboring Purkinje cells are inhibited. Since all excitatory inputs to the cerebellum *give rise to inhibition after at most two synapses,* the effect of any input is extinguished after about 100 ms, and the site concerned again becomes avaiable for processing a new input. It can be assumed that this automatic "erasing" is crucial to the participation of the cerebellum in rapid movements.

The physiology of the cerebellum can serve as an example of the extent to which brain physiology has resolved the "How" and "Why" of central nervous activity. Despite our detailed knowledge of the cerebellar circuits and their inputs and outputs, we still cannot state much more precisely than has been done here how these circuits carry out the functions of the cerebellum—which, indeed, we have also been able to infer. The task now is to improve this unsatisfactory situation by continued reflection and experimentation. The advances that have been made in recent decades in many areas of neurophysiology suggest that there will be critical breakthroughs in our understanding of central nervous activity. In any case, there are at present no grounds for believing the assertion one sometimes hears, that the brain is incapable of "understanding itself."

Q 6.21 The majority of all afferents in the basal ganglia come from
a. the brainstem.
b. the motor cortex.
c. the associative cortex.
d. the thalamus.
e. the cerebellum.

Q 6.22 The cells of origin of the corticospinal tract lie
a. exclusively in the postcentral gyrus.
b. predominantly in the postcentral gyrus and adjacent cortical areas.
c. in the basal ganglia, especially the pallidum.
d. exclusively in the precentral gyrus.
e. predominantly in the precentral gyrus and the adjacent (frontal) cortex.

Q 6.23 Which of the following symptoms are characteristic of the simultaneous interruption of pyramidal and extrapyramidal tracts in the left internal capsule?
a. Tremor at rest.
b. Intention tremor.
c. Flaccid paralysis on the left.
d. Adiadochokinesis.
e. Parkinsonism on the right.
f. None of these symptoms is characteristic.

Q 6.24 Inhibitory synapses on the dendrites of the Purkinje cells of the cerebellum are formed by
a. climbing fibers directly.
b. mossy fibers, via granule cells and parallel fibers.
c. climbing fibers, via basket cells.
d. mossy fibers, via granule cells and basket cells.
e. All the above are correct.
f. All the above are false.

Q 6.25 Which of the following symptoms are characteristic of the absence of cerebellar function?
a. Tremor at rest.
b. Intention tremor.
c. Asynergy.
d. Hypotonus.
e. Parkinsonism.
f. Hemiplegia.

7

REGULATORY FUNCTIONS OF THE NERVOUS SYSTEM, AS EXEMPLIFIED BY THE SPINAL MOTOR SYSTEM

M. Zimmermann

Many functions of living organisms are *regulatory processes*—a certain state, characterized by a measurable quantity, is kept constant. Examples are the regulation of body temperature, of blood pressure, and of the position of the body in the gravitational field. In many cases of regulation, the nervous system plays a central role.

Biologic regulatory processes lend themselves to description in terms of *control theory*, a discipline that was originally developed in a technologic framework. In a similar way, communication in biology has been systematized by the application of information theory (see *Fundamentals of Sensory Physiology*, Chapter 2).

Control theory and information theory are, of course, but two specialties within the general area of applied mathematics. Physiologists are increasingly using mathematical and computer methods to study the consequences of their necessarily complex hypotheses, and to compare these with new experiments. Especially in Europe, application of control and information theory is often lumped under the term *biocybernetics* (*cybernetics* itself usually connotes both control and information theory).

Neuroscientists, in particular, have a natural interest in control theory, for its methods—or some equivalent—are often the only way one can work out what will happen in a (real or hypothesized) complex circuit of dynamic elements, especially one including feedback. As an introduction to some of the elementary ideas of control theory, we shall in this chapter show how the spinal motor functions discussed in Chapter 6 can be represented in the language of that field.

7.1 The Stretch Reflex as a Length-Control System

The Concept of Regulation. The two experimental results of Fig. 7-1 illustrate the fundamental observation that has led to the interpretation of the stretch reflex as a *control circuit*. A muscle preparation isolated from its afferent and efferent connections to the spinal cord shows a passive length–tension behavior (Fig. 5-5) as illustrated by the black curve in Fig. 7-1. If a force (ΔF) is applied to this resting (unstimulated) preparation, the muscle is seen to be stretched (by ΔL_0).

Now consider the same length–tension measurement applied to a muscle *in situ*. In this case, the connections with the spinal cord are intact; the experiment is best done with a decerebrate animal (see Sec. 6.4). The result is illustrated by the red curve in Fig. 7-1. Here application of the same force (ΔF) produces a much smaller change (ΔL_1) in length; it is as though the muscle had become "stiffer." Evidently, with the inclusion of the spinal control, the system "spinal cord and muscle" responds actively to the disturbance so as to oppose changes in length. That is, muscle length tends to be kept approximately constant (see Fig. 6-3), or "regulated."

The Notion of a Closed Loop. If we now say that the "stretch reflex" (involving the spindle afferents and α motoneurons) underlies the

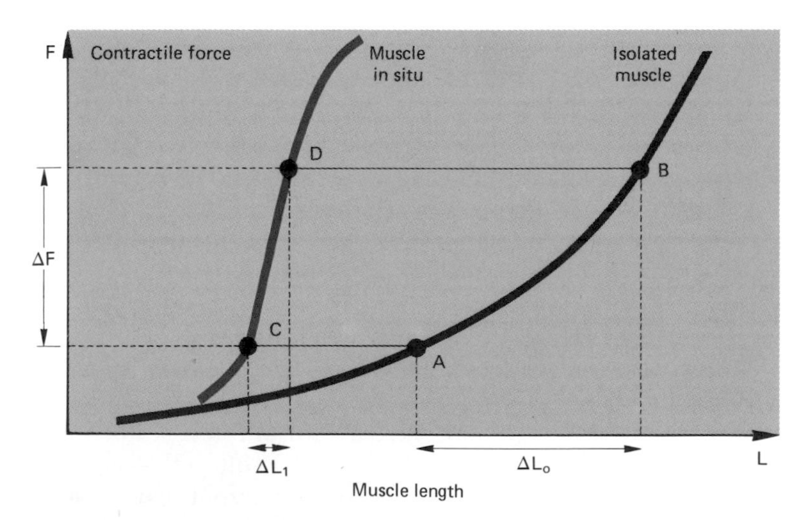

Fig. 7-1. Representative relationships between force and length in a muscle. The increase in force ΔF (ordinate) causes a large change in length ΔL_0 (abscissa) in an isolated muscle preparation, while in a muscle with nervous connection to the spinal cord intact (*red curve*) the change is small (ΔL_1).

above result (increase of static stiffness), a conceptual problem is likely to arise: ordinarily, such a reflex is described simply in terms of an initial "stimulus" (stretch) and a later response (contraction)—as in Fig. 7-2A (black). This description is adequate for a clinical test, but it runs into trouble as soon as one wants to analyze the reflex and its components. That is, if length changes induce tension changes, one can hardly ignore the fact that (depending on load) the resulting tension changes then cause further length changes, which in turn . . . and so on (Fig. 7-2A, red). In general, if the response affects the stimulus (as it does in the stretch reflex because of the parallel arrangement of the extrafusal fibers and spindles in the muscle; Fig. 7-2B), then we are dealing with a *closed loop*. This is the situation in nearly all reflexes. Because of this fact, it is difficult to intuit directly just what the overall performance of such a system (as in Fig. 7-1) tells us about, for example, the components and their connections. And it is precisely here that control theory helps—engineers have worked out standard experimental and mathematical methods for treating such tricky situations. In this book we shall not go into the mathematical methods. We emphasize, though, that there is nothing mysterious about them—a full analysis of at least the static behavior of the closed-loop systems to be mentioned here requires only high-school algebra (see any introductory text on control-system analysis).

One of the most useful and common instances of such closed loops is the negative-feedback control system. Here a controller influences some controlled system, and there is feedback from some variable in

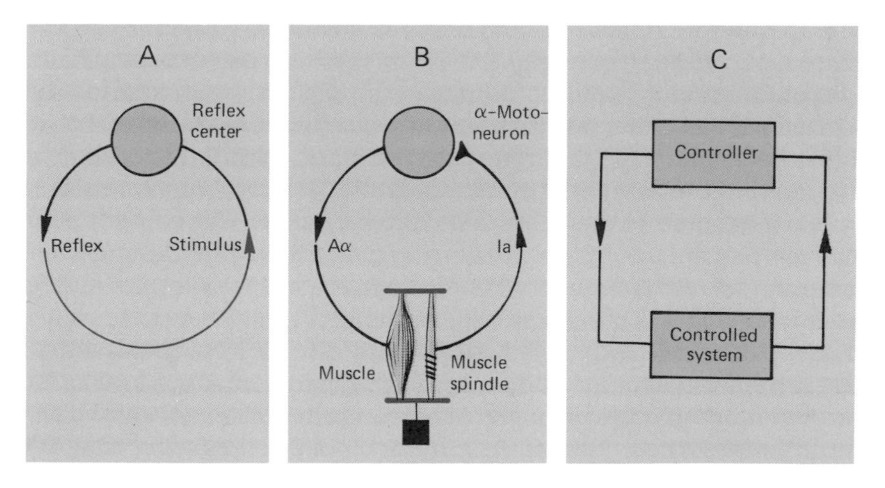

Fig. 7-2. Closed-loop situations. **A**, Diagram of a generalized reflex with feedback (red); **B**, stretch reflex, in which negative feedback is brought about by the parallel arrangement (red) of muscle and muscle spindle; **C**, simplest block diagram of a control circuit.

the controlled system, which acts back on the controller, affecting the control. The analogy with our interpretation of the stretch reflex as a length-control system is illustrated in Figs. 7-2B and C. Before returning to our discussion of the stretch-reflex system, however, we introduce the elements and function of a control system more generally, by comparison with an automatic central-heating system.

Structure of a Control Circuit. Suppose you wish to keep the temperature of a room as constant as possible year-round, near some prechosen level. There are many influences on room temperature that you can anticipate in advance, and compensate for by certain strategies. For example, you might simply turn all your heaters on from November to March, when you know the outside temperature will tend to be low. This kind of control, based on expectation but without feedback concerning the current room temperature, is called *open-loop control.* The disadvantages are obvious in this case, though occasionally the principal is useful (for example, in systems processing very rapid events).

Clearly, for room temperature control and in many other cases, it is more sensible to measure the room temperature itself, and adjust the heaters accordingly. This process is called *closed-loop* (or *feedback*) *control;* it is, of course, usually automated so that once set it runs by itself. The main elements of such a system are labeled in the block diagram of Fig. 7-3B, and illustrated by corresponding components of the room-heating system in Fig. 7-3A. The *controlled variable* is the thing being kept constant (room temperature). The physical structures within which regulation occurs are the *controlled system* (room with heater). A *sensor* (thermometer) monitors the current value of the controlled variable, and its output is the *feedback* signal (electrical signal from thermometer). In the *controller* (thermostat), the feedback signal is compared with a reference signal or *set point* (the selected temperature on the thermostat dial). If the feedback signal reflecting the current value of the controlled variable differs from the set point, this comparison results in an *error signal.* From this the controller generates a *control signal* (vent sitting), which acts (via the fuel-feed rate of the heater) on the controlled system in such a way as to change the controlled variable in the appropriate direction, so that feedback and set point coincide. The closed-loop control thus works continuously to minimize the error signal, even in the presence of unforseen perturbations (heat-loss rate, e.g. by opened doors or windows, changes in outside temperature).

The critical property of such a regulator is the comparison that generates the error signal. In its simplest form, the error signal is the result of *subtracting* (mechanically or electrically) the feedback signal

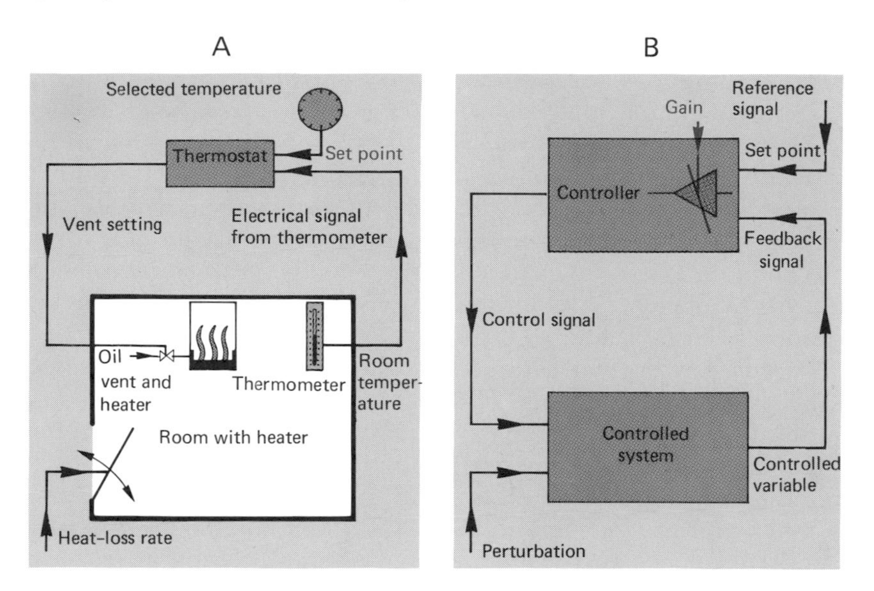

Fig. 7-3. A. Illustration of the control circuit, using the example of a heating system to regulate the temperature in a room. Each element in the physical control system is associated with an equivalent general designation, in control-systems terminology, as in *B*. **B.** Block diagram of a generalized feedback control system. The terms are explained in the text. The gain of the controller (output per unit deviation of the feedback from the set point) is assumed to be variable (*red*).

from the reference signal. Hence such an arrangement is called *negative-feedback* regulation. As we shall see, however, the crucial change of sign producing the regulatory effect can be at other points in the loop.

Regulators and Servomechanisms. Closed-loop control circuits like the above temperature control system are designed to hold a controlled variable at a fixed preset level, and are usually called *regulators*. Note, however, that we can at any time adjust the set point (thermostat dial) to a new desired level. This will cause an error signal in the controller, and the system will change the controlled variable to match the new set point. To go further, if one continuously changes the set point, such a system becomes a device causing one physical variable (the controlled variable) to *follow* another. Closed-loop control systems designed to do this are called *servomechanisms*. For example, a servomechanism makes the angle of a ship's rudder follow the setting of the wheel.

Control of Muscle Length. Returning to our discussion of the stretch reflex in terms of control theory, it should be clear from the physiologic properties described in Chapter 6 that there are many points of correspondence with the elements of the above generalized control circuit. Taking muscle length to be the *controlled variable,* the muscle spindle amounts to a *sensor* which generates *feedback* (activity in the Ia afferent fibers). *Control signals*—based on evaluation of both the feedback and other influences ("reference signals" from higher motor centers) on the reflex—are sent out ultimately as activity of the α motoneurons, which act on the controlled system (muscle with tendons and joints) so as to change length. Finally, *perturbations,* which the system operates to counteract, exist in the form of unforeseen variation of load or fatigue of the muscle.

There is, however, an interesting difference of design with respect to the origin of the change of sign required for negative feedback control. In the central-heating system, we ascribed this for simplicity to the controller's comparison of feedback with set point. In the stretch-reflex loop, on the other hand, increasing length increases Ia afferent activity, which in turn increases α-motoneuron activity—both "positive" effects. In fact, the change of sign actually arises in the muscle; increased α-motoneuron activity tends to *reduce* length, reducing the discharge rate of the afferents from the spindles. In both cases, then, a disturbance increasing the value of the controlled variable elicits negative feedback effects—effects that cause the controller to counteract that increase.

Q 7.1 The stretch reflex can be regarded as a control circuit because
- a. the muscles make circular movements.
- b. the afferents involved come from the annulospiral endings of the muscle spindles.
- c. an isotonic muscle contraction is enhanced by the stretch reflex (positive feedback).
- d. disturbances affecting muscle length are compensated by a reflex contraction.
- e. the muscle length is kept essentially constant.

(more than one answer is correct)

Q 7.2 In a refrigerator (considered as a control circuit), which of the following are true?
- a. The thermostat is the controlled system.
- b. The temperature of the interior is the controlled variable.
- c. The cooling apparatus (motor and compressor) is the controller.
- d. Heat conduction through the wall and the entry of warm air through the opened door are disturbances.
- e. With the thermostat set at a fixed value, the refrigerator is a servomechanism.

(more than one answer is correct)

Q 7.3 What is the "master/servant" relationship to which the term "servomechanism" refers?
- a. The controlled variable changes in obedience to an arbitrarily varied reference signal.

b. The feedback signal changes in obedience to the changeable controlled variable, with some delay.
c. The controlled variable changes in obedience to a perturbation.
d. Any disturbance of the controlled variable is compensated for.

Q 7.4 The following biologic and technical functions can be regarded as control circuits:
a. the mating drive of animals.
b. maintenance of a constant body temperature in mammals and birds.
c. the adjustment of breathing to meet changing needs.
d. power steering in an automobile.
e. the production of either light or dark toast by an automatic toaster.
(more than one answer is correct)

7.2 Static and Dynamic Properties of Control Systems

Above we have considered only the static (or steady-state) properties of the elements of control circuits. The situation is usually more complicated, for such elements often show an initial transient behavior following a perturbation (see the "phasic" and "tonic" aspects of the afferents and efferents illustrated in Figs. 7-4B and C). Accordingly, the closed-loop control circuit also exhibits important transients.

Here again control theory is useful, for it offers standard methods of description of these dynamic properties, and keeps track of their interactions when such elements are interconnected. A basic mathematical tool in such analysis is the *transfer function*—a function that allows one to calculate the time-varying *output* of an element or a circuit in response to *any input waveform* (in its range of validity). Here, for simplicity, we illustrate the dynamics in terms of the graphic *step response* of these elements and circuits—the response in time following a sudden step change of an input variable. The experimentally measured step response in principle contains the same information as the transfer function used in control theory to characterize a control circuit.

Because, as we have seen, the closed-loop behavior of a circuit can obscure the properties of the individual elements, a standard technique is to open the loop at some point, and then to measure the response at various points to a perturbation. One can do this with the stretch reflex by cutting the dorsal or ventral roots.

Figure 7-4A, B, and C illustrate *open-loop responses* at two points in the stretch reflex. The muscle length (A) has been caused to undergo a step increase. The spindle afferents (B) and the motor efferents (C) both respond with an overshoot before settling down to the steady-state response associated with the new length. The dotted lines show

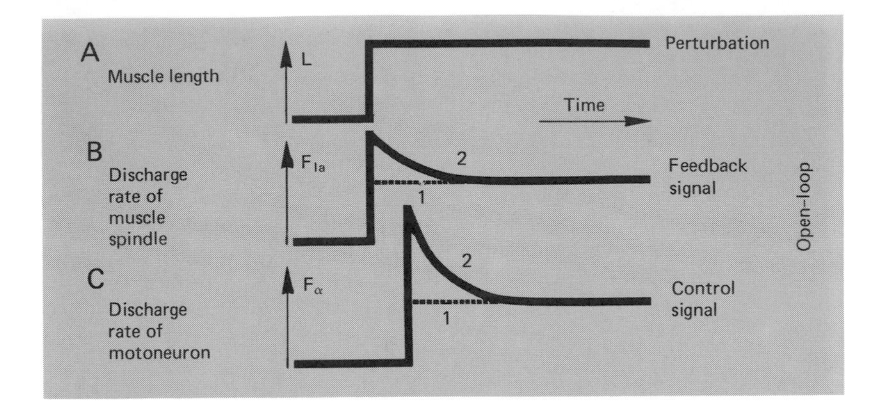

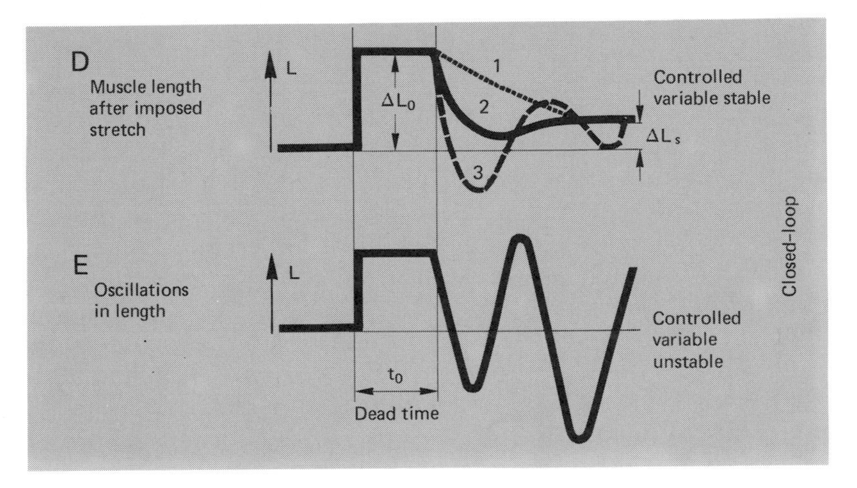

Fig. 7-4. Dynamics of length-control circuit, illustrated by the step response. **A**, Step change in muscle length L induced by a change in load; **B**, response of the sensor (frequency F_{Ia} of the muscle-spindle discharge); **C**, delayed response of the controlling element (frequency F_{α} of the motoneuron discharge) to the same disturbance. The responses in **A**, **B**, and **C** are represented under open-loop conditions. **D**, Response of the controlled variable (muscle length L) under closed-loop conditions for the case of proportional (tonic) circuit elements (1) and with elements having initial transients (2, 3); for 2 the net closed-loop gain is smaller than for 3. **E**, Undamped oscillations; t_0 = dead time. See discussion in text.

the responses expected if there were no dynamic effects. The overshoot—even more pronounced in the delayed motoneuron discharge—is evidently caused by the sudden change in length. These are not defects; rather, the reflex has evolved to deal effectively with rapid changes.

In Fig. 7-4D, the corresponding dynamic *closed-loop* behavior of the reflex is illustrated in terms of the changes of length. First a force is applied so as to cause a step change in length ΔL_0. After a *delay*, or *dead time*, t_0 (ca. 30 ms for conduction and synaptic delay in the human calf-muscle reflex), the reflexly altered motoneuron activity begins to compensate for the disturbed muscle length. The dotted curve 1 shows the kind of time course that might be expected if the control signals had no overshoot—the slow change would be determined only by the muscle dynamics and load inertia. Curve 2 shows that the overshooting dynamics of the control signal can produce more rapid recovery of length; curve 3 illustrates that even larger initial transients in the motoneuron response could lead to an overshoot in the controlled variable.

Effectiveness of Regulation and Stability. Following the transients in Fig. 7-4D, muscle length settles down to a steady value ΔL_s. Perfect regulation of length would have compensated entirely for the perturbations and brought length back to its initial value. ΔL_s is sometimes called the "steady-state error"; in regulation of this type (proportional control), it increases with the magnitude of the perturbation being compensated. In another type of control circuit—integral control—the steady-state error is zero. A fundamental way of reducing this steady-state error in proportional control is to increase the *open-loop gain* (or amplification factor) of the controller. The gain of the stretch reflex control, for example, is the change in discharge rate of motoneurons (mediated by Ia afferents) per unit change in length of the muscle (see below, and Fig. 7-7).

However, the price paid for the improved regulation due to high gain is the danger of *instability*. Especially because of the overshooting dynamic behavior of the feedback and control signals and the conduction delay times, it is easy to picture a situation in which the loop—at very high gain—*overcompensates* briefly. This overcompensation is then reversed by the reflex but again with a delay. That is, the system can get out of step with itself, and uncontrolled oscillations can result (Fig. 7-4E).

Certain pathologic changes in the motor system are characterized by oscillations of muscle length. These may be caused by an increase in the facilitation by the brain of the spinal stretch reflexes. These phenomena occur in neurologic disturbances such as clonus (ongoing rhythmic reflex twitches when a stretch reflex is triggered by a hammer) or the tremor (trembling) of the hands as in Parkinsonism. In both cases, there are undesirable oscillations of muscle length.

Modification of the Stretch Reflex from Higher Centers; Servocontrol of Muscle Length. So far we have considered only the reflex mechanisms tending to hold muscle length near a given value. In movement, however, length is changed in a directed manner. As described in Chapter 6, these effects are brought about by motor centers in the brain that act on the spinal control systems via descending pathways.

In the temperature controller of Fig. 7-3, we noted that not only perturbations of the controlled variable but also changes of the set point caused an error signal, and the latter caused the regulator to control temperature at a new operating point. In an analogous way, descending commands in the spinal cord can alter the thresholds and discharge rates of the α motoneurons (and γ motoneurons as well; see

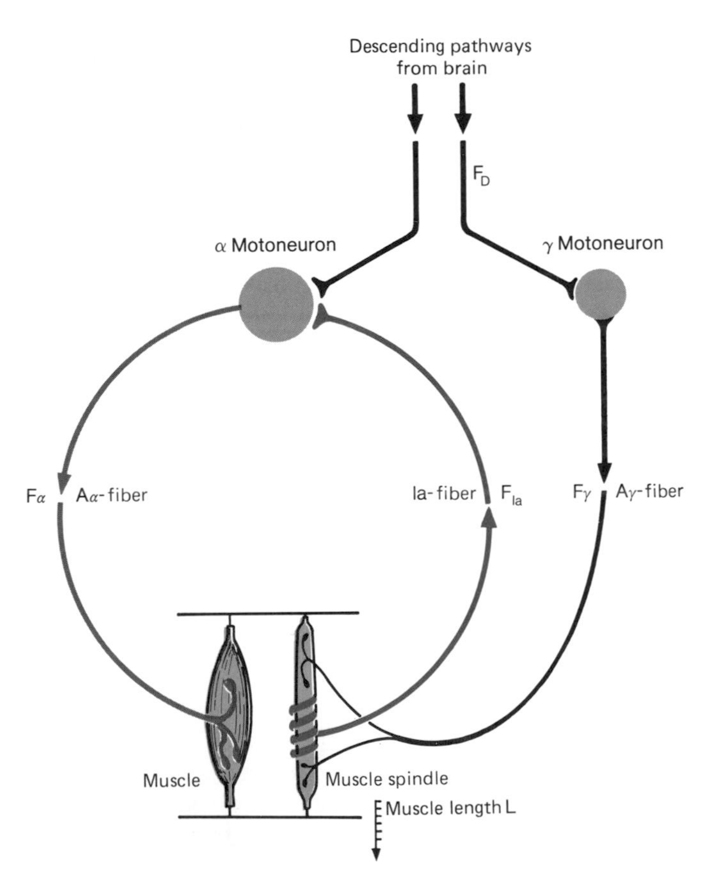

Fig. 7-5. The spinal stretch reflex and the descending pathways that mediate supraspinal influences.

below) changing the effective set point about which the reflex regulates length.

In Fig. 7-5 the diagram of the stretch reflex (Fig. 7-2B) is complemented by addition of the pathways descending from the brain. To understand the following discussion, you should make sure that you can see the qualitative points of correspondence between this diagram and the block diagram of the control circuit shown in Fig. 7-3B.

Figure 7-6 schematizes the time course of events in such a situation. In analogy with the format of Fig. 7-4, we postulate a step change (A) in the discharge rate of fibers in a descending motor tract. The dynamics of the *open loop*-induced changes (for example, with

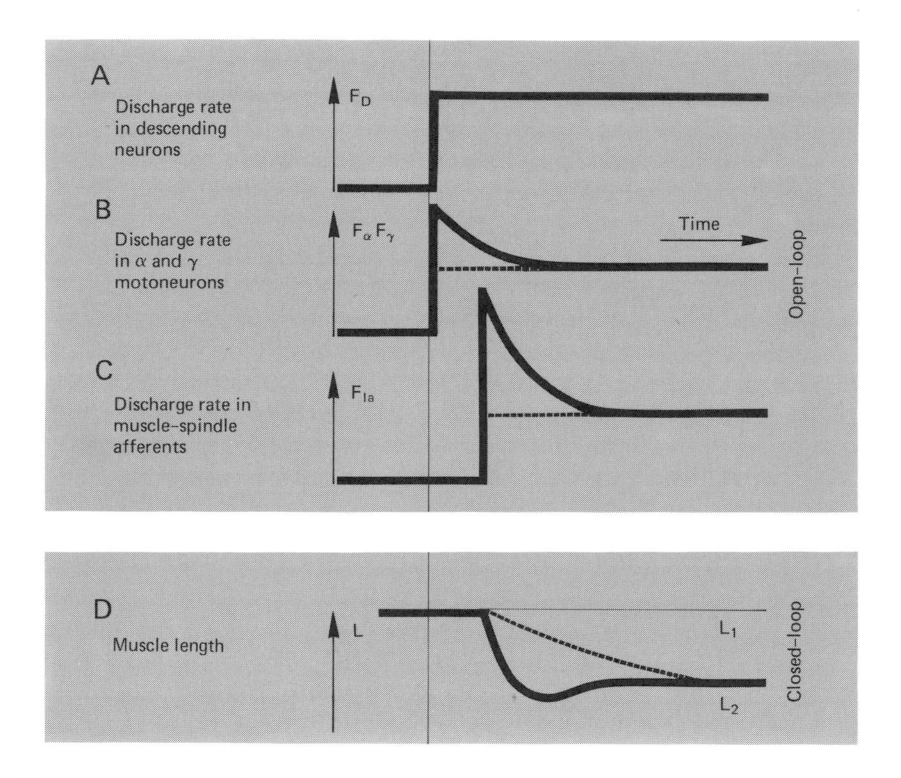

Fig. 7-6. Alteration of the set point of a reflex control circuit. **A**, Hypothetical step change in the frequency F_D of descending excitation; **B**, corresponding response of the α motoneuron activity; **C**, later response of the sensor (muscle spindle, affected via the γ motoneuron). In **B** and **C** the loop is assumed open. **D**, Transient response of the controlled variable after the same step change in the descending control, under closed-loop conditions. The *dotted curve* is the response assuming purely proportional response of the circuit elements and the *solid curve*, that with transient response as indicated in **B** and **C**.

dorsal roots cut) in α and γ fibers are similarly shown in B, and the subsequent open-loop response of the spindle afferents in C. In the *closed-loop* (intact reflex) case (Fig. 7-6D), the net result of the imposed activity of higher centers has been to shift the controlled variable from L_1 to a new value L_2. The dynamics of the transition are similar to those in Fig. 7-4. The spinal control circuit now continues to act as a fixed-point regulator, tending to hold length near L_2.

A remarkable property of the control of movement is that the descending control signals enter the stretch reflex circuit at two different points: at the α motoneuron and at the muscle spindle, via the γ motoneuron. This situation is called *α–γ coactivation,* and is discussed in Chapter 6. These complex combined effects can be analyzed in detail by control theory, but here we shall simply note some of the qualitative advantages. For example, when the muscle shortens due to descending α activation, the concomitant γ activity ensures that the state of stretch of the spindles, and hence their sensitivity, is maintained. That is, the *working range over which reflex regulation of length can operate* is extended. Moreover, the fact that the descending influences via α and γ motoneurons affect the reflex differently allows more flexibility for the higher centers to manipulate the limbs rapidly and at the same time to avoid instability. We can also infer that α–γ coactivation reflects a sophisticated form of movement control from the fact that the phylogenetically lower amphibians have no separate γ system; in these animals the spindles share efferent innervation with the extrafusal muscle fibers.

Steady-State Operating Characteristics. In Figs. 7-4 and 7-6, after the transients have settled down, the system is said to be in a *steady state.* We will use the operation of the reflex in this state to illustrate further some of the principles of control—the advantage being that we can ignore for the moment the transients. In particular, in the steady state we can easily describe the properties of the elements over a *wide range of inputs* by drawing *characteristic curves*—curves showing the steady outputs resulting from a given steady input.

A steady-state characteristic curve of the muscle spindle (Ia discharge rate versus length) has already been shown in Fig. 6-4. Similarly, one can define a curve for the relation between Ia discharge and α-motoneuron discharge. If the above two (open-loop) relations are combined, one gets an overall curve relating length change of the muscle to the α-motoneuron discharge rate. Open-loop curves of this kind are shown in Fig. 7-7A. For simplicity, here all these effects have been assumed to be linear.

The curves of Fig. 7-7A illustrate two kinds of effects that can be

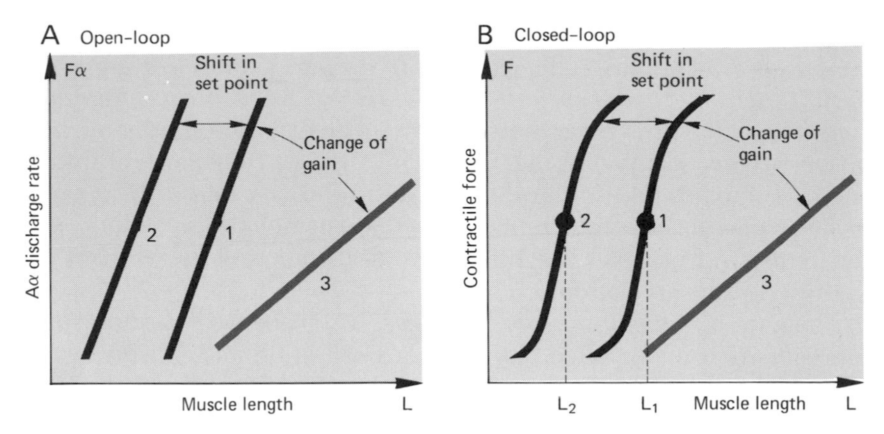

Fig. 7-7. Steady-state characteristic curves. **A**, Input–output relation between length and controller output under open-loop conditions. When the effective set point is changed the characteristic curve shifts along the abscissa (*1*, *2*); when the closed-loop gain changes (*red*) the slope of the curve is altered (*1*, *3*). **B**, Characteristics of the closed-loop control circuit, in a force-length diagram. Other symbols as in **A**. See discussion in text.

imposed upon such a control circuit: *change of set point* and *change of gain*. In principle, influences upon either the α or the γ motoneurons by supraspinal centers could act to change the α-activity-versus-length curve as shown by the black curves of Fig. 7-7, though experiments imply that usually both mechanisms operate (see α-γ coactivation, Chapter 6). The shift (for example, from 1 to 2) without change of slope denotes the result that the control circuit operates at a new set point but with unchanged gain (and hence with similar effectiveness of regulation).

The corresponding *closed-loop* case in Fig. 7-7B, in which tension versus length is shown, is analogous to the experimental situation of Fig. 7-1. The changes in set point described in Fig. 7-7A are here reflected in equivalent tensions at different lengths (1 and 2); again these represent a shift of the characteristic curve along the length axis, with regulation about the new set length.

By contrast with these changes of set point, we illustrate (in red in Fig. 7-7) the effect of *change of gain* in the transfer properties between length and α-motoneuron discharge. As already mentioned, high gain can improve the speed of regulation and reduce steady-state error, but too high gain can lead to instabilities such as undamped oscillations. In a control system adapted to a large repertoire of subtle movements, one can imagine that *modulation of gain* would be an important feature. Indeed, a number of neuronal effects can be interpreted as modifying gain from moment to moment and in various situations as needs arise. These include the various cases of segmental *inhibition of motoneurons involving activity in inhibitory spinal neurons impinging onto the* α motoneurons—for example, autogenic inhibition (triggered by activity in tendon organs), Renshaw inhibition, and antagonist inhibition. Autogenic and Renshaw inhibition account in large part for the transient in the response of the α motoneuron, illustrated in Fig. 7-4C; the initially high gain is rapidly reduced due to these inhibitory actions.

The gain of the spinal stretch reflex is also modified by supraspinal elements that give rise to descending facilitation and inhibition. Supraspinal influences can thus act in two ways: they can *shift* the curve along the abscissa (servoregulation, see above; transition 1–2 in Fig. 7-7) and they can *change the slope* of the curve (gain change; transition 1–3 in Fig. 7-7). Experiments on animals in which the activity of the muscle spindle is modified by way of the γ innervation have provided an especially good example of this twofold action.

Interaction of Length Control During Coordinated Movement. In this chapter, we have tried to illustrate that closed-loop situations, especially where dynamic effects are fundamental to their operation, can be analyzed in terms of control theory. It is, however, important to stress that, where behavior is concerned, many such loops interact with one another to form the system of motor control.

To see how control theory can help to sort out such a complex system, let us consider an example. We know that in the case of the movement of a joint, at least two length-control circuits are involved, namely the stretch reflexes of the two antagonist muscles of that joint. These two length-controls are interdependent in several ways: by way of the mechanical coupling of the muscles at the joint, by the reciprocally acting antagonist inhibition, and by coordination via the supraspinally generated movement program. Thus, when the flexor of that joint contracts so as to cause a movement, the gain of the extensor stretch reflex is reduced appropriately to yield a controlled yet stable change in the overall mechanical situation.

Feedback signals arise from muscle receptors as well as from cutaneous and joint afferents. Feedback loops extend to several levels of the CNS: to the spinal cord, the cerebellum, and the motor cortex. Higher-level sensory monitoring of the effects of ongoing motor commands also occurs via the visual system and the labyrinths.

If one considers the several joints that participate in a movement (such as walking or grasping), the order of complexity is still considerably higher. In principle, even systems of this complexity are ultimately analyzable by the methods of control theory. But painstaking experimentation and a greater fund of neurophysiologic data will be required before such a complete analysis can be achieved.

Q 7.5 Which of the following statements apply to the description of the dynamic behavior of a control circuit?
a. One can measure the step responses of the individual elements of a control circuit when the circuit is open.
b. When a control loop is closed, no dynamics can be measured, for all the disturbances are completely compensated.
c. When a control circuit is unstable, the step response may show undamped oscillations under closed-loop conditions.
d. The dynamic behavior of a linear control system can be derived from the change of the controlled variable in time following a step change of an input.
e. A dead time in a control circuit has no effect on its dynamic behavior.
(more than one answer is correct)

Q 7.6 In the example of the stretch reflex, the effective readjustment of the set point of muscle length in order to produce a movement is by
a. the excitatory input via the Ia afferents from the muscle spindles.
b. the excitation of the total population of homonymous α-motoneurons (integrated frequency F_α) which elicits muscle movement.
c. stretch ΔL of a muscle by variable loading, as a result of which there is compensatory regulation via the stretch reflex.
d. the change in gain of the controlling elements by inhibition (for example, Renshaw inhibition, antagonist inhibition).
e. the change in discharge rates in descending pathways, which can bring about a change in the activity and excitability of α and γ motoneurons.

Q 7.7 The effectiveness of regulation depends upon several properties of a control circuit; which of the following properties have the indicated effects?
a. Oscillations in the circuit improve the precision with which the controlled variable moves to a new set point.
b. If the gain of the controlling elements is too high and there is a dead time in the circuit, regulation is more likely to become unstable.
c. Dead time and inertia in the controlled system can be counteracted by introduction of transient components in the control circuit—for example, in the feedback and controlling signals.
d. The shape and the slope of the steady-state characteristic curve are irrelevant to the precision of regulation by a control circuit.
(more than one answer is correct)

Q 7.8 The stretch reflex, under supraspinal control, can be regarded as a servomechanism because
a. muscle forces counteract gravity forces.
b. the muscle length is regulated at a new set point when the excitation of motoneurons is changed via descending pathways.
c. antagonist inhibition and reciprocal interactions with contralateral stretch reflexes facilitate walking movements.
d. the steady-state characteristics of the stretch-reflex control circuit are shifted by supraspinal influences.
e. the set point for muscle length is variable and is programmed by the brain.
(more than one answer is right)

8

THE AUTONOMIC NERVOUS SYSTEM

W. Jänig

The organism communicates with its environment by means of its somatic nervous system: the sensory system receives and processes information from the environment (see Chapter 1 and *Fundamentals of Sensory Physiology*), and the motor system provides the means for getting about in the environment (see Chapters 4, 6, and 7). Many processes in the somatic nervous system are subject to conscious and voluntary control.

The autonomic nervous system behaves in a quite different way. It innervates the smooth musculature of all organs and organ systems, the heart, and the glands. It regulates breathing, circulation, digestion, metabolism, secretions, body temperature, and reproduction and coordinates all these vital functions. As its name implies, the autonomic nervous system—sometimes also called the *vegetative* or the *involuntary* nervous system—is ordinarily not subject to direct voluntary control.

In functional terms, the actions of the autonomic and the somatic nervous systems are ordinarily interlinked. The two systems are integrated centrally, so that their central neuronal structures are also often inseparable.

8.1 Functional Anatomy of the Peripheral Autonomic Nervous System

The autonomic nervous system consists of three functionally different parts: the *sympathetic, parasympathetic,* and *enteric* systems. The terminal neurons of the sympathetic and parasympathetic systems, which correspond to the motoneurons in the somatic nervous system, are located outside the CNS. The aggregations of cell bodies of such neurons are called autonomic ganglia. The neuron that has its cell body in the CNS and terminates with its axon in such a ganglion is called a *preganglionic* neuron. The neuron that has its cell body in the ganglion and terminates with its axon on effectors is called a *postganglionic* neuron (see Figs. 8-1B and C). The neurons of the gastrointestinal system, situated in the walls of the gastrointestinal tract, are in part identical with postganglionic parasympathetic neurons.

The Peripheral Sympathetic System. The cell bodies of all preganglionic sympathetic neurons are located in the *thoracic* and the *upper lumbar spinal cord* (gray shading in Fig. 8-1A). The axons of these

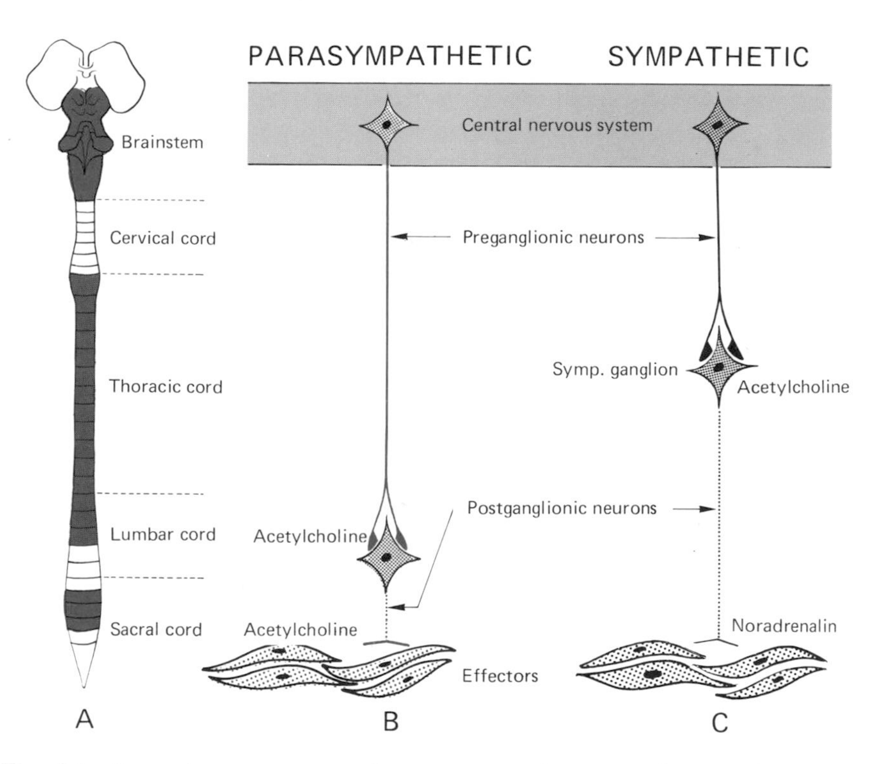

Fig. 8-1. General arrangement and transmitter substances of preganglionic and postganglionic neurons. **A.** Locations of the cell bodies of preganglionic neurons of the sympathetic (*gray*) and parasympathetic (*red*) systems in brainstem and spinal cord. **B, C.** Schematic comparison of the preganglionic and postganglionic neurons in the two systems, and of the synaptic transmitter substances in the ganglia and at the effectors.

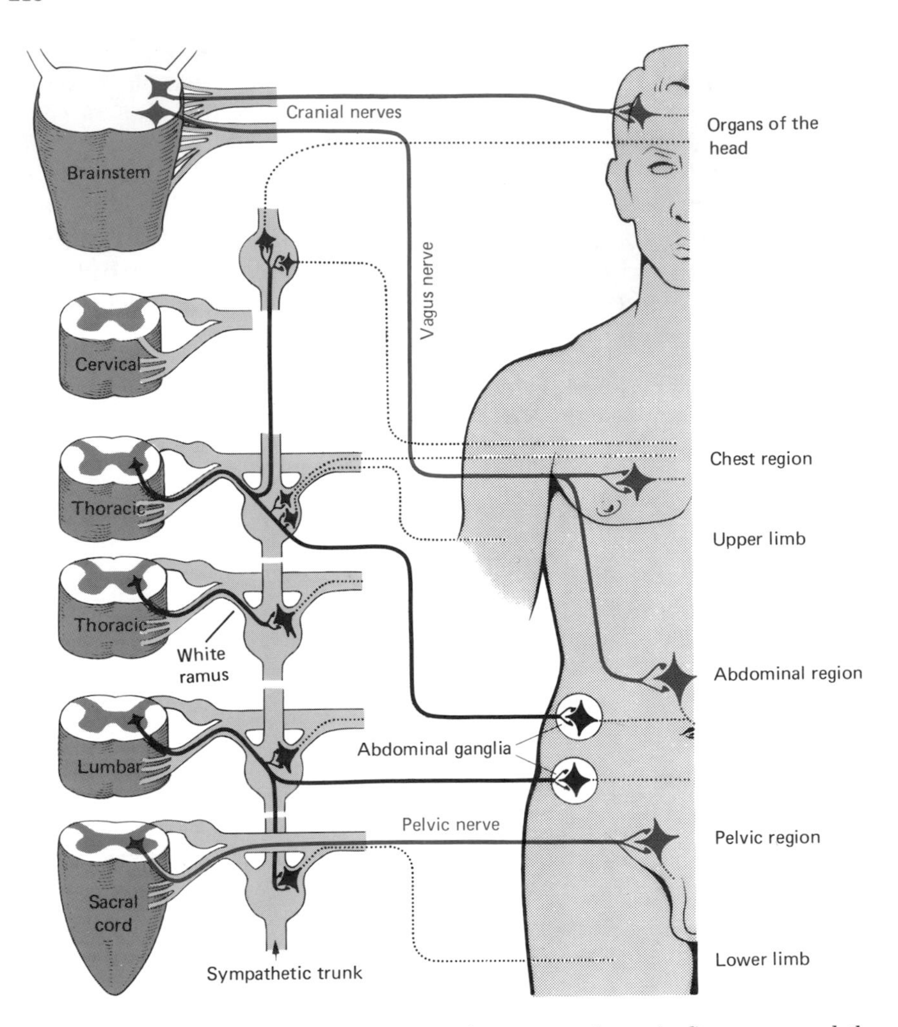

Fig. 8-2. Origins of sympathetic (*black*) and parasympathetic (*red*) neurons and the regions of the body they innervate. Postganglionic axons are dotted. The autonomic ganglia and nerves are drawn too large in relation to the spinal-cord segments. (Modified from Netter (1972) *Ciba Collection of Medical Illustrations, Vol. 1, Nervous System,* CIBA, Summit, N.J.)

neurons (solid lines in Fig. 8-2) leave the spinal cord in the ventral roots and run through the white rami to the *autonomic ganglia* situated *outside the CNS*. The axons of the preganglionic neurons synapse in the sympathetic ganglia with the cell bodies of the postganglionic neurons. The sympathetic ganglia in the region of the thoracic, the lumbar, and the sacral spinal cord are arranged segmentally to the left and the right of the spinal column (*Th, L, S* in Fig. 8-3A); in the region of the cervical cord (*Cv* in Fig. 8-3A) there are only two paired ganglia. The ganglia on each side of the vertebral column are connected from top to bottom by nerve trunks. These chains of ganglia are called the right and the left *sympathetic trunks* (Fig. 8-3A). In addition to these paired ganglia, there are unpaired ganglia in the abdominal and the pelvic regions in which the axons of the preganglionic neurons from both halves of the spinal cord terminate (Fig. 8-2). The preganglionic axons of these unpaired ganglia do not end in the sympathetic trunk ganglia, but run through them directly to the unpaired ganglia.

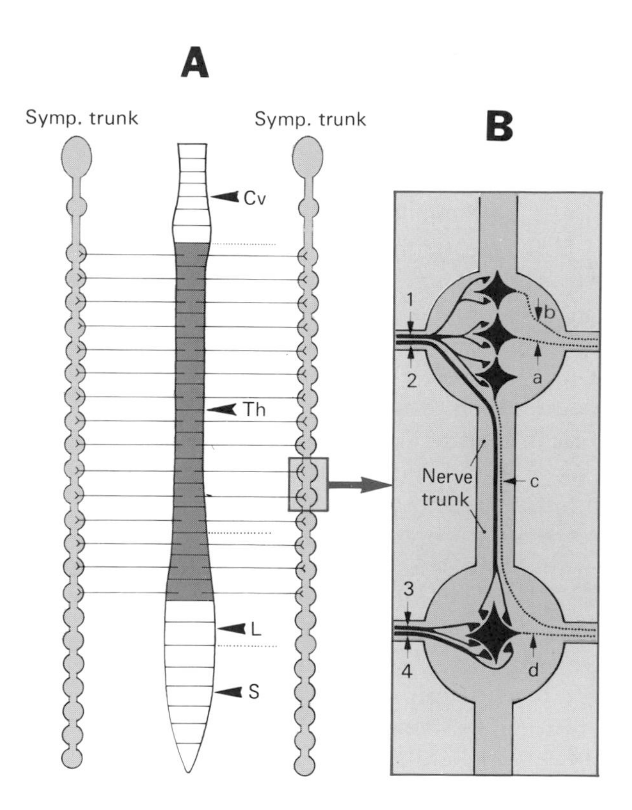

Fig. 8-3. Sympathetic trunks. **A.** Position of the left and right sympathetic trunks in relation to spinal cord and brainstem. The ganglia are drawn too large, as compared with the spinal-cord segments, and the lumbar and sacral regions of the cord are too long as compared with the sympathetic trunks. Cv, cervical cord: Th, thoracic cord; L, lumbar cord; S, sacral cord. **B.** Divergence (*1* onto *a*, *b*, and *c*) and convergence (*2*, *3*, and *4* onto *d*) of preganglionic axons onto postganglionic neurons in the sympathetic trunk ganglia.

In the paired and the unpaired ganglia, a preganglionic axon diverges on to many postganglionic cells, and many preganglionic neurons converge on a postganglionic cell. *Divergence* and *convergence* are illustrated in Fig. 8-3B, in which the interconnections of four preganglionic axons and four postganglionic neurons in two ganglia are shown. The preganglionic axon 1 in the upper ganglion diverges onto the postganglionic cells *a*, *b*, and *c*. In the lower ganglion the three preganglionic axons 2, 3, and 4 converge on the postganglionic neuron *d*. As a result of this circuitry of the preganglionic and the postganglionic neurons, the activity of a few preganglionic neurons is transmitted to many postganglionic neurons, and an individual postganglionic neuron receives the activity of many preganglionic neurons. This type of circuitry is an important safety factor guaranteeing ganglionic synaptic transmission.

Most preganglionic sympathetic fibers are myelinated. They are less than 4 μm in diameter (B fibers; see Table 2-2a, p. 67) and propagate excitation at speeds below 20 m/s. The postganglionic fibers are very thin and unmyelinated. Their propagation speed is about 1 m/s (C fibers; see Table 2-2a).

The axons of the postganglionic neurons (black dotted lines in Fig. 8-2) emerge from the ganglia and innervate the effectors of the sympathetic system. The postganglionic neurons on which preganglionic neurons from the thoracic spinal cord converge innervate the head, the thorax, the abdomen, and the upper extremities. The postganglionic neurons on which preganglionic neurons from the lumbar spinal cord converge innervate the pelvis and the lower extremities (Fig. 8-2). The ganglia of the sympathetic system are usually relatively far removed from the target organs; therefore, the postganglionic sympathetic axons are often very long (Figs. 8-1 C and 8-2).

The effectors of the sympathetic system are the *smooth muscles* of all the organs (blood vessels, viscera, secretory organs, hairs, and pupils), the *cardiac muscle,* and *glands* (sweat, salivary, lacrimal, digestive). The sympathetic nervous system has an inhibitory effect on the non-sphincter smooth muscles of the gastrointestinal and excretory organs and bronchi, and probably also on the digestive glands and those in the bronchi, while on all other effectors it has an excitatory effect.

The Peripheral Parasympathetic System. The cell bodies of the preganglionic neurons of the peripheral parasympathetic nervous system are located in the *sacral cord* and the *brainstem.* The preganglionic parasympathetic fibers are, for the most part, unmyelinated and, as indicated in Fig. 8-1C, are very long in contrast to the preganglionic sympathetic fibers because the parasympathetic ganglia are located in the vicinity of the effectors. The parasympathetic axons from the brainstem run in the *vagus nerve* to the organs in the thorax and the abdomen (solid red lines in Fig. 8-2) and in other cranial nerves to the organs of the head. The fibers from the sacral cord run in the pelvic nerve to the organs in the pelvis (Fig. 8-2). The autonomic ganglia in which preganglionic and postganglionic parasympathetic neurons are interconnected are distributed in the walls of the effectors or in the vicinity of the effectors. The postganglionic parasympathetic fibers (red dotted lines in Fig. 8-2) are, therefore, very short in comparison with the corresponding sympathetic fibers (black dotted lines in Fig. 8-2). All the parasympathetically innervated organs, such as the bladder, the rectum (pelvic region), the gastrointestinal tract (abdomen), the heart, the lungs (thorax), and the lacrimal and salivary glands (head region), are also innervated by sympathetic fibers. However, not all sympathetically innervated organs are innervated by the parasympathetic system. In particular, this is true—with a few possible exceptions—of the entire vascular system (arteries, veins).

Enteric Nervous System. In a fundamental sense, the enteric nervous system is the real *autonomic* nervous system. It can function without central inputs, either sympathetic or parasympathetic, and it is capable of controlling the variety of movements of the intestinal tube that serve to mix and propel its contents, as well as regulating some of the secretory processes. The enteric nervous system consists of groups of nerve cells (small ganglia) situated between the smooth longitudinal musculature and the smooth circular musculature in the myenteric plexus, and below the circular musculature in the submucosal plexus. The neurons of the enteric nervous system are (1) sensory neurons excited by stretching and contraction of the intestinal wall, (2) motor neurons that innervate the smooth circular and longitudinal musculature, and (3) interneurons that couple the afferent and motor elements to one another. Indeed, the enteric nervous system could be called the *brain of the gut*.

The Visceral Afferents. So far, we have discussed the efferents of the autonomic nervous system. There are, however, also afferents that are sometimes assigned to the autonomic nervous system. These come from the visceral region and are, therefore, called visceral afferents (see also Fig. 1-8). In the present state of our knowledge, we cannot subdivide these afferents into a sympathetic and a parasympathetic group, and the functional validity of such a subdivision is doubtful. Neither the visceral nor the somatic afferents have effects limited to the corresponding efferent system; that is, visceral afferents can also affect the somatic efferent system, and vice versa.

The *receptors of the visceral afferents* are situated in the organs of the thorax, the abdomen, and the pelvis, and also in the walls of the blood vessels. These receptors measure the intraluminal pressure (for example, in the arterial system) or the degree of fullness of the hollow organs (for example, the bladder, the veins, the intestine) indirectly by the stretching of their walls. They may also record the acidity and the electrolytic concentration of the contents of the hollow organs (for example, of the blood or the stomach contents), as well as noxious stimuli in the visceral region. Some of the visceral afferents, like somatic afferents, enter the spinal cord in the dorsal roots, and have cell bodies in the spinal ganglia. Many of these afferents encode noxious events in the viscera. A large number of the visceral afferents from the abdomen and the thorax run in the vagus nerve. Their cell bodies are situated in the sensory ganglion of the vagus (the nodose ganglion), below the base of the skull.

Q 8.1 Which of the following statements apply to the peripheral parasympathetic nervous system?
- a. It controls the hormonal balance.
- b. It has long preganglionic neurons.
- c. It innervates only the organs in the head, the thorax, the abdomen, and the pelvis.
- d. It consists of preganglionic and postganglionic neurons that are connected in the sympathetic trunk.
- e. It innervates all organs that are also innervated by the sympathetic system.

Q 8.2 In which of the following are the cell bodies of the preganglionic neurons of the sympathetic nervous system situated?
- a. Effectors.
- b. Sympathetic trunk.
- c. Thoracic spinal cord.
- d. Sacral spinal cord.
- e. Lumbar spinal cord.
- f. Mesencephalon (midbrain).

Q 8.3 The visceral afferents
- a. have cell bodies in the sympathetic trunk.
- b. may enter the CNS with the somatic afferents.
- c. come from the intestinal and the vascular region.
- d. have cell bodies in the spinal ganglia or in the sensory ganglion of the vagus nerve.
- e. make synapses outside the CNS.

8.2 Acetylcholine, Noradrenaline, and Adrenaline

Synaptic transmission from the preganglionic axons to the postganglionic neurons in the parasympathetic and sympathetic systems is *cholinergic* (Fig. 8-1B,C). Most of the postganglionic sympathetic neurons transmit their activity to the effectors by releasing *noradrenaline*, whereas most of the postganglionic parasympathetic neurons release *acetylcholine* (Fig. 8-1B,C).

Acetylcholine; Nicotinergic and Muscarinergic Transmission. The membranes of the postganglionic neurons and of the cells of the effector organs contain molecular structures with which acetylcholine reacts. The chemical composition of these so-called *cholinergic receptors* is not known in detail. When they react with acetylcholine, the result is an increase in the conductance for small ions through the membrane, which generates postsynaptic potentials (see also neuromuscular end plate, Sec. 3.1).

Nicotine has the same effect as acetylcholine on the postganglionic neurons in the autonomic ganglia. At the effector organs (smooth muscles, glands), however, it cannot simulate the action of acetylcholine; here the action of acetylcholine is simulated by muscarine, a poison derived from the mushroom *Amanita muscaria.* Both cholinergic transmission in the autonomic ganglia and the action of nicotine on the postganglionic neurons can be selectively blocked by quaternary ammonium bases (ganglion-blocking agents). The cholinergic transmission to the effector organs and the action of muscarine can be selectively blocked by atropine, a poison from the deadly nightshade. From these pharmacologic observations it is inferred that the cholinergic receptors of the autonomic effector organs are different from those in the postganglionic neurons. They are distinguished by the terms *nicotinergic* and *muscarinergic* acetylcholine receptors; correspondingly, the cholinergic actions of substances and cholinergic transmission are classified as either nicotinergic or muscarinergic.

Noradrenaline, Adrenaline; α/β Receptor Concept. Noradrenaline circulating in the bloodstream comes from the postganglionic adrenergic sympathetic neurons and from the adrenal medulla. Adrenaline comes exclusively from the adrenal medulla (see p. 225). As in the case of cholinergic transmission, the actions of adrenaline and noradrenaline on the effector organs are mediated by an interaction between these adrenergic substances and specific molecular structures in the cell membranes of the organs, the adrenergic receptors. By pharmacologic criteria, a distinction is made between α- and β-adrenergic receptors and hence between α-and β-adrenergic actions of adrenaline and noradrenaline. Many of these actions can be selectively prevented by substances called α and β blocking agents.

The membranes of most of the organs and tissues affected by adrenaline and noradrenaline contain both α and β receptors. In reacting with adrenaline and noradrenaline, α and β receptors usually mediate opposite (antagonistic) effects. Under physiologic conditions the response of an organ to the adrenergic substances depends on which of the two predominates, the α-receptor or the β-receptor action. Table 8-1 shows the responses of various organs to noradrenaline and to the electric stimulation of adrenergic postganglionic axons innervating the organs, as well as the adrenergic receptors that mediate these actions.

By systematically altering the structure of the noradrenaline molecule, a great variety of substances have been synthesized, each of which elicits preferentially α-receptor or β-receptor effects in particular organs or groups of organs. These drugs have proved very useful in therapeutic medicine. For example, when the methyl residue at the nitrogen of the adrenaline molecule (see Fig. 8-4) is replaced by a propyl group, a substance is produced that has only β-adrenergic actions. In an aerosol spray, this substance assists the asthmatic patient by relaxing the smooth tracheal musculature (see Table 8-1). A further modification of this purely β-adrenergic substance, by replacing the two OH groups on the benzene ring (Fig. 8-4), produces a substance that selectively blocks the β-receptor action.

Table 8-1.

Organ	Action of Noradrenaline or Stimulation of Postganglionic Adrenergic Neurons	Receptor
Heart	Increased heart rate and contractile force	β
Most blood vessels	Vasoconstriction	α
Muscle arteries	Vasoconstriction	α
	Vasodilation (only to circulating adrenaline)	β
Gastrointestinal tract		
Longitudinal and circular musculature	Relaxation	α and β
Sphincters	Contraction	α
Bladder		
Detrusor vesicae	Relaxation	β
Trigonum vesicae (internal sphincter)	Contraction	α
Seminal vesicle	Contraction	α
Vas deferens	Contraction	α
Dilator of pupil	Contraction (mydriasis)	α
Tracheal and bronchial musculature	Relaxation	β

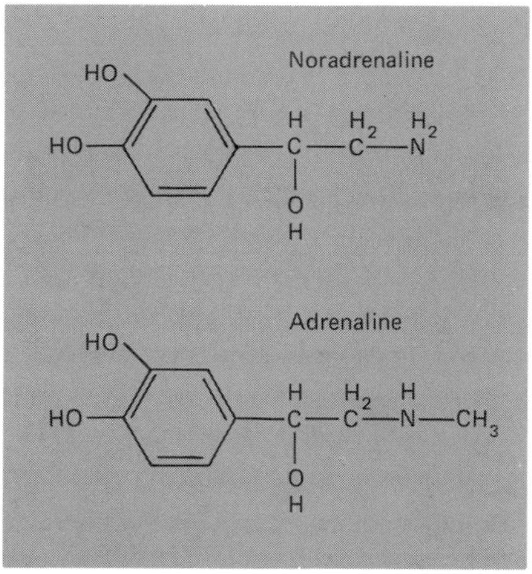

Fig. 8-4. Catecholamines: noradrenaline and adrenaline.

The Adrenal Medulla. The medulla of the adrenal gland plays a special role in the body. It is a transformed sympathetic ganglion and consists of modified postganglionic neurons. When the preganglionic neurons that innervate the adrenal medulla are excited, these postganglionic neurons release a mixture of about 80% *adrenaline* and 20% *noradrenaline,* which diffuses into the bloodstream. These adrenergic substances may reinforce the effects of the sympathetic neurons on organs, but they are primarily to be regarded as *metabolic hormones.* That is, their liberation results in the mobilization of oxidizable substances like glucose and free fatty acids from the glycogen and fat stores. The adrenergic substances from the adrenal medulla thus ensure that when the sympathetic system is activated fuel is quickly made available. This process is especially significant when severe demands are made on the body, as under conditions of extreme physical exertion, exhaustion, or mental strain.

Other Transmitter Substances in the Peripheral Autonomic Nervous System. Acetylcholine and noradrenaline are probably not the only transmitter substances in the peripheral autonomic nervous system. Recent physiologic and pharmacologic studies have shown that even after complete pharmacologic blocking of cholinergic (nicotinic and muscarinic) and adrenergic transmission, it is still possible to elicit responses of many autonomic effector organs by electric stimulation of the postganglionic innervation. Furthermore, pre- and postganglionic neurons as well as neurons of the enteric nervous system have been shown, experimentally and histochemically, to contain substances that could either be synaptic transmitters or influence synaptic (cholinergic and adrenergic) transmission. Examples of these substances include dopamine, serotonin, adenosine triphosphate (ATP), and neuropeptides. None of these candidates, however, has yet been proved beyond doubt to function as a transmitter substance in the peripheral autonomic nervous system.

Q 8.4 Cholinergic synaptic transmission in the peripheral autonomic nervous system
- a. is limited to the parasympathetic system.
- b. is muscarinergic in the autonomic ganglia and nicotinergic at the autonomic effectors.
- c. is nicotinergic in the autonomic ganglia and muscarinergic at the autonomic effectors.
- d. can be blocked by atropine at the effectors.
- e. is a characteristic that distinguishes the enteric nervous system from the sympathetic and parasympathetic.

Q 8.5 Adrenergic synaptic transmission in the peripheral autonomic nervous system has the following characteristics:
- a. It is limited to the sympathetic nervous system.
- b. When they are excited, preganglionic sympathetic axons release noradrenaline.
- c. When excited, postganglionic axons release adrenaline.
- d. The adrenaline circulating in the bloodstream comes from the adrenal medulla.
- e. Noradrenaline and adrenaline are released from the postganglionic sympathetic axons in the ratio 4 to 1.

Q 8.6 The action of noradrenaline on peripheral effector organs
- a. is always excitatory.
- b. can be prevented at the tracheal and bronchial musculature by a β blocking agent.
- c. can be mediated only in the presence of acetylcholine.
- d. is of the β-receptor type on the heart muscle and of the α-receptor type on the blood vessels.
- e. causes dilation of most blood vessels.

8.3 Smooth Muscle: Myogenic Activity and Responses to Stretching, Acetylcholine, and Adrenaline

The autonomic nervous system innervates nearly all the smooth muscles of the body. Therefore, some of the identifying features of smooth muscle based on the special characteristics of the cell membrane and the structure of the contractile mechanism must first be described. By reference to these features, it is possible to explain the way in which many autonomically innervated organs function.

The smooth muscles of an organ consist of single fusiform cells that are about 20 to 200 μm long and 2 to 10 μm thick. The cells are interconnected in reticular fashion. Smooth muscle cells, like skeletal muscle cells, contain myofilaments, although in far smaller quantities. These myofilaments are not regularly arranged as in the skeletal muscle, so it is not possible to detect any striation in the smooth muscles (hence the term "smooth"). Smooth muscles shorten like skeletal muscles, by means of a *sliding-filament mechanism* (see Chapter 5). But the thick and thin filaments of smooth muscle slide past one another much more slowly than those of skeletal muscle, so that smooth muscles are particularly well suited for functions in which force must be maintained for a long time with little expenditure of energy.

Myogenic Activity. Many smooth muscles (for example, those of the gastrointestinal tract, the blood vessels, and the bladder) can contract spontaneously, in the absence of neuronal activation. This property can be demonstrated in an experiment like that of Fig. 8-5. A smooth-muscle preparation is mounted in a bath of physiologic saline so that it is about at its natural length in the organ. Both the force developed by the preparation (left in Fig. 8-5) and the membrane potential of one of the muscle cells are measured. After a short time, the preparation begins to depolarize spontaneously and to develop force (Fig. 8-6A). These depolarizations may be *phasic-rhythmic* and/or *tonic*. They occur at intervals of seconds, minutes, or hours; accordingly, they are called second, minute, or hour rhythms. If the neurons in the prepara-

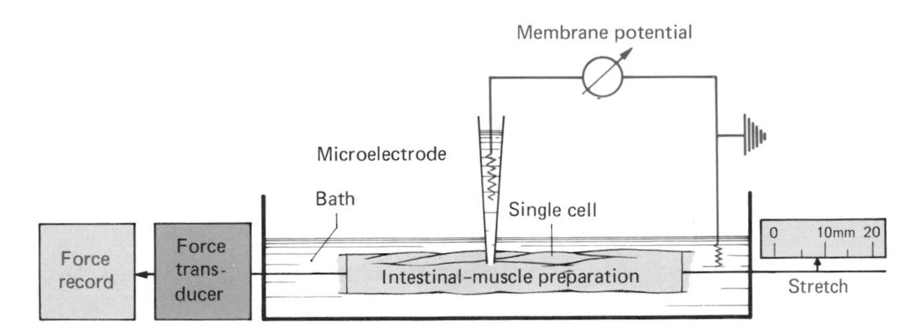

Fig. 8-5. Diagram of an experimental arrangement for recording active development of force by smooth muscles and the membrane potential of a muscle cell under passive stretch. The arrangement is in principle the same as in Fig. 3-3. On the *right*, the muscle preparation is prestretched; on the *left*, the force is measured isometrically. Preparation: guinea-pig intestinal muscle.

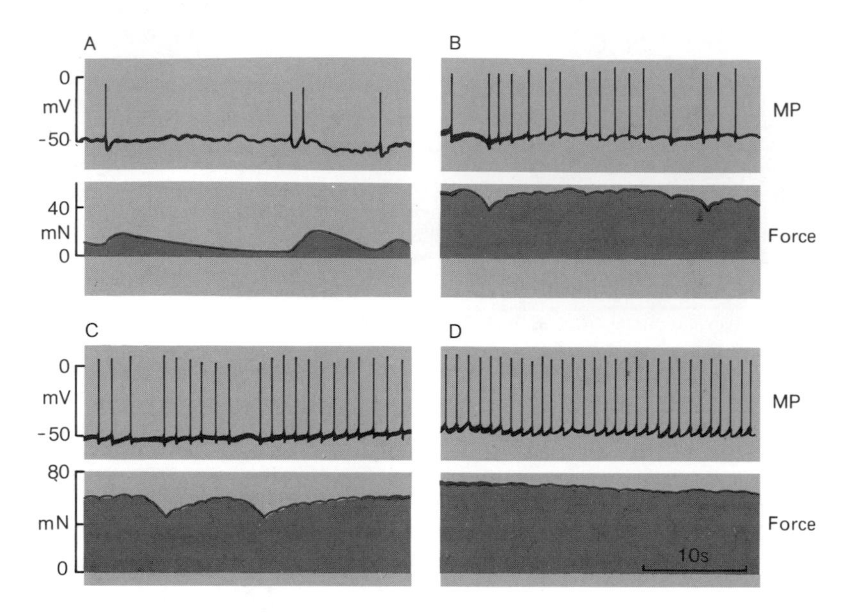

Fig. 8-6. Force development in a multicellular strip of smooth muscle with increasing stretch. The membrane potential of a single smooth muscle cell was measured with a microelectrode. Experimental arrangement as in Fig. 8-5. Intestinal muscle preparation from guinea pig. Modified from Bülbring (1962) *Physiol. Rev.* 42, Suppl. 2:160.

tion are poisoned, the electrical and mechanical activity of the preparation persists. Therefore, this activity must be muscular in origin, and is called *myogenic activity*.

The depolarization (which would elicit contraction in isometric conditions) consists of pulselike discharges ("spikes") resembling in many ways the action potentials of nerve and striated-muscle cells. These discharges arise in a group of smooth muscle cells in the preparation, which have an especially low threshold for the production of action potentials; the potentials spread from cell to cell throughout the preparation. The cells in which the activity originates are *pacemakers* for the surrounding tissue. The spread of excitation from cell to cell occurs at contact points between the cell membranes (so-called nexus), which present low electrical resistance to conduction. In this way, the electrical activity of many adjacent smooth muscle cells is synchronized, so that they behave as a *functional unit (functional syncytium).*

Note, however, that contractions are not triggered by conducted action potentials in all smooth muscles. An equally important mechanism in some smooth muscles is the elicitation of contraction by membrane depolarization without production of action potentials. In both cases the contractions are controlled by an *increase in the intracellular concentration of calcium ions* (see Sec. 5.3).

Apart from these spontaneously active smooth muscles, there are some smooth muscles with cells that generally show no spontaneous activity—for example, the smooth muscles of the hairs and the smooth musculature that adjusts the lens of the eye. These muscles can be activated only by way of their autonomic nerves.

Time Course of Smooth-Muscle Contraction. When a skeletal muscle is excited by a single suprathreshold electric stimulus, the result is a brief action potential in the skeletal muscle fibers. A few milliseconds after the onset of this action potential, contraction begins. The contraction rises to a maximum within about 60 ms and then falls back to the baseline in 200 ms (see Fig. 5-2). The rather different contraction of a smooth muscle is illustrated in Fig. 8-6A. This smooth muscle is a strip of intestinal musculature from a guinea pig. The contractile force developed by the preparation was measured isometrically, while the membrane potential (MP) of a single cell in the strip was recorded intracellularly with a microelectrode.

After suprathreshold excitation of the intestinal muscle the preparation contracts. Each action potential is followed by a phasic contraction (Fig. 8-6A). In this muscle the rise time of a single contraction is approximately 1–2 s, and the time to return to baseline is about 5–10 s. In comparison with the skeletal muscle, then, the contraction of the smooth muscle is about 20 to 50 times slower. This slow time course results mainly from the slow speed at which the thick and thin myofilaments slide past one another (see also Sec. 5.1 and 5.3).

To produce a maintained, uniform contraction (tetanus) of the skeletal muscle, one must excite it at a rate of about 50 to 125 stimuli/s (see Sec. 5.2). Because of the *slower contraction* of smooth muscle, considerably lower pulse frequencies—in the range 0.5 to 3 Hz—suffice to produce a steady contraction. The smooth muscle shown in Fig. 8-6D discharges one action potential per second. At this rate the individual contractions of the smooth muscle are almost entirely fused. We can infer from this experiment that even relatively low frequencies, one action potential per second or less, in the excitatory efferent postganglionic fibers to the smooth muscles elicit a sustained, uniform contraction.

Force Development by Smooth Muscles in Response to Stretch. Many smooth muscles are very sensitive to stretch, responding with a depolarization of the fiber membrane and with contraction. In the experiment shown in Fig. 8-6, the intestinal preparation was progressively stretched with the apparatus diagrammed in Fig. 8-5. When the preparation is slightly stretched, the discharge rate of the smooth muscle cells is low and the force developed by the whole muscle strip is slight (Fig. 8-6A); stronger stretching (increasing from B to D in Fig. 8-6) causes an increase in both the frequency of the action potentials and the force the preparation develops.

The increased excitability due to stretching of the membranes of the smooth muscle cells is of great importance in the hollow organs of the body, such as the intestine, blood vessels, and urinary bladder. Any increase in the filling of a hollow organ generates increased activity in its wall muscles. Thus, for example, if nervous control of the bladder is lost as a result of destruction of the sacral spinal cord, the bladder will, nevertheless, release its contents spontaneously (though it does not empty completely) once it has filled to a certain point. The intrinsic excitability of the smooth muscles and its modification by mechanical stretching enable the hollow organs to perform their functions to a limited extent without nervous control. Thus, the organs innervated by the autonomic system may also exhibit *autonomy* in a different sense—because of their myogenic activity.

Direct Effects of Acetylcholine, Noradrenaline, and Adrenaline on Smooth Muscle Cells. Smooth muscles can be influenced directly by a large number of drugs and hormones. It is for this reason that smooth muscles are often used as biological test objects in pharmacologic assays. In the following, the effects of acetylcholine and adrenaline on a resting (prestretched) preparation of intestinal smooth muscle will be described.

The experimental arrangement is the same as in Fig. 8-4. A preparation of intestinal muscle is immersed under tension in a physiologic saline solution. The membrane potential of a smooth muscle cell and the force developed by this preparation are measured.

The upper trace in Fig. 8-7A shows the membrane potential (MP) of a single prestretched smooth muscle cell from the intestine. The MP in this cell is about −50 mV. The beginning of the intracellular recording shows that as a result of the prestretching the cell depolarizes to the threshold and thus continuously triggers action potentials. The lower trace in Fig. 8-7A shows the force developed by the entire preparation. At the start of the measurement with the preparation at rest, the force developed amounts to about 10 mN. When a very small quantity of ACh is added to the bath (black bar in Fig. 8-7A), the membrane of the muscle cell depolarizes, and the frequency of the action potentials along the muscle fiber increases. At the same time, the force developed by the preparation increases to 30 mN. When the

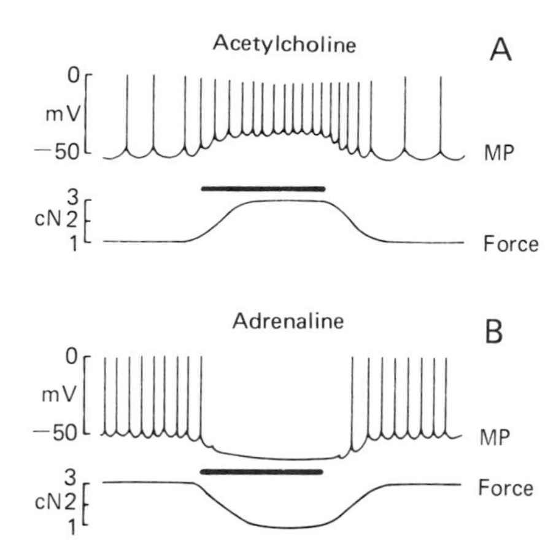

Fig. 8-7. Responses of smooth muscle cells to application of acetylcholine (**A**) and adrenaline (**B**), measured as diagrammed in Fig. 8.4. The intestinal muscle preparation is prestretched. The *upper curves* in **A** and **B** depict the membrane potential of a muscle cell of the preparation, and the *lower curves* show the force developed by the entire preparation. During the time denoted by the thick black line (ca. 10 s), the preparation was bathed in a solution of acetylcholine or adrenaline. In **B** the preparation was prestretched more strongly than in **A**.

ACh solution is replaced by a normal bath solution (last third of the trace in Fig. 8-7A), the MP returns to its initial value. The tension of the preparation decreases again as a result of the reduced frequency of the propagated action potentials.

The effect of a dilute solution of adrenaline on a preparation of intestinal muscle is illustrated in Fig. 8-7B. As a result of the prestretching, the preparation develops a force of 30 mN. When a small amount of adrenaline is added to the bath, the potential of the muscle cells becomes more negative. The cell from which the recording is made in Fig. 8-7B hyperpolarizes, and action potentials are no longer generated. As a result, the force developed by the preparation declines, and the preparation becomes flaccid. During stretching, adrenaline impedes the depolarization of the intestinal muscle cell membranes and thus the contractions of the muscle cells. In many smooth muscle cells on which adrenaline has an inhibitory action, its administration gives rise to no discernible hyperpolarization.

Noradrenaline and adrenaline have an *inhibitory* effect on the nonsphincter smooth muscles of the gastrointestinal tract, bladder, and lungs. Other smooth muscles—for example, the vascular muscles—are excited by noradrenaline and adrenaline. In low (physiologic) concentration adrenaline has a relaxing effect on the arterial vessels of skeletal muscle (see Table 8-1). *ACh* has an *excitatory* effect on the smooth muscles of the gastrointestinal tract, lungs, and secretory organs.

Neuromuscular Transmission in Smooth Muscle. Electrophysiologic experiments and electron-microscope observations have shown that neuromuscular transmission in smooth muscle qualitatively resembles that in skeletal muscle. There are, however, certain differences of detail. The autonomic nerve fibers do not terminate on the smooth-muscle cells in morphologically distinct neuromuscular synapses; rather, the axons travel past the smooth-muscle cells at varying distances from them. The transmitter substance is released by these axons and diffuses to the smooth-muscle cells. It is thought that the transmitter substance from one fine axonal branch affects many smooth-muscle cells in its vicinity. The *excitatory postsynaptic potentials* measurable in muscle cells following nerve stimulation last about 10 to 20 times as long as the EPP in skeletal muscle fibers. In some muscle cells (for example, those of the intestine), *inhibitory postsynaptic potentials,* in the hyperpolarizing direction, can also be recorded after stimulation of the autonomic nerves. These inhibitory postsynaptic potentials cause relaxation of the smooth musculature.

In summary, the smooth musculature can be characterized as follows. Smooth muscle cells have a membrane potential and are electrically coupled with one another to form functional syncytia. Force development in smooth musculature depends on a number of factors (Fig. 8-8): myogenic activity, mechanical factors (stretch), local metabolic influences (e.g., P_{O_2}, P_{CO_2}, pH, osmolarity of the surrounding milieu), hormonal influences (such as adrenaline), and neuronal influences. Each of these influences is involved in the regulation of force development to an extent that depends on the function of the organ of

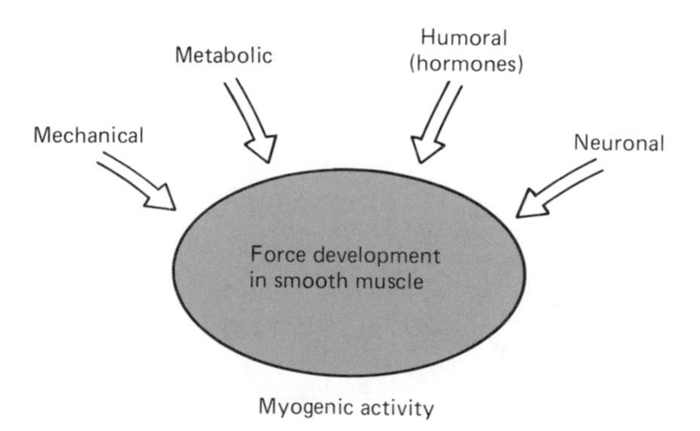

Fig. 8-8. The influence of various factors on the myogenic activity of smooth musculature.

which the muscle forms a part. The smooth musculature of the hair follicles and of the eye is entirely under neuronal control, whereas the smooth musculature of the uterus is essentially regulated by the mechanical, myogenic, and hormonal components, with only a very minor contribution of its innervation.

Smooth musculature can be roughly categorized according to the presence or absence of myogenic activity. Smooth muscle cells *with myogenic activity* have unstable membrane potentials, which spontaneously give rise to action potentials. The action potentials spread over the entire tissue, triggering the contraction mechanism. The smooth muscles of the gastrointestinal and urogenital tracts are in this category.

Smooth muscle cells *without myogenic activity* are usually electrically inexcitable and have stable membrane potentials. Contraction of smooth musculature of this type is in most cases triggered only neuronally. This category includes the smooth musculature of certain arteries, the smooth eye muscles, and the anococcygeal muscle.

Q 8.7 Stretching a preparation of smooth muscle from the intestine
- a. results in depolarization of the membrane of the smooth muscle cells.
- b. brings about a reduction in the discharge frequency of the smooth muscle cells.
- c. leads to relaxation of the muscle preparation.
- d. causes the muscle preparation to develop force.
- e. increases the membrane potential.

Q 8.8 The graded development of force by a smooth muscle (intestine)
- a. is controlled by the depolarization of the fiber membranes of the muscle cells.
- b. can be triggered by adrenaline solutions of various concentrations.
- c. is impossible because the contraction of the smooth muscle is an all-or-none event.
- d. can be triggered by ACh solutions of various concentrations.
- e. can be generated by mechanical stretching of the muscle.

Q 8.9 The contraction of a smooth muscle after neural stimulation
- a. is an event lasting _____ (tens of milliseconds; 100 ms; seconds).
- b. lasts about _____ (2; 5; 50) times longer than the contraction of a skeletal muscle.
- c. increases _____ (more quickly than; at the same rate as; more slowly than) that of a skeletal muscle.
- d. is caused by _____ (the same; different) basic mechanisms.

8.4 Antagonistic Effects of Sympathetic and Parasympathetic Activity on Autonomic Effectors

Most autonomically innervated organs are independently active; that is, they function even in the denervated state. Since the organs in question are almost exclusively hollow organs, such as the gastrointestinal tract, the bladder, and the blood vessels, their function is regulated by the degree of filling or the internal pressure. Increased internal pressure stretches the smooth muscles in the walls of the organ and depolarizes the membrane of the smooth muscle cells. This in turn leads to force being generated by the smooth musculature and thus to displacement of the contents of the hollow organ. This autonomous function of the organs can be traced back to the characteristics of the smooth musculature (see Sec. 8.3).

Most of these organs are innervated by sympathetic and parasympathetic fibers and visceral afferent fibers. The activity in the efferent autonomic fibers is superimposed on the autonomous activity of the organs. The sympathetic and the parasympathetic activities are mostly antagonistic in their effect on the organs. In the following, we will describe this antagonistic action on two preparations—an intestinal muscle and a frog's heart. These examples have been chosen to demonstrate nervous control of the digestive, the excretory, and the cardiovascular systems.

Autonomic Influences on the Heart. An isolated frog's heart continues to beat *spontaneously* without any connection whatsoever with the body. Like almost all other autonomically innervated organs the heart is *independently active*. The heart rate is controlled by a group of modified muscle cells situated at the entrance to the heart. These cells depolarize spontaneously and generate propagated action potentials that are transmitted by other specialized muscle cells to the muscles of the chambers of the heart, which pump the blood into the arterial system. In this way the contractions of separate areas of the heart are coordinated with one another. These spontaneously depolarizing cells at the entrance are called *pacemaker cells*. The autonomic nervous system acts on both the pacemaker cells and the other cardiac muscle cells.

Figure 8-9A (left) shows an isolated frog's heart, the beating rate and the contractile force of which are recorded mechanically at the apex of the heart by a pointer. The heart rate is given by the *frequency* of the pointer deflections, and the force of contraction is given by the *magnitude* of the deflection (Figs. 8-9B and C). The two autonomic nerves that innervate the heart, the sympathetic cardiac nerve and the parasympathetic cardiac nerve, are connected to stimulating electrodes. A solution with composition similar to blood runs from the reservior, *R*, into the heart and is pumped out again through the tube *T*.

At the start of the recordings in Figs. 8-9B and C (upper records), the heart beats spontaneously at a rate of about 18 beats/min. When the

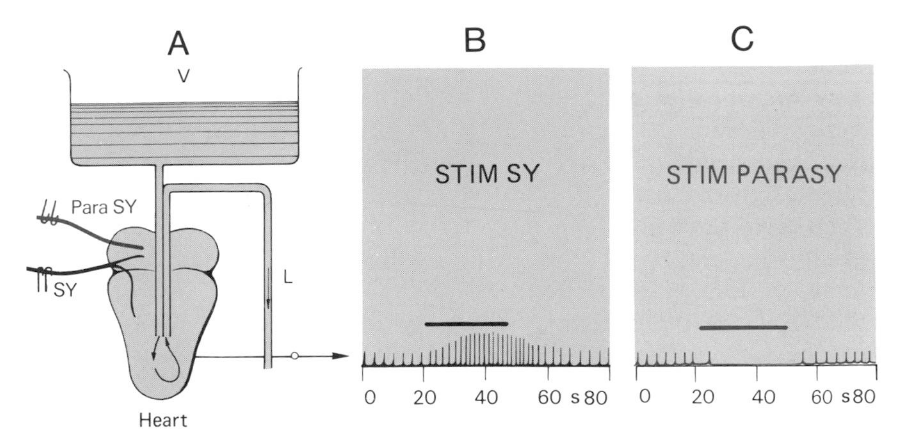

Fig. 8-9. Nervous control of the heart by stimulation of the autonomic cardiac nerves. **A.** Diagram of an experiment with a frog heart. Saline solution from the reservoir, *R*, passes through the heart and is pumped out through the tube, *T*. The contractions are recorded mechanically (lever with arrow, attached to the heart). The sympathetic (SY) and parasympathetic (PARA SY) cardiac nerves lie across stimulating electrodes. **B, C,** diagrams of the contractile force (amplitude of deflections) and frequency (rate of deflections) of the heartbeat as recorded before, during, and after stimulation (horizontal bar) of the cardiac nerves. Modified from Bain (1932) *Quart. J. Exp. Physiol.* **22:** 269–274.

sympathetic nerve is stimulated electrically (Fig. 8-8B) the time between the pointer deflections becomes shorter, and the magnitude of the deflection increases. This means that the *heart rate* and also the *contractile force* of the heart both *increase.* Once stimulation of the sympathetic nerve stops, the heart resumes its spontaneous rhythm. Excitation of the *parasympathetic nerve* (Fig. 8-9C) results in quite different changes in the activity of the heart. The time between the contractions of the heart increases until the heart comes to a standstill; in other words, the parasympathetic nerve *lowers* the *heart rate.* The magnitude of the deflections remains constant during electrical stimulation of the parasympathetic nerve. We can conclude from this that the parasympathetic activity does not influence directly the contractile force of the heart. The regions of the heart innervated by the parasympathetic and the sympathetic nerves correspond to the effects these nerves have on contractile force and heart rate. The sympathetic nerve innervates the pacemaker cells in the atrial wall, the atria and the muscles of the ventricles; the parasympathetic nerve innervates only the pacemaker cells and the atria of the heart. When the transmitter substances of the sympathetic and parasympathetic cardiac nerves, noradrenaline and acetylcholine, are applied to the heart directly in low concentrations, the effect is the same as that of electrical stimulation of the cardiac nerves.

The amount of blood pumped by the heart per unit time (*cardiac output*) depends on the heart rate and on the contractile force of the heart. Sympathetic activity increases the cardiac output; parasympathetic activity reduces it. The two parts of the autonomic nervous system thus have *antagonistic* effects on the spontaneously active heart. Of course, regulation of the cardiac output by the autonomic nervous system never occurs in the organism in just the way diagrammed in Figs. 8-9B and C, because the sympathetic and the parasympathetic systems influence the heart simultaneously. The heart is subjected continuously to inhibitory parasympathetic and excitatory sympathetic influences. Each change in activity in one or the other of the two autonomic systems results in changes in the heart rate and/or the contractile force. Thus, the cardiac output (amount of blood pumped by the heart per unit time) increases when there is a rise in sympathetic activity and/or a drop in parasympathetic activity. Conversely, the cardiac output declines when there is a drop in sympathetic activity and/or a rise in parasympathetic activity. Since the CNS has these means at its disposal to regulate the cardiac output through the autonomic nervous system, the body can *adapt* the operation of the cardiovascular system to the requirements of the moment.

Autonomic Control of the Intestinal Musculature. Pronounced antagonistic effects of the sympathetic and parasympathetic systems can also be observed in the entire digestive system. Figure 8-10 illustrates these, for a strip of intestinal muscle; the force developed by the preparation (lower records in parts A and B) and the membrane poten-

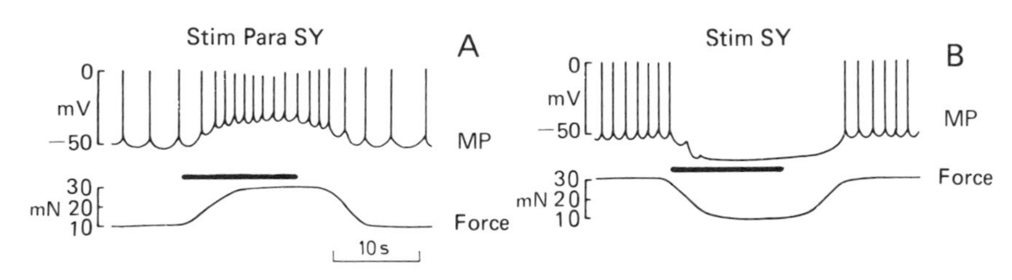

Fig. 8-10. The actions of parasympathetic (**A**) and sympathetic (**B**) nerves on the smooth muscle cells of the intestine; experimental arrangement as diagrammed in Fig. 8-5. The parasympathetic and sympathetic nerves of the intestinal muscle preparation have been left intact. The *upper records* in **A** and **B** show the membrane potential (MP) of a single muscle cell of the preparation; the *lower records* show the force developed by the whole preparation. During the period marked by the *heavy line* (duration ca. 10 s) the parasympathetic (**A**) or sympathetic (**B**) nerve was stimulated.

tial of a single smooth-muscle cell (upper records) are shown. The experimental arrangement is the same as in Fig. 8-5. The autonomic nerves to the preparation have been left intact.

When the parasympathetic nerve is stimulated, the membrane of the smooth-muscle cell is depolarized (upper record in A). The frequency of the action potentials propagating along the fiber increases. As a result, the smooth-muscle cells contract. The force developed by the entire preparation increases (lower record in A). Stimulation of the sympathetic nerve produces the opposite effects (B). The membrane of the cell hyperpolarizes, action potentials are no longer generated, and, consequently, the intestinal muscle relaxes. These antagonistic effects of the sympathetic and the parasympathetic nerves on the intestinal musculature parallel the effects of noradrenaline and ACh, the autonomic transmitter substances, applied directly to the intestinal muscles (see Fig. 8-7).

The above statements must be qualified as follows: (1) The decrease in activity of the intestinal musculature after stimulation of sympathetic nerves is caused mainly by inhibition of the transmission from the preganglionic to the postganglionic parasympathetic neurons. (2) The smooth musculature of the intestinal sphincters is excited by sympathetic postganglionic axons. (3) The parasympathetic innervation of the intestine also includes preganglionic axons, excitation of which immobilizes the intestine (by inhibition of the smooth musculature). The inhibition is induced by postganglionic neurons that are neither adrenergic nor cholinergic (see p. 225).

A transmitter of the autonomic nervous system can have inhibitory or excitatory effects, depending on the effector. It is probable that such transmitters can increase the conductances of the membranes in a relatively selective fashion for either potassium or sodium ions, depending upon properties of the subsynaptic membrane. These changes in conductance lead to shifts in the membrane potential of the cells in the direction of the potassium or the sodium equilibrium potential, producing hyperpolarization or depolarization, respectively.

Q 8.10 In Sec. 8-1, you learned that after excitation of the sympathetic system, mainly adrenaline is released into the bloodstream by the cells of the adrenal medulla. What effect does this adrenaline produce?

a. It increases the movement of the intestine.
b. It reduces the contractile force of the heart.
c. It inhibits intestinal function.
d. It increases the cardiac output.
e. It facilitates cholinergic transmission from the parasympathetic system to the intestinal muscles.

Q 8.11 Which of the following statements are correct?
a. ACh reduces cardiac output.
b. Noradrenaline increases intestinal motility.
c. ACh decreases intestinal motility.
d. Noradrenaline increases the cardiac output.

Q 8.12 In which way can the CNS increase the volume of blood pumped per minute by the heart? (Think of the heart rate and the contractile force of the heart.)
a. Increasing the activity in the sympathetic cardiac nerves.
b. Decreasing the activity in the sympathetic cardiac nerves.
c. Increasing the activity in the parasympathetic cardiac nerves.
d. Decreasing the activity in the parasympathetic cardiac nerves.

8.5 Central Nervous Regulation: Spinal Reflex Arc, Bladder Regulation

As we have seen, the independent activity of autonomically innervated organs can be inhibited or enhanced by the sympathetic and parasympathetic systems. These effects are involved in vital functions such as regulation of digestion, circulation, and body temperature and control of the bladder. The regions of the brainstem and spinal cord in which these regulatory signals arise are called *centers* (for example, *vasomotor and cardiac center, micturition center, respiratory center*). A center is a localized assemblage of neurons with a specific effect on a certain organ or organ system. The various autonomic centers are intimately interlinked; thus far, they have been identified by stimulation, recording, and ablation experiments but not morphologically. For this reason, we speak of autonomic centers only in a *functional* sense.

In the following paragraphs a few reflex circuits underlying autonomic regulation are described.

The Spinal Autonomic Reflex Arc. The simplest connection between afferents and autonomic efferents is situated at the segmental level in the spinal cord. This neuronal circuit is called the *autonomic reflex arc*. Fig. 8-11 shows a cross section through the spinal cord with the autonomic reflex arc on the left and the simplest somatic reflex arc (monosynaptic stretch reflex) on the right. The efferent neuron of the autonomic reflex arc that transmits its activity to the autonomic effectors is the postganglionic neuron. Its soma is located outside the spinal cord in an autonomic ganglion. The cell body of the efferent neuron of the somatic reflex arc (the soma of the motoneuron) lies in the ventral horn of the spinal cord.

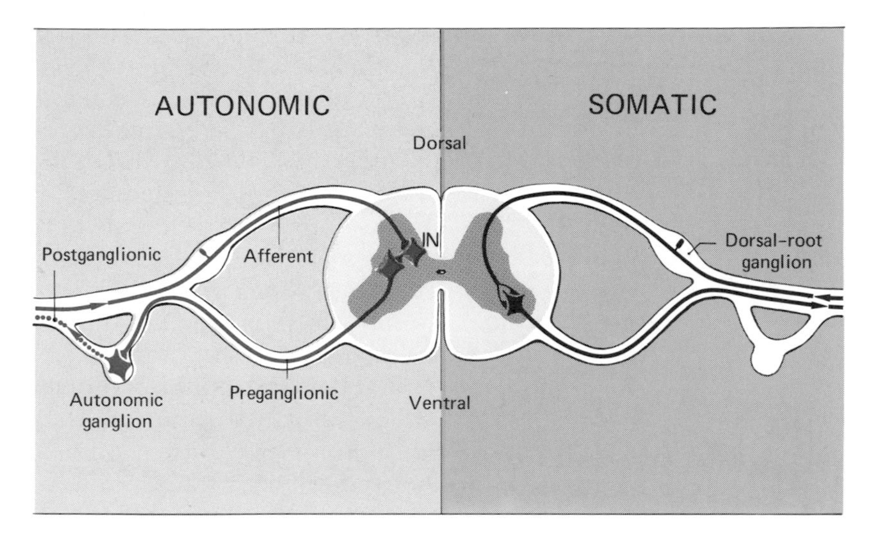

Fig. 8-11. Autonomic reflex arc (*red*) compared with the monosynaptic reflex arc. IN, interneuron.

The afferent fibers of the autonomic reflex arc are both *visceral and somatic,* and they enter the spinal cord in the dorsal roots. At least two neurons are interposed between the afferent neuron and the postganglionic neuron: an interneuron (*IN* in Fig. 8-11) and the preganglionic neuron. The monosynaptic reflex arc, on the other hand, contains no neuron between afferent fiber and motoneuron. The *simplest autonomic reflex arc* thus has at least two synapses in the gray matter of the spinal cord and one synapse in the ganglion between preganglionic neuron and postganglionic neuron. The *simplest somatic reflex arc* has only one synapse between afferent and efferent neuron.

The Segmental Connection of Autonomic Efferents with Visceral and Somatic Afferents. When pathologic processes occur in the viscera (for example, cholecystitis or gastritis), the muscles over the site of the disorder may be taut. The skin area (dermatome) that is innervated by afferents and efferents of the same spinal-cord segment as the affected organ may be reddened. The "stomach ache" caused by convulsive contractions of the intestines can be eased or even eliminated by changing the skin temperature (for example, by applying poultices) of the dermatome which is innervated by the same spinal-cord segment as the affected intestine.

From these observations we must conclude that the visceral and the somatic afferents are connected with autonomic and somatic efferents at the *segmental level* of the spinal cord. Figure 8-12 is a cross section through the spinal cord with the reflex arcs that can explain these observations. The *reddening* of the skin area is caused by dilation of the blood vessels in the skin. Thus, visceral afferents of the intestines must be connected with the autonomic (sympathetic) efferents to the cutaneous blood vessels (*viscero-cutaneous* reflex; Reflex Pathway 1 in Fig. 8-12). If, at the same time, the *abdominal muscles* over the affected intestines are *taut*, then we must further conclude that the visceral afferents from the intestines are also connected with the motoneurons whose axons innervate the abdominal muscles (*viscero-somatic* reflex; Reflex Pathway 2 in Fig. 8-12). Warming the skin results in *inhibition of the intestinal motions* and thus reduces the pain. This effect is most certainly not direct but is transmitted as a nervous reflex. It is based on the segmental connection of the afferents of the thermoreceptors in the skin with the autonomic efferents to the intestine (*cutaneo-visceral reflex*; Reflex Pathway 3 in Fig. 8-12). The sympathetic efferents to the intestine are also excited by the visceral afferents from the intestine itself (*intestino-intestinal reflex*; Reflex

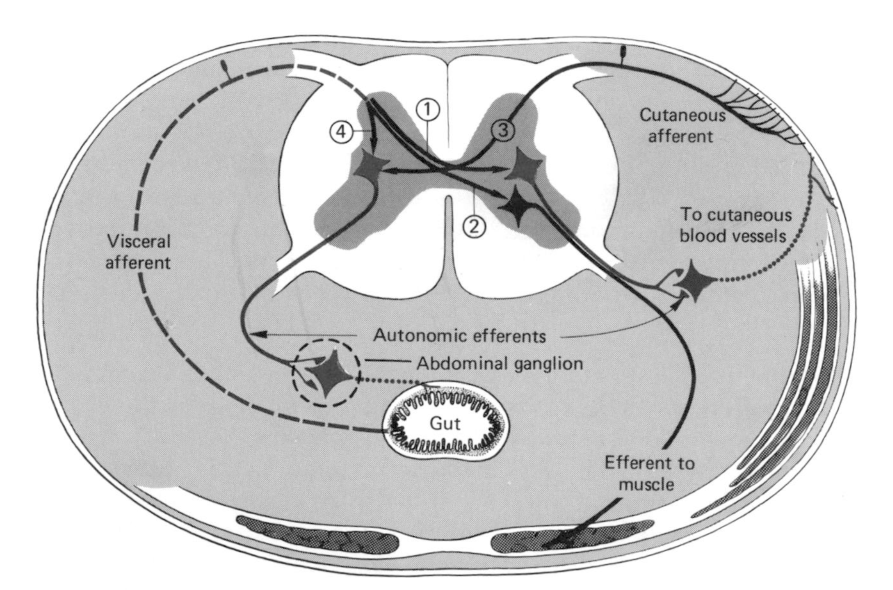

Fig. 8-12. Synaptic connections joining autonomic and somatic efferents and somatic and visceral afferents in the spinal cord to form reflex arcs. (*1*) viscero-cutaneous reflex; (*2*) viscero-somatic reflex; (*3*) cutaneo-visceral reflex; (*4*) intestino-intestinal reflex. Interneurons in the spinal cord have been omitted.

Pathway 4 in Fig. 8-12). This last reflex is of special concern following abdominal surgery, since it may bring about an undesirable postoperative immobilization of the intestine.

These spinal autonomic reflexes are particularly marked in patients whose spinal cords have been transected as a result of an accident (paraplegics; see Sec. 6.2). About two months after the accident, mechanical stimulation of the skin can provoke massive sweating and violent vascular reactions in the skin. The relay stations in the spinal cord that transmit these reactions are continuously inhibited in healthy persons by descending pathways.

Neuronal Regulation of Micturition. The wall of the urinary bladder and its internal sphincter consist of smooth muscle. In addition, the bladder has a voluntarily controllable striated external sphincter muscle (see Fig. 8-13). In the denervated state the bladder partially empties itself once a certain degree of filling has been reached. The basic mechanism behind this autonomy of the micturition process, which takes place in the smooth muscles, was dealt with in Sec. 8.3 (see Fig. 8-6). Central nervous control of the bladder is superimposed on this autoregulation process. Nervous regulation is mediated by the parasympathetic system, from centers situated in the *sacral cord* and in the *anterior pontine region* of the brainstem (see Fig. 6-11). Spinal control probably still predominates at the beginning of life, in infants. As the CNS matures, neuronal regulation of micturition is transferred to the anterior pontine region.

In Fig. 8-13 the pathway for reflex micturition in the adult human is shown in red. In the wall of the bladder are mechanoreceptors that measure stretching of the wall. These send out visceral afferents which conduct excitation reporting filling of the bladder to the *sacral cord*. In turn, an ascending spinal tract transmits excitation from the sacral cord to the "micturition center" in the *anterior pons*. A descending spinal tract from this region excites the preganglionic parasympathetic neurons in the sacral cord. These latter neurons excite the postganglionic neurons that terminate in the musculature of the bladder wall. As a result, the smooth muscles of the bladder wall contract, the internal sphincter muscle expands by shortening of the urethra, and at the same time there is a tendency for the external sphincter to relax because of inhibition of the motoneurons innervating it.

If the spinal cord of a human is accidentally severed above the sacral level, so that *paraplegia* or *quadriplegia* results, the bladder is initially paralyzed. One to five weeks after spinal transection, the bladder begins to empty automatically after it is filled. That is, mictu-

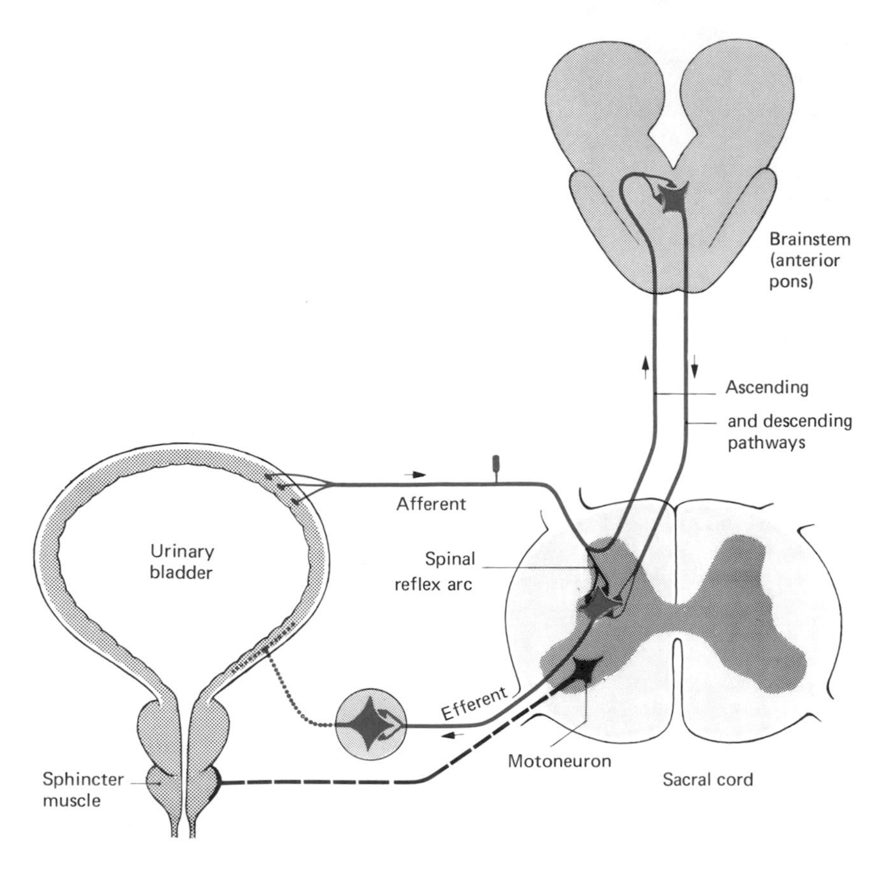

Fig. 8-13. Nervous control of micturition. The parasympathetic reflex arc in an animal with intact brain is drawn in *red*. In a chronically spinal animal or paraplegic human, the urinary bladder is controlled via the spinal reflex arc. For greater clarity, the interneurons have been omitted. The sympathetic innervation of the bladder musculature, which originates in the first two lumbar segments, is also not shown. Modified from de Groat (1975) *Brain Res.* **87**: 201–213.

rition is thereafter brought about reflexly by way of the spinal cord (*spinal reflex arc* in Fig. 8-13).

The smooth muscles of the bladder are also innervated by sympathetic fibers originating in the lumbar cord (not shown in Fig. 8-13). These fibers have an inhibitory effect on the smooth musculature of the bladder. The extent to which this sympathetic innervation is of functional significance is still disputed.

Voluntary control of micturition is exerted by way of descending inhibitory and excitatory pathways from the *cortex*, which act on the pontine micturition center, the sacral preganglionic neurons, and the motoneurons to the external sphincter. We must view the control of micturition as being *hierarchically* organized (organ level, segmental level, brainstem level, cortical level). The higher the level, the better bladder control is adapted to the needs of the organism. For example, control at the level of the pons would invariably cause complete emptying when the bladder is full. Higher centers conveniently intervene in this regulatory process, postponing or accelerating micturition.

Q 8.13 Sketch the autonomic spinal reflex arc and label the afferent, preganglionic, and postganglionic neurons.

Q 8.14 "Stomach aches" are usually caused by cramplike movements of the intestine. These pains are relieved by warming the skin of the abdomen because
- a. the intestine is warmed directly.
- b. the afferents from warm receptors in the stimulated area of the skin are connected in the spinal cord to sympathetic efferents that innervate the intestine.
- c. the parasympathetic centers in the medulla oblongata are reflexly excited.
- d. the cutaneous vessels are constricted by the heat.
- e. a cutaneovisceral reflex arc exists at the segmental level.

Q 8.15 The reflex arc by which micturition is controlled in the adult human
- a. has its reflex center in the lumbar cord.
- b. passes through the lumbar sympathetic nerves.
- c. has its reflex center in the pontine region.
- d. is purely somatic.
- e. has visceral afferents from the bladder to the sacral cord forming its afferent pathway.

8.6 Genital Reflexes

The response cycle during human copulation can be subdivided into four phases: the excitation, plateau, orgasm, and resolution phases. The time course of the cycle varies greatly among individuals. The excitation and resolution phases last the longest, whereas the plateau and orgasm phases are usually relatively brief. In men the response cycle is ordinarily stereotypic, and the resolution phase is followed by a refractory time during which orgasm cannot be reached. The response cycles of women are very variable by comparison, and women are capable of multiple orgasms.

The spinal neuronal processes underlying this cycle consist of complex series of reflexes involving parasympathetic, sympathetic, and motor efferents as well as visceral and somatic afferents. We have only a fragmentary understanding of these reflexes.

Genital Reflexes in the Man. The physiologic phases of the sexual response cycle are as follows, in order of occurrence: erection of the penis, emission of sperm and glandular secretions into the urethra, and ejaculation. Orgasm begins with or before emission and ends with ejaculation.

The *erection* of the penis is brought about by dilation of the arteries in the erectile tissue (corpora cavernosa), which fills the venous sinuses maximally so that the pressure within them rises. Drainage from the erectile tissue is cut off by the strong connective-tissue sheath of the penis. The arterial dilation is an active process, induced by parasympathetic efferents from the sacral cord (vasodilators) (Fig. 8-14). The parasympathetic neurons themselves are activated by two routes: (i) afferents from the penis and surrounding tissues, and (ii) a psychogenic route involving higher brain structures and spinal descending pathways. The density of mechanoreceptors on the penis is highest on the glans. The adequate stimulus for these receptors is gliding and massaging shear motion.

Emission and ejaculation are the climax of the sex act (orgasm) in men. Stimulation of the afferents from the internal and external genitalia (see Fig. 8-14) during sexual activity reflexly, by way of the thora-

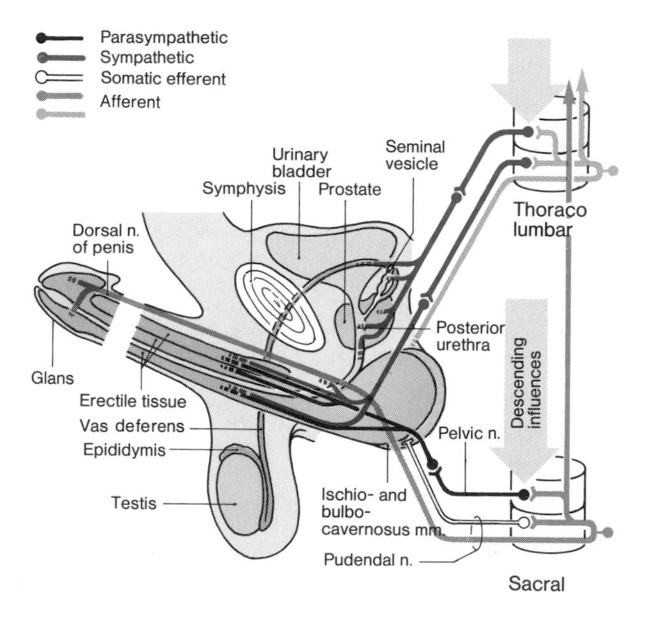

Fig. 8-14. Innervation of the male genitalia. Interneurons between the afferents and the efferent neurons in the spinal cord are not shown. (From *Human Physiology*, p. 128, Eds. Schmidt and Thews (1983). Springer-Velag Berlin Heidelberg).

columbar cord, excites sympathetic efferents. Their activity causes contractions of the epididymis, vas deferens, seminal vesicle, and prostate. The semen, a mixture of sperm and glandular secretions, is propelled into the posterior urethra. It is prevented from flowing back into the bladder by constriction at the base of the urethra. After emission, excitation of the afferents from the genitalia induces rhythmic contractions of the musculature of the pelvic floor and the skeletal muscles enclosing the proximal erectile tissue, which expel the semen from the urethra (ejaculation). This process is a reflex involving the sacral cord (Fig. 8-14). It is accompanied by rhythmic contractions of the pelvic and trunk musculature, which cause jerky movements during intercourse. These transport the semen into the proximal vagina. During the ejaculation phase parasympathetic and sympathetic neurons to the genitalia are maximally excited.

Genital Reflexes in the Woman. There is great interindividual variation in the duration and intensity of the various phases of the response cycle in women. Stimulation of the mechanoreceptors in and around the female genitalia, the axons of which run in the pudendal nerve to the sacral cord, causes changes in both external and internal genital organs. The same changes can also be psychogenically induced.

The labia majora spread apart, moving forward and to the sides, and as excitation continues they become congested with venous blood and enlarged. The labia minora become so filled with blood that they double or triple in size, pushing out through the labia majora and changing color from pink to bright red. The clitoris swells, increases in length, and is pulled up to the edge of the pubic symphysis. Again, the swelling results from filling of the organ with blood. The engorgement of the external genitalia may be produced by vasodilator parasympathetic efferents from the sacral cord that run in the pelvic nerve (Fig. 8-15).

The internal genitalia also undergo remarkable changes during the excitation, plateau, and orgasm phases. The vagina elongates and expands. A mucous fluid appears on the surface of the squamous epithelium of the vagina; it lubricates the vagina and is necessary for adequate stimulation of the mechanoreceptors of the penis during copulation. As excitation builds up, the *orgasmic platform* is produced by venous congestion in the outer third of the vagina (Fig. 8-15). This thickened region contracts during orgasm. During sexual excitation the uterus becomes more nearly vertical, so that the cervix moves away from the posterior wall of the vagina, leaving room in the inner third of the vagina for the reception of the sperm. At the same time, the uterus enlarges, and during orgasm it contracts. All these

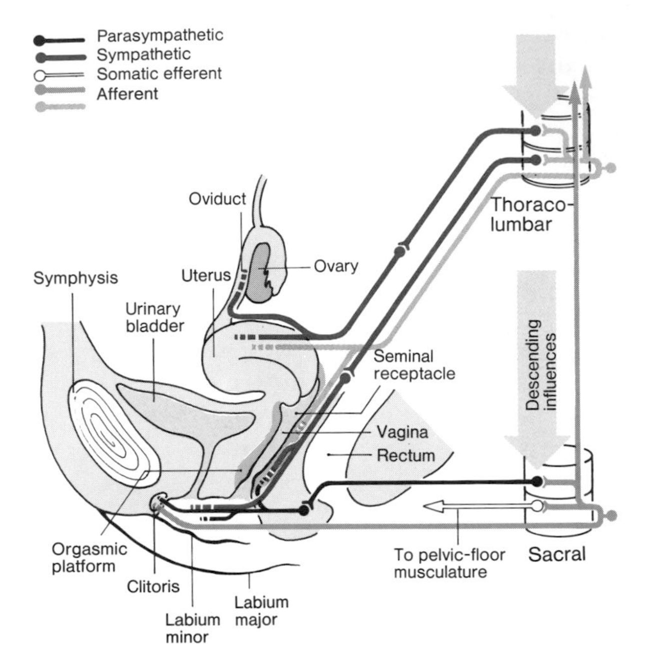

Fig. 8-15. Innervation of the female genitalia. Interneurons between the afferents and the efferent neurons in the spinal cord are not shown. (From *Human Physiology*, p. 129, Eds. Schmidt and Thews (1983). Springer-Verlag: Berlin Heidelberg).

changes are probably induced reflexly by excitation of parasympathetic neurons from the sacral cord or sympathetic neurons from the thoracolumbar cord (Fig. 8-15).

Extragenital Changes. Orgasm is a response of the entire body. It consists of the reactions of the genitalia produced by the autonomic nervous system, general autonomic reactions, and central nervous excitation, usually intense, which enhances sexual sensations.

During the sexual response cycle heart rate, blood pressure and respiratory frequency all increase, sometimes very greatly. Because of vasocongestion the breast of the woman becomes larger and the pattern of veins is more distinct. The nipples are erected and the areola swollen. These breast reactions can also occur in the man, but they are far less conspicuous. In many women and some men a "sexflush" of the skin can be observed. Typically, it begins, in the late excitation phase, on the upper abdomen, and as excitation increases it spreads over the breasts, shoulders, abdomen, and under some circumstances the whole body. The skeletal musculature undergoes both voluntary and involuntary contractions. There are nearly convulsive contractions of the facial, abdominal and intercostal musculature. During orgasm voluntary control of the skeletal musculature is often almost completely lost.

Q 8.16 The genital reflexes in the man have the following characteristics:
a. The erection is produced entirely by excitation of sacral motoneurons.
b. The reflexes are mediated only by way of the thoracolumbar spinal cord.
c. Filling of the penis with blood during erection is caused by excitation of parasympathetic efferents from the sacral cord.
d. Erection, emission, and ejaculation are a sequence of events independent of the autonomic nervous system.
e. Erection can be elicited both by stimulation of genital afferents and purely psychogenically.

Q 8.17 Responses of the genital organs during sexual excitation of the woman
a. appear only during orgasm.
b. are elicited by activity in afferent and efferent nerve fibers in the pelvic nerves and in visceral nerves from the upper lumbar cord.
c. are regulated only by hormones.
d. are probably mediated chiefly by way of the sacral spinal cord.
e. are limited to the external genitalia.

8.7 Central Nervous Regulation: Arterial Blood Pressure, Regulation of Muscle Perfusion

Control of Arterial Blood Pressure. Blood circulation is the *transport system of the body*. The circulatory system carries oxygen and energy-rich substances to the organs (the CNS, internal organs, musculature, and so on) and removes waste products. Two subsystems are involved, the *pulmonary circulation* and the *systemic circulation*.

The *arterial* part of the systemic circulation consists of the left heart and the arteries and arterioles; the heart is the pump that drives the blood through the arteries, from which it flows into the arterioles and out into the capillary bed. In a young, healthy human the mean blood pressure in this system is about 100 mm of mercury, which amounts to about one-seventh of an atmosphere. This pressure is necessary to supply the *tissues* with sufficient blood by way of the capillaries. In the *veins* the blood is carried back to the right heart. Here the pressure is about one-tenth of the arterial pressure. The walls of the veins are very soft and elastic; as a result, the venous system contains about 80% of the total blood volume. The right heart pumps the venous blood into the *lungs* through the pulmonary arteries; there it is reloaded with oxygen and releases carbon dioxide, before returning through the pulmonary veins to the left heart. Heart, arteries, and veins are autonomically innervated and are the effectors of the neuronal regulatory system governing circulation.

Figure 8-16 diagrams the most important components of the system controlling arterial blood pressure in the systemic circulation. The brain region governing blood pressure is in the lower brainstem, in the medulla oblongata (see Fig. 6-13). It is also called the *cardiovascular center* (vasomotor and cardiac center). This center functions even in the absence of modifying influences from higher regions of the brain—for example, in decerebrate animals (see p. 185). It receives information about the blood pressure from the pressoreceptors in the arteries carrying blood away from the heart (on the left in Fig. 8-16). These receptors monitor the degree of stretching of the vessel walls and thus measure both the mean blood pressure and the pulsatile variations in pressure. When either the mean arterial pres-

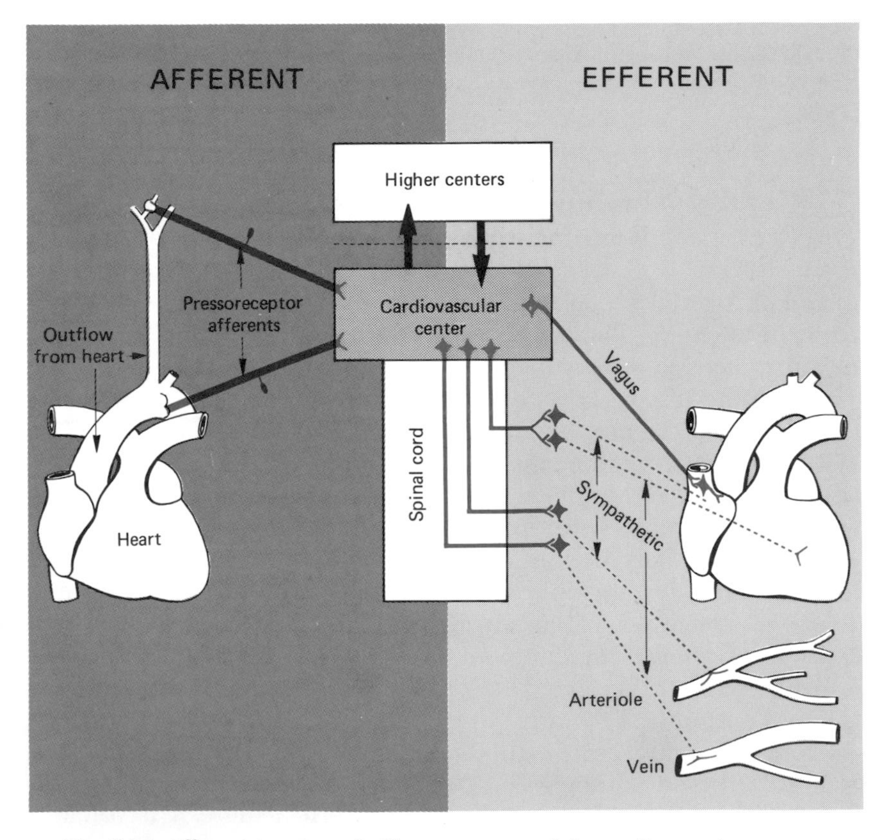

Fig. 8-16. Afferent inputs and efferent outputs of the cardiovascular center.

sure or the pulse amplitude increases, the discharge rate in the afferents of the pressoreceptors increases; lowering the mean pressure and pulse amplitude leads to a decrease in discharge rate. The fibers of the pressoreceptors are visceral afferents.

The efferents most important in regulating the arterial blood pressure innervate the *heart* and the *arterioles* (right side in Fig. 8-16). These efferents send a steady barrage of impulses to their effectors—they are *tonically active*. The heartbeat frequency and the contractile force of the heart are increased by activity in the sympathetic fibers (Fig. 8-9). Parasympathetic fibers running to the heart in the vagus nerve (Figs. 8-2 and 8-16) affect only the frequency of beating. The arterioles are innervated only by sympathetic fibers; these are called *vasoconstrictors*. The diameter of the blood vessels is under control of activity in these fibers. An increase in sympathetic activity constricts the vessels, and a reduction results in expansion. Fig. 8-17A shows the most important elements in arterial blood-pressure regulation as they are arranged in a control circuit. With this diagram, you can work out the principle of arterial blood-pressure regulation. Figure 8-17B shows the (compensatory) changes in discharge rate of a presso-affer-

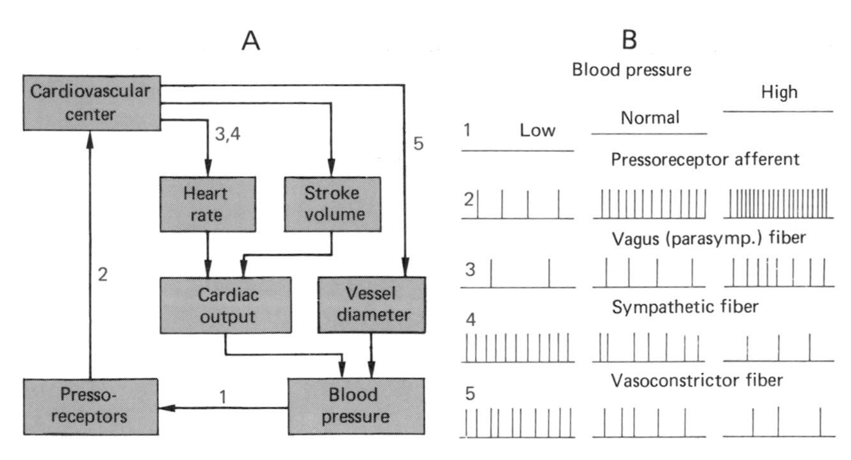

Fig. 8-17. A. Block diagram of the regulation of blood pressure. "Stroke volume" denotes the amount of blood pumped by the heart at each contraction. "Cardiac output" is the amount of blood pumped by the heart in a certain time (for example, a minute). The red numbers refer to the five graphs in B. Modified from Ruch and Patton (1965) *Physiology and Biophysics*, Philadelphia and London, Saunders. **B.** Diagram of the discharge in a typical pressure afferent (*2*), an efferent fiber of the cardiac vagus nerve (*3*), an efferent fiber of the sympathetic cardiac nerve (*4*), and a postganglionic vasoconstrictor neuron (*5*) during normal, increased, and lowered mean arterial blood pressure (*1*). Modified from Rushmer (1972) *Structure and Function in the Cardiovascular System*, Philadelphia, London, Toronto, Saunders.

ent (2), a vagus fiber to the heart (3), a sympathetic fiber to the heart (4), and a vasoconstrictor fiber (5) when the mean arterial blood pressure (1) is reduced (left column in Fig. 8-14B) or increased (right column).

An *increase in blood pressure* is signaled to the cardiovascular center by increased discharge in the presso-afferents (Fig. 8-17B, 2). The cardiovascular center counteracts this increased pressure in the following way: it increases the impulse activity in the parasympathetic fibers to the heart (Fig. 8-17B, 3) and reduces the activity in the sympathetic fibers (Fig. 8-17B, 4). As a result, both the heart rate and the volume of blood expelled in a single contraction (stroke volume) are reduced. Thus, the amount of blood pumped out by the heart per unit time (the cardiac output) is decreased. In addition, the impulse activity in the vasoconstrictors to the arterioles declines (Fig. 8-17B, 5). This causes expansion of the arterioles—a reduced peripheral resistance to flow—and, thus, an increase in the rate of blood flow out of the arterial system. The changes in all three parameters—heart rate, stroke volume, and diameter of the arterioles—counteract the increase in blood pressure.

When the *blood pressure falls,* the neuronal parameters (Fig. 8-17B, left column) and the responses of the effectors change oppositely. The activities of the presso-afferents and the parasympathetic fibers to the heart decrease, and those of the sympathetic fibers and the vasoconstrictors increase. Consequently, heart rate, stroke volume, and cardiac output increase, and the arterioles constrict. These changes counteract the fall in pressure. These regulatory events occur within a few seconds following a disturbance in pressure.

Neuronal control of the cardiovascular system comprises many specific functions in addition to regulation of arterial blood pressure, such as the control of the volume of extracellular fluid (p. 259), the control of body temperature (p. 254ff), and the control of digestion. Therefore, other afferents (for example, those from volume receptors, thermoreceptors, chemoreceptors, and receptors in the gastrointestinal tract), other effectors (for example, the venous sytem, see Fig. 8-16, and the kidney) and other regions of the brain (for example, the upper brainstem and the hypothalamus) play a role in these processes. In the following paragraphs, an example of these complex interactions—the changes in blood flow through organs and in cardiac output during muscular exertion—is discussed.

Control of Blood Flow Through Organs During Muscular Exertion. In a person at rest, about 20% of the cardiac output flows through the skeletal musculature, 40% through the visceral region, 14% through

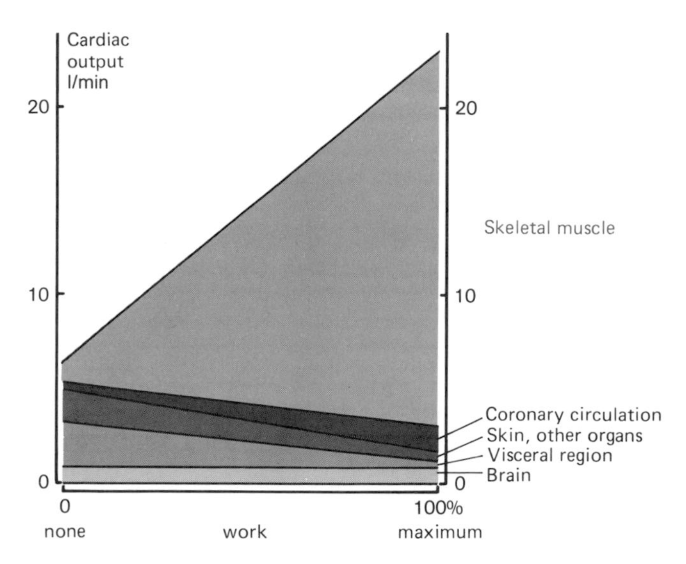

Fig. 8-18. Blood flow through various organs and organ systems, before and during physical work. The thermoregulatory increase in cutaneous blood flow during light to moderately hard work is not shown. Modified from Best and Taylor's *Physiological Basis of Medical Practice,* ed. by Brobeck JR (1979) 10th ed., Williams & Wilkins, Baltimore.

the brain, 6% through the coronary circulation, and 20% through the skin and other tissues (see Fig. 8-18, left). During muscular work the following changes occur (Fig. 8-18, right). The cardiac output rises by a factor of four at most, blood flow through the visceral region, skin, and other organs decreases (except that for purposes of thermoregulation, at intermediate levels of work, flow through the skin increases; this effect is not shown in Fig. 8-18), blood flow through the brain does not change, and there are increases in flow through the heart (coronary circulation) by as much as a factor of four and through the skeletal musculature by as much as 20-fold.

Of all these changes, the increases in muscle blood flow and cardiac output are by far the largest. It is noteworthy that the onset of these *adjustments of the cardiovascular system to muscular work* occurs within a few seconds after the work is begun.

Although these important fast adaptive responses have long been known, we are still not sure just how they are brought about. It is possible that the circulatory changes described above are triggered by the *CNS, simultaneously* with the contractions of the skeletal muscles. It is also possible that the changes are in part responses to central nervous signals elicited by muscle afferents that monitor the changes in the muscle as it works.

Q 8.18 In response to a sudden decrease in blood pressure,
a. the heart rate increases/decreases.
b. the diameter of the peripheral vessels increases/decreases.
c. the activity in the sympathetic cardiac nerves increases/decreases.
d. the activity in the pressoreceptors increases/decreases.

Q 8.19 During physical work the circulation is adjusted in the following ways:
a. blood flow through the organs in the abdominal and pelvic regions increases.
b. the cardiac output increases.
c. blood flow through the brain decreases.
d. the heartbeat rate decreases.
e. blood flow through the muscles increases.

8.8 The Hypothalamus. The Regulation of Body Temperature, Osmolarity of the Extracellular Fluid, and the Endocrine Glands

In a very real sense, the complex behavior of vertebrates is made possible because conditions within the body (the so-called *internal milieu*) are maintained within narrow limits by mechanisms not requiring conscious control. Among these internal conditions are, for example, the temperature of the body, the concentration of ions and the volume of fluid in the extracellular space, and the concentration of glucose in the blood. The relatively stable state of equilibrium achieved among all these interdependent factors is called *homeostasis*. Blood and extracellular fluid are kept at constant ion concentrations, carbon-dioxide and oxygen tensions, and so on. The body carries its "internal environment" about with it; these mechanisms might be compared with the provisions made for an astronaut, who in his space suit or the cabin of his vehicle carries controlled terrestrial conditions (oxygen, carbon dioxide, pressure and so on) with him into outer space.

The part of the brain most important in maintaining homeostasis is the *hypothalamus*. This is a phylogenetically old part of the brain; as animals evolved, its structure has remained relatively unchanged. It lies approximately in the middle of the brain and is the center for all regulatory processes in the body. Hypothalamic functions integrate spinal reflexes and autonomic regulation originating in the brainstem. These *integrative functions of the hypothalamus* involve not only the autonomic nervous system, but the somatic nervous system and the endocrine system as well. An animal lacking the cerebrum is, thus, not particularly difficult to keep alive, whereas one without a hypothalamus requires elaborate care if it is to survive.

The functions of the hypothalamus are ordinarily treated in the context of a number of areas of physiology—for example, the regulation of temperature or electrolyte balance, control of the endocrine organs, and the physiology of the emotions. This fact reflects the *diversity of hypothalamic functions*. But all these functions have one property in common—they serve to keep internal conditions constant.

In the following section, the position of the hypothalamus in the brain and its most important afferent inputs and efferent outputs are described. The regulatory nature of its operation is illustrated by examples; these include the regulation of body temperature, osmolarity of the extracellular fluid, and the endocrine glands.

Anatomy of the Hypothalamus. Figure 8-19A will be familiar to you from your study of Fig. 6-13. It shows the medial aspect of the brain; the plane of the section is oriented vertically and from front to back. The site of the hypothalamus is shown in red in Fig. 8-19A. Together with the thalamus, it lies between cerebrum and midbrain, in a region of the brain therefore called the *diencephalon*. As its name implies, the *hypo*thalamus is situated below the thalamus.

The hypothalamus has a special relationship to the *hypophysis*, or *pituitary gland* (Fig. 8-19A). This gland has two parts, an anterior and a posterior lobe. The anterior pituitary produces hormones, one of the functions of which is to control secretion by other endocrine glands in

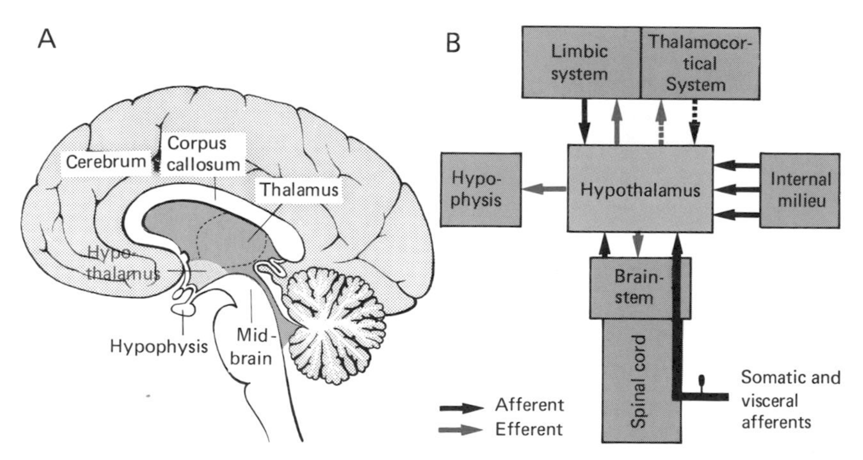

Fig. 8-19. A. Position of the hypothalamus (*red*) in the brain; **B.** Afferent and efferent neuronal connections and humoral influences of the hypothalamus. The afferent connections are drawn in *black*, the efferent connections in *red*.

the periphery of the body, such as the thyroid and sexual glands. In the functional hierarchy of the brain, the pituitary is subordinate to the hypothalamus.

In Fig. 8-19B, the most important afferent (black arrows) and efferent (red arrows) connections of the hypothalamus are diagrammed. Examination of this diagram will make the central position of the hypothalamus in the brain even more obvious than did the anatomical relationships shown in Part A of the figure.

The hypothalamus makes neural connections, both afferent and efferent, with all superordinate and subordinate regions of the CNS. The two major superordinate regions are the *limbic system* and the *thalamo-cortical system*. Those regions of the CNS that are in part subordinate to the hypothalamus are the *brainstem* and the *spinal cord*. The hypothalamus also receives crucial afferent information from the *environment* by way of peripheral sense organs (auditory, olfactory, gustatory, and somatic receptors) and from the *viscera* by way of the visceral afferents. Moreover, it receives special inputs regarding the *internal milieu*—for example, from neurons in the hypothalamus itself, which monitor the temperature of the blood, the salt concentration in the extracellular fluid, and the concentrations in the blood of hormones released by the endocrine glands. The efferents by which the hypothalamus sends outputs to the pituitary are of particular importance. Those to the anterior lobe are *hormonal* in nature, and those to the posterior lobe are *neuronal* (hence the alternative terms "adenohypophysis"—from the Greek for "gland"—and "neurohypophysis" for the anterior and posterior pituitary, respectively). These connections control secretion of hormones from the pituitary.

The Regulation of Body Temperature. We will now describe the maintenance of constant body temperature in detail, as an example of this higher level of hypothalamic control. In mammals it is essential for the functioning of the organism that body temperature be kept constant; the rates of all chemical reactions in the body are temperature dependent, and the system has evolved to work optimally at 37° to 38°C.

When we consider the body temperature of warm-blooded animals, a distinction must be made between the temperature inside the body (for example, in the thorax or the brain), the *core temperature*, and that at the periphery of the body (for example, the limbs or the skin), the *body shell temperature*. The body shell temperature fluctuates considerably as a function of the ambient temperature (think of your cold fingers in winter), whereas the core temperature is held almost constant. The organism regulates its core temperature by two main mechanisms: the control of *heat production* and the control of *heat loss.* Thermoregulatory production of heat in adult humans is achieved primarily via the somatomotor system—by *shivering,* which in turn accelerates metabolism; another mechanism operating especially in the newborn is acceleration of the breakdown of fats by activation of the sympathetic nervous system (nonshivering thermogenesis). Heat loss is regulated by control of the *cutaneous circulation.* The heat generated in the body is transported by the blood stream to the skin and given off to the environment. The blood flow through the fingers, for example, can be varied by a factor of 1 : 600, and the amount of heat transported can be varied accordingly.

An important mechanism of cooling, particularly when the ambient temperature is high, is the *evaporation* of actively produced *sweat* at the surface of the body. Each liter of sweat that is completely evaporated removes a total heat energy of 580 kcal from the body—that is, approximately a quarter of the amount of energy that you take in each day with your food. In addition to these mechanisms, warm and cold sensations elicit certain behavior patterns, such as the avoidance of extreme ambient temperatures, the choice of particular clothing, and so on; these can also be regarded in the broadest sense as control mechanisms for maintaining heat balance.

If the organism is to "know" when to remove heat and when to produce it, it must have receptors (sensors) to measure the temperature. Sensors of this sort are located in the anterior portion of the hypothalamus and the skin of the organism. The sensors in the *anterior part of the hypothalamus* are specialized neurons (Fig. 8-20A) that measure a rise in the core temperature (*warm neurons*). The sensors in the *skin* are *cold receptors* (Fig. 8-20A), which record a lowering of the body shell temperature. Thus, cooling is reported to the center by these cold receptors before any drop in the core temperature can occur. Cutaneous warm receptors probably are unimportant in thermoregulation.

The posterior section of the hypothalamus controls, in particular, the production and the loss of heat (*control center* in Fig. 8-20A). Information from the warm neurons in the anterior part of the hypothalamus and from the cold receptors in the skin converges upon this center. If this part of the hypothalamus is destroyed, the organism may become "poikilothermal"; that is, it can no longer keep its core temperature constant and independent of the ambient temperature.

Figure 8-20B shows a block diagram of the core temperature control system of the body (terms in red) as compared with a modern thermostatically controlled central-heating system (terms in black). The body core temperature is maintained at 37°C, and the temperature in the

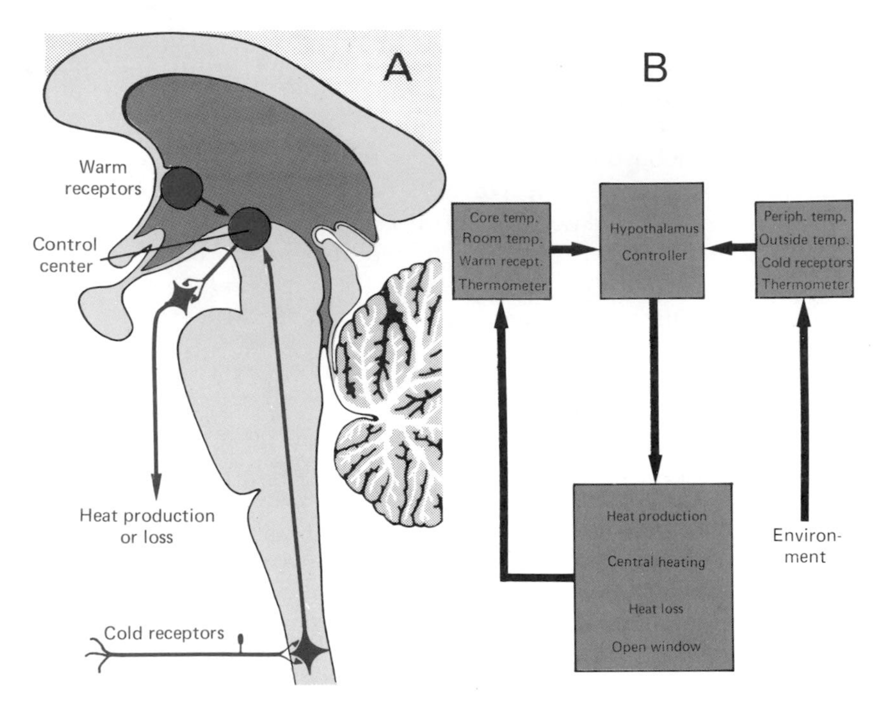

Fig. 8-20. Regulation of body temperature. **A.** Semianatomic diagram of the most important elements in temperature regulation. **B.** Block diagram showing the analogy between regulation of body temperature (terms in *red*) and that of the temperature of a room (*black*).

room is held at 21° to 22°C. The inside thermometer in the house thus corresponds to the warm neurons in the hypothalamus. This inside thermometer measures the room temperature, and, similarly, the neurons in the anterior part of the hypothalamus measure the core temperature. The information from the warm neurons goes to another group of neurons in the posterior region of the hypothalamus (Fig. 8-20A, control center) that controls the production and loss of heat. In analogous fashion, the information from the room thermometer is transmitted to the thermostat of the central-heating system. If the room is too warm, the heat output from the boiler is cut back. However, to lower the temperature in the room one can also open a window. In a very similar manner, when the core temperature is raised, the blood flow in the skin is increased, and sweat is produced. Both mechanisms of increasing the rate of heat loss—the increase in cutaneous blood flow and the secretion of sweat—are controlled by the sympathetic system. An increase in activity in the sweat gland fibers increases the secretion of sweat. A decrease in the activity in the fibers that innervate the cutaneous blood vessels produces dilation of the vessels and increases the blood flow through the skin.

As mentioned above, the control center in the posterior portion of the hypothalamus also receives information from the cold receptors in the skin. When these receptors are excited, by a drop in the peripheral zone temperature, more heat is produced by *increasing the metabolic rate,* and heat loss is diminished by a *reduction in cutaneous circulation.* Some central-heating systems make use of an outdoor thermometer (Fig. 8-20B), which operates in a fashion analogous to these cold receptors. The information from the outdoor thermometer is transmitted to the thermostat of the central-heating system. If the ambient temperature drops, then the outdoor thermometer senses this, and the boiler heat output is set higher as a precaution because the heat losses from the room are greater when the ambient temperature is low.

We must emphasize that this analogy between biological thermoregulation and the control of room temperature is very incomplete. For example, other brain structures such as the spinal cord and brainstem also have thermosensitive and thermoregulatory functions. Biological thermoregulation is considerably more complicated than is implied in the block diagram of Fig. 8-20.

Control of the Endocrine Glands by the Hypothalamus. Some glands in the body have no ducts through which the substances they produce are sent out; their secretions enter the circulating blood directly. These are called *endocrine glands,* and the substances they secrete are called *hormones.* Hormones affect organs and organ systems and have a fundamental influence on physical, sexual, and mental development; they enhance the ability of the body to adjust its performance as required, and aid in stabilizing certain physiologic variables. This hormonal communication system is coupled to the CNS via the hypothalamus. The following discussion concerns the principle of control of the endocrine glands by the *hypothalamus* and the *anterior lobe of the pituitary, or adenohypophysis.*

The adenohypophysis releases specific hormones that control the production and secretion of hormones by the endocrine glands (adrenal cortex, thyroid gland, and gonads in Fig. 8-21A) or which affect organs directly (control of growth and of the mammary glands in Fig. 8-21A). Secretion of these *adenohypophyseal hormones* is itself controlled by hormones from the hypothalamus. These hypothalamic hormones are called *releasing hormones* (RH). They are produced by neurons in the hypothalamus and secreted into a special vascular sys-

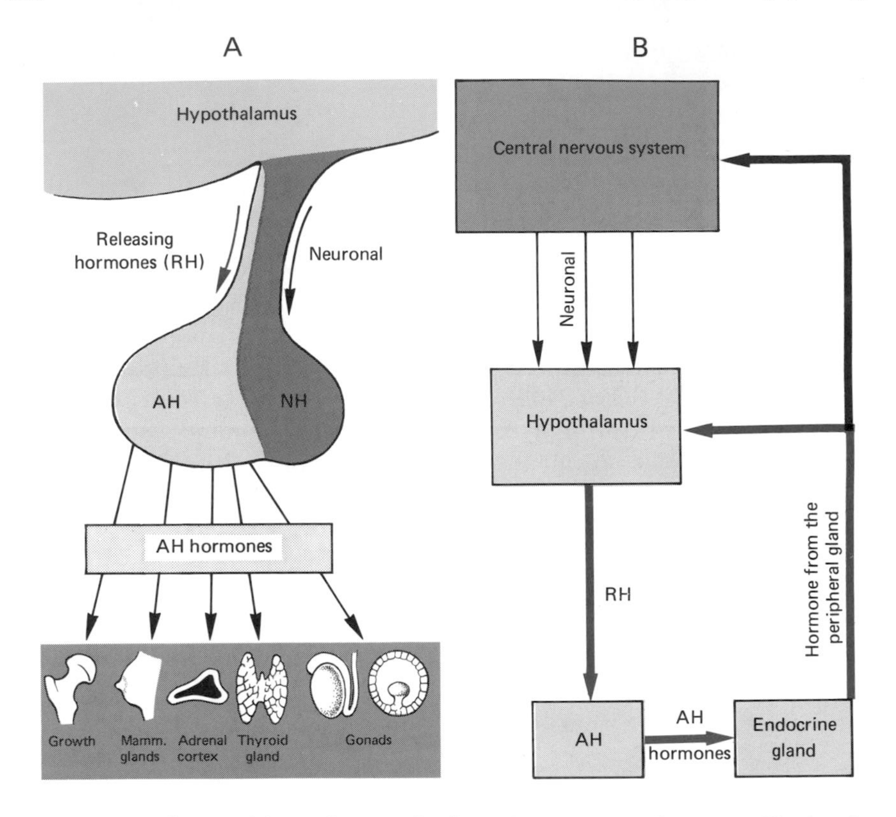

Fig. 8-21. Regulation of the endocrine glands. **A.** Semianatomic diagram. AH, adenohypophysis; NH, neurohypophysis. **B.** Regulation of the concentration of an endocrine hormone in the blood by the hypothalamus (*red*); the diagram also reflects the influence exerted on this system by other regions of the CNS (*black*).

tem, the *hypothalamic portal vessels*. Through these vessels, the releasing hormones pass to the adenohypophysis. There is a special releasing hormone for each adenohypophyseal hormone. *Inhibitory releasing hormones* (IH) are also known to affect certain adenohypophyseal hormones, acting to prevent their release.

Regulation of the endocrine glands by the hypothalamus takes the form of a *control circuit with negative feedback* (red in Fig. 8-21B; see Chapter 7). The hormone of an endocrine glands acts upon specific hypothalamic neurons in such a way that lowering of the concentration of the hormone in the blood causes, in the hypothalamus, increased production and secretion of the associated releasing hormone. This in turn brings about an increased liberation of the respective adenohypophyseal hormone and thus raises the concentration of the hormone of the endocrine gland in the blood. The feedback systems between hypothalamus, adenohypophysis, and endocrine glands (red in Fig. 8-21B) can operate in the absence of neuronal influences from other regions of the brain—for example, in animals in which the hypothalamus has been surgically isolated from all other parts of the brain.

But normally, neuronal influences from the CNS can affect the hormonal control circuits, adjusting them to momentary internal and external requirements (Fig. 8-21B). These *adaptive processes* become evident, for example, in the increased secretion of thyroid hormone that occurs with prolonged exposure to cold, in the activation of the adrenal cortex associated with any sort of physical or mental strain, and in the changes in the gonads that accompany sexual maturation. The details of the neuronal control of these adaptive processes in the body are not known. It can be shown that neurons in various regions of the CNS respond specifically to certain endocrine hormones; we may, therefore, assume that the CNS regions that control the hypothalamic–hypophyseal system receive messages from the endocrine glands by way of the hormones in the blood (Fig. 8-21B, black arrow).

Regulation of the Osmolarity (Water Content) of the Extracellular Space. The ingestion of excessive amounts of water leads very rapidly to the production of urine. This rapid excretion of fluid demonstrates that the control system regulating the body's water content is functioning successfully; this system protects the blood and the tissue fluids from becoming diluted. On the other hand, if nothing is drunk for a long time, urine production drops to a low level. The body tends to lose as little water as possible.

These processes are also controlled by the hypothalamus. In the anterior part of the hypothalamus are specialized neurons very sensitive to changes in the salt concentration (chiefly that of sodium chloride) or—equivalently—the "water content" of the blood and the surrounding extracellular space. The total concentration of dissolved molecules (mainly salts) in these fluid-filled spaces determines their osmolarity; hence these neurons are called *osmoreceptors*. An increase or decrease in salt concentration (osmolarity)—produced, for example, by consumption of table salt or water—is accompanied by an increase or decrease in the activity of these neurons. The activity of the neurons is conducted along their axons to the posterior lobe of the pituitary, or *neurohypophysis* (NH in Fig. 8-21A). In the terminals of these axons in the neurohypophysis an antidiuretic hormone (ADH, or *vasopressin*) is stored. When the neurons are excited, this hormone is released into the bloodstream from the axonal endings. In the case of the neurohypophysis, then, communication between hypothalamus and pituitary does not involve release of hormones like those that affect the adenohypophysis, but is *neuronal.*

The *concentration of ADH in the blood* is a signal to the effector organ, the kidney, concerning the amount of water to be excreted. When the ADH concentration is high, little water is excreted, and when it is low much water is lost. Thus, if the water content of the body is high, or (equivalently) the salt concentration low, little ADH is released from the pituitary and a great deal of dilute urine passes out through the kidney (Fig. 8-22A). It is likely that the hypothalamic osmoreceptors are also involved in producing the *sensation of thirst* (see *Fundamentals of Sensory Physiology*).

Figure 8-22B is a block diagram of the regulation of osmolarity just described; using the example of water consumption, the diagram shows how the water content of the body is held constant. This hormonal control is very rapid, coming into play within 15 min. Use the

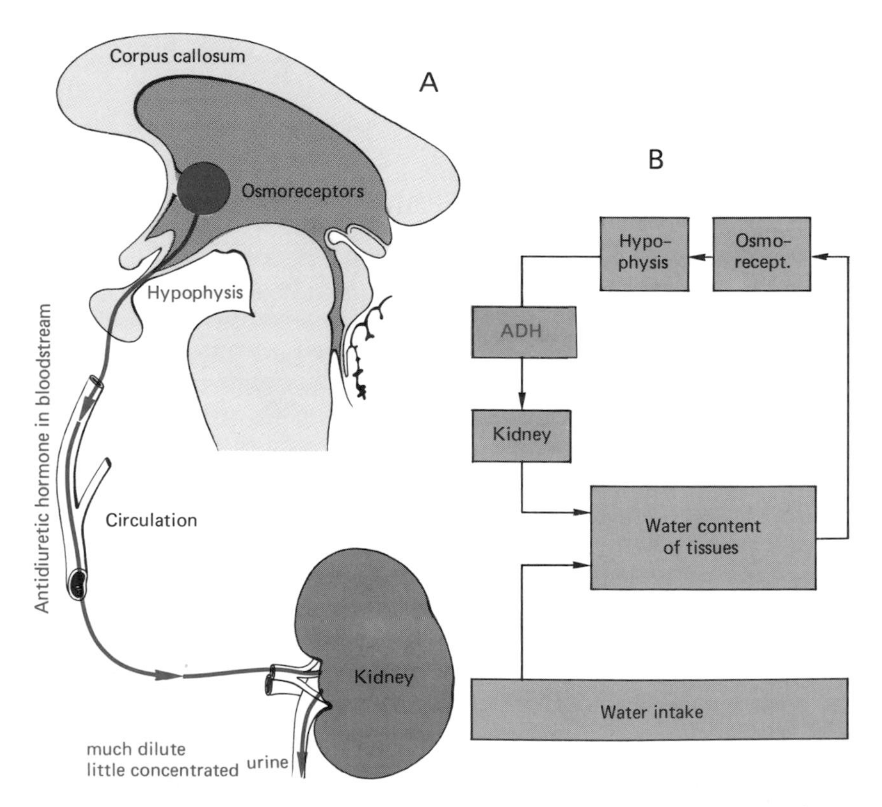

Fig. 8-22. Regulation of osmolarity of the extracellular fluid. **A.** Semianatomic diagram; **B**, block diagram. ADH, antidiuretic hormone.

block diagram to follow the changes in water content (or osmolarity) of the extracellular space, activity of the osmoreceptors, ADH concentration in the blood, and water excretion by the kidneys; then repeat the process under the assumption that the lowest box in the diagram represents water loss (for example, by profuse sweating in hot weather) rather than water intake.

Finally, it should be mentioned that the situation is complicated by the close link between control of osmolarity in the extracelluar space and *control of the volume of extracellular fluid.* For example, the excitation of receptors that monitor the volume of fluid circulating in the pulmonary part of the circulatory system results in a reduction in ADH release from the posterior pituitary, and thus to an increased excretion of dilute urine.

Q 8.20 The hypothalamus is located
a. beneath the thalamus.
b. in the brainstem.
c. between the medulla oblongata and the cerebellum.
d. in the cerebrum.
e. between cerebrum and mesencephalon.

Q 8.21 During muscular exercise the body's heat production is raised. Which two ways are available to the body to keep its core temperature constant?
a. Increase in sweat production.
b. Reduction in cutaneous blood flow.
c. Release of ADH into the blood stream.
d. Increase in cutaneous blood flow.
e. Release of adrenalin from the adrenal cortex.

Q 8.22 When it is overheated, the body must give off heat to the environment. This emission of heat is controlled from the hypothalamus by
a. neuronal mechanisms.
b. hormonal mechanisms.
c. neuronal and hormonal mechanisms.
d. the sympathetic system.
e. the parasympathetic system.

Q 8.23 Someone excretes 10 liters of urine per day and must, therefore, drink 10 liters of water per day. What could the fault in the salt—water balance control system be?
a. The kidney can no longer excrete concentrated urine.
b. The hypothalamus produces too much vasopressin.
c. The water is being resorbed too quickly in the intestines.
d. The ADH production is restricted as a result of a tumor in the hypothalamus.
e. The releasing hormone that liberates ADH from the posterior lobe of the pituitary is not being produced.

Q 8.24 The following factors participate in control of the level of thyroid hormone in the blood:
a. pre- and postganglionic fibers to the thyroid gland.
b. all the releasing hormones from the hypothalamus.

c. releasing hormones from the posterior lobe of the pituitary.
d. a specific hormone from the anterior lobe of the pituitary.
e. osmoreceptors in the hypothalamus.
f. a specific releasing hormone from the hypothalamus.

8.9 Integrative Functions of the Hypothalamus. Limbic System

In the preceding section, the hypothalamus was shown to be the control center for many of the homeostatic processes that take place in the organism. The internal conditions of the organism are held constant by hypothalamic control. As a result, the organism is relatively independent of changes in the external environment. These homeostatic processes involve corresponding *behavioral patterns* in man and animals—for example, thermoregulatory behavior and drinking. Such behavior is also extensively, or at least in part, controlled by the hypothalamus. The hypothalamus is also the coordinating center for *defense behavior* (which comprises attack, self-defense, and flight), *eating behavior* (which regulates food intake), and simple *reproductive (sexual) behavior.*

These aspects of behavior can be observed in an animal with an intact hypothalamus, but from which the entire cerebrum has been removed. In such cases, the behavior is stereotyped and not properly related to environmental conditions. These attitudes may also be regarded as *homeostatic processes in a wider sense*, in that they enable the individual to exist in a hostile environment (defense behavior), guarantee intake of food (eating behavior), and serve to ensure propagation of the species (sexual behavior). In this section we shall take the examples of eating and defense behavior to show that the elementary patterns consist of the coordination of individual somatic and autonomic reactions and that the hypothalamus integrates these reactions into certain behavior patterns. Finally the limbic system, which is closely related to the hypothalamus in both anatomy and function, will be described.

Eating Behavior. The animal used in the following experiments is a conscious cat able to move about without restriction. Prior to the experiments a metal electrode was attached firmly to its head, under anesthesia. The tip of the electrode was positioned in the hypothalamus. If a certain group of cells in the *lateral portion* of the hypothalamus is stimulated through the electrode (100 stimuli/s for a duration of 10 s), eating behavior is triggered. At the beginning of the experiment, the animal is lying quietly, not paying attention to anything in particular. When the stimulus is applied the cat raises its head and looks alert. It stands up and starts to walk slowly around the room as if looking for something. It sniffs at the floor, goes to the food trough, and starts eating. This sequence of actions, from the first reaction to the final act of eating, is usually carried out within the 10-s-long series of stimuli. If the series of stimuli is interrupted, the cat stops eating. Repeating the stimulation produces the same behavior pattern.

In order to check whether the eating behavior observed after electrical stimulation of the lateral hypothalamus also comprises autonomic responses, the following were measured: blood pressure, intestinal movement, intestinal blood flow, and muscular blood flow (see Fig. 8-23, eating). Since it is difficult to measure the effect of hypothalamic stimulation on these parameters in the unrestrained animal, the cat was anesthetized. The four variables were continuously recorded in graph form in Fig. 8-23. Electrical stimulation (heavy black line over time scale) of the region in the hypothalamus from which eating behavior can be triggered causes the following changes in these variables: the blood pressure increases, the movements of the intestine increase, the blood flow through the intestine increases, and the blood flow through the skeletal muscles decreases. When the stimulation of the hypothalamus is repeated, the same autonomic changes are observed.

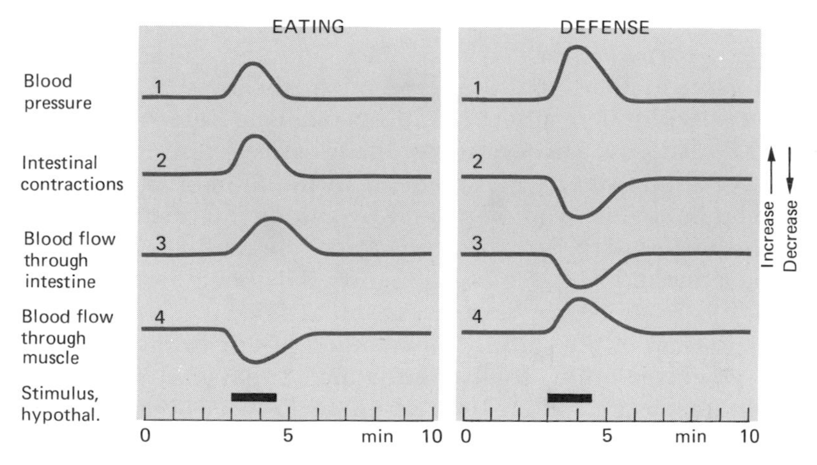

Fig. 8-23. Autonomic responses during eating and defensive behavior. The variables *1* to *4* were monitored in a cat under anesthesia, while small regions of the hypothalamus associated with the indicated behavior were stimulated electrically (black bars) with fine electrodes. Modified from Folkow and Rubinstein (1966) *Acta Physiol. Scand.* 65: 292.

This redistribution of the blood flow through the organs in favor of the intestines is brought about by a lowering of the activity in the sympathetic fibers that innervate the intestinal blood vessels and an increase in the activity in the sympathetic fibers that constrict the blood vessels of the muscles (vasoconstrictors). The intestinal movements are induced by parasympathetic fibers that run in the vagus nerve.

These experiments show that the behavioral pattern "eating" comprises both somatomotor and autonomic responses. The autonomic responses to a certain extent put the animal in a state of readiness for the processes "food intake" and "food digestion." This *coordinated sequence of somatomotor and autonomic responses* can be elicited electrically only from very restricted areas in the left and right lateral hypothalamus; this region of the hypothalamus is therefore called the "eating" or "hunger" center. Destruction of this center in an animal causes it to refuse food (aphagia) and can result in starvation. In the medial hypothalamus there is another neuronal region, electrical stimulation of which causes inhibition of eating behavior. This area is also called the "satiety center." When it is destroyed, hyperphagia (compulsive eating) results.

Defense Behavior. If the position of the tip of the electrode in the hypothalamus is advanced by about 2 mm, then a completely different behavioral pattern can be triggered from this new area, namely, defense behavior. When the hypothalamus is stimulated, the animal, which was lying peacefully, suddenly becomes very alert. It gets up, arches its back, and starts to snarl, hiss, and paw. The digits of the paws are spread apart, and the claws are unsheathed. All these reactions occur within a few seconds of hypothalamic electrical stimulation. The response can terminate either in a violent attack on the person conducting the experiment or in the animal's trying to run away. The behavior is accompanied by autonomic reactions such as the secretion of saliva, urination, dilation of the pupils, erection of the hair, and a marked increase in respiration. The behavior we have just described is triggered by an increase in activity in the sympathetic system (dilation of the pupils, fur standing up) or in the parasympathetic system (secretion of saliva, urination). Behavioral attitudes with these autonomic characteristics are generally ascribed to the emotional expressions of *anger and fear* in man.

In order to measure the autonomic parameters shown in Fig. 8-23, the animal was again anesthetized. Figure 8-23 (defense) shows that the measurements change as follows during electrical stimulation (thick black line) of the region of the hypothalamus from which defense behavior can be triggered: the blood pressure increases, the movements of the intestine decrease, the blood flow through the intestines decreases, and the blood flow through the skeletal muscles increases. If we disregard the change in blood pressure, the measured autonomic reactions in the body change in opposite directions during the behavioral responses of "eating" and "defense." All the autonomic reactions during defense behavior can be explained by the increase in the activity in the sympathetic system. The intestinal movements and the blood flow through the wall of the intestine both decrease as a result of the increase in the sympathetic activity. The blood vessels of the skeletal musculature dilate as a result of decreased activity in the sympathetic vasoconstrictor fibers. In addition, it is thought that there are special sympathetic fibers to blood vessels in the skeletal muscles that produce an active dilation, so that the blood flow through the musculature is increased. Further changes accompanying defensive behavior are *activation of the adrenal medulla*, which leads to the release of adrenaline and noradrenaline into the blood (see p. 225), and the release of *adrenocortical hormones* because of activation of the hypothalamic-hypophyseal system (see p. 257). These autonomic and endocrine responses enable an animal to respond optimally to threats from the environment.

Experiments like that of Fig. 8-23 indicate that the hypothalamus contains a large number of different neuron populations that are responsible for activating the somatomotor, autonomic, and endocrine responses in particular *patterns*. The animal's behavior is an expression of these patterns. The excitatory and inhibitory synaptic connections between the various hypothalamic neuron populations and other groups of neurons in the hypothalamus, brainstem, and spinal cord could be termed "programs." One such *neuronal program*, for example, ensures that the flow of blood through intestines and skin decreases during defensive behavior (Fig. 8-20). These programs operate, to a limited extent, even in the absence of the cerebrum—as can be demonstrated in animals from which all the cortical structures have been removed. The application of nociceptive and non-nociceptive stimuli to the skin of such animals elicits behavior patterns that resemble defensive behavior, eating behavior, sexual behavior, and so on. If the hypothalamus of such a decorticate animal is destroyed, these behavior patterns are produced only in fragments, or not at all, by either central electrical stimuli or natural (food and so on) stimuli.

The Limbic System. The behavioral patterns triggered by electrical stimulation of the hypothalamus of a decerebrate animal, or by natural stimulation, are stereotyped. When stimulation stops, they are abruptly halted. The same forms of behavior can be triggered repeatedly by the same stimuli. Behavior produced in this way in decorticate or hypothalamic animals is not directed toward specific environmental situations, and it seems likely that such behavior is not correlated with particular feelings or emotional "sets" within the animals.

In the discussion of Fig. 8-19B, you learned that the hypothalamus is under the control of the limbic system. This system is the phylogenetically older part of the cerebrum; in man it is entirely covered by the more recently evolved neocortex (Fig. 8-24A). The various structures of the limbic system are arranged in a ring around the upper end of the neuraxis, and it is from this arrangement that the name is derived (*limbus* means hem or border).

The limbic system comprises a number of cortical structures, nuclei, and fiber tracts. The cortical structures have three or five layers, in contrast to the six-layered structure of the neocortex (see p. 275). The fiber connections between the limbic system and nonlimbic areas of the brain (Fig. 8-24B) are remarkable in their extent. Massive tracts mediate reciprocal communication with the hypothalamus and the upper part of the brainstem. The connection between limbic system and neocortex involves chiefly the frontal cortex. Information from a great variety of sensory areas is probably obtained indirectly, by way of the neocortical association areas.

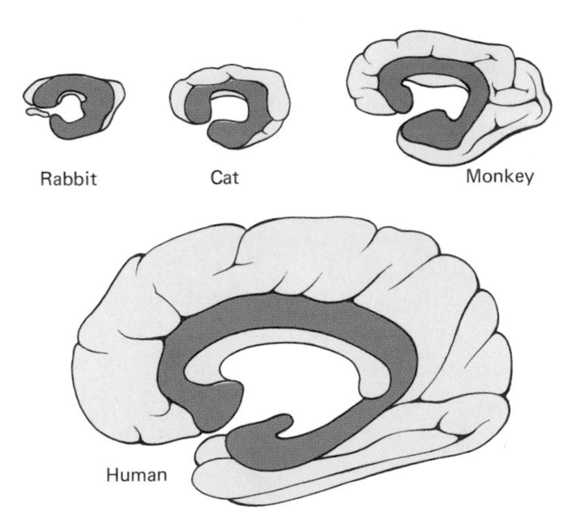

Fig. 8-24.A. Dimensions and arrangement of the limbic system (*red*) relative to the neocortex in several mammalian species including man, as seen in the medial aspect of the brain. Modified from MacLean (1954) in Wittkower and Cleghorn, *Recent Developments in Psychosomatic Medicine*, London, Pitman Medical Publ. Co. Ltd.

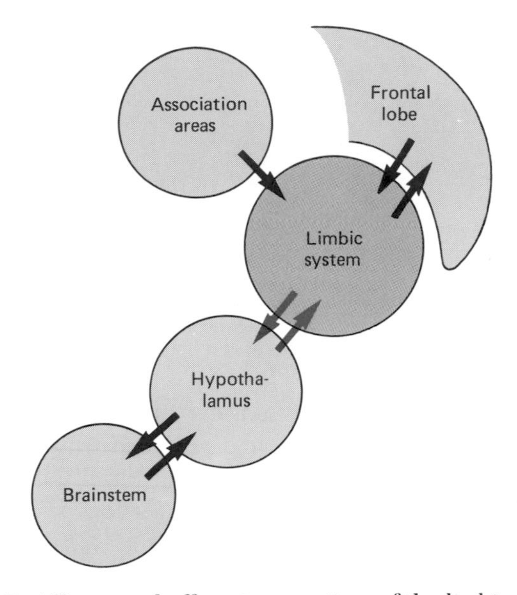

Fig. 8-24.B. Afferent and efferent connections of the limbic system.

Because most structures of the limbic system are derived phylogenetically from structures serving an olfactory function in primitive animals, it was originally thought that this system was involved in analysis of olfactory information; it was, therefore, given the name *rhinencephalon* (olfactory brain). We are now certain that only very few structures, located basolaterally in the limbic system, function in olfaction. The current view of the limbic system is that it is the substrate for *species-specific behavior*. Experiments involving electrical and chemical intracranial stimulation, ablations, and self-stimulation have produced very different results in different species of animals; interpretations of these results have, therefore, been within the framework of behavior specific to the particular species.

If one tries to extend to humans the concept of the limbic system as the substrate for species-specific behavior, the first question concerns the definition of such behavior in humans. It is likely that the phenomena we call "emotions," "affective behavior," "feelings," and so on can be categorized as species-specific behavior. In this sense, one function of the limbic system would be to control the *"expression" of the emotions*. Such control involves the somatomotor, autonomic, and endocrine systems and takes place by way of the hypothalamus and the upper part of the brainstem. This situation is reflected in the massive reciprocal fiber connections between the various structures of the limbic system and the hypothalamus and brainstem (Fig. 8-24B). On the other hand, the limbic system also regulates the *"affective aspect" of the emotions,* which is experienced subjectively. The critical connections in this case are probably those between limbic system and neocortex, for it is in the neocortex that environmental events receive their *affective coloration* and thus their "meaning" to the organism.

From a biological viewpoint, the "emotions" have the character of *signals.* The expression of emotions conveys the state of an individual to its conspecifics—for example, during rage behavior—and thus brings about corresponding behavioral changes in them. Internally, the emotions cause changes in the behavior of the individual itself. For example, sexual feelings, which can be produced by sexual hormones, result in quite specific behavioral patterns.

Such control of behavior is achieved, on the one hand, by means of complete *programs developed during evolution,* the neuronal correlates of which have been laid down in the limbic system and are *genetically* inherited. On the other hand, they also involve *learning,* by which programs are stored in *memory.* Since an animal can learn only that which is significant in the context of behavior, this learning is also intimately associated with inherited behavior patterns. The functional and anatomic proximity of the neuronal structures controlling species-specific behavior and those involved in memory (the hippocampal formation; see p. 310) is understandable in this light.

As the cerebrum became progressively more elaborate during mammalian evolution, the neocortex expanded relative to the limbic system and eventually enclosed it almost entirely (Fig. 8-21). Parallel with this development, there appears in man, and to a limited extent in anthropoid apes, language and higher forms of mental performance. At this level of behavior, the formation of concepts and strategies results in modification and concealment of the inherited species-specific behavioral repertoire.

Q 8.25 During defense behavior (flight, attack),
a. the cardiac output is increased.
b. the digestion is promoted.
c. the blood flow through the skeletal muscles is increased.
d. the blood flow through the intestine is lowered.
e. ACh is released from the adrenal cortex.

Q 8.26 A decorticate dog (cerebrum removed)
a. can no longer maintain a constant body temperature.
b. can remain alive if given the proper care.
c. can still regulate its feeding behavior to some extent.
d. reacts defensively to strong cutaneous stimuli.
e. dies because it can no longer regulate its circulation.

Q 8.27 Draw a diagram showing the changes, during eating behavior and defense behavior, in
- a. blood pressure.
- b. intestinal contractions.
- c. intestinal blood flow.
- d. muscle blood flow.

Q 8.28 Which of the following associations between structure and global function are approximately correct?
- a. Cerebrum—control of species-specific behavior.
- b. Hypothalamus—standing reflexes (extensor rigidity).
- c. Hypothalamus—transmission of information from specific sensory systems.
- d. Precentral gyrus—motor activity.
- e. Limbic system—emotions.

9
INTEGRATIVE FUNCTIONS OF THE CENTRAL NERVOUS SYSTEM

R.F. Schmidt

The integrative functions of the central nervous system include quite fundamental processes, operations that cannot be classified directly as processing of sensory inputs or motor and autonomic activity. Essentially, such neuronal mechanisms underly the sleeping–waking cycle, consciousness, language, and memory, as well as learning. Other integrative mechanisms such as the bases of elementary behavior patterns and the emotions have been treated in the preceding chapter (see Secs. 8.5 and 8.6). Still others—such as the neurophysiologic bases of complex behavior patterns—will be omitted or mentioned only briefly because of the inadequacy of present knowledge (see Sec. 6.5, paragraph headed *Impulse to act and design of the motor pattern,* p. 196).

An essential—but not the only—part of the nervous system involved in these integrative tasks is the *cerebral cortex,* the phylogenetically newest development; in man, this has become so extensive that it is fitted into the skull only by considerable folding. The structure, connections, and general physiology of the cerebral cortex will thus be described first in this chapter.

Without the *sensory systems*—the peripheral and central mechanisms of the nervous system that receive signals from the environment and the interior of the body (receptors), conduct them (via afferents), and process them (in sensory centers)—integrative operations of the CNS would be inconceivable, just as without the effector (autonomic and motor) systems. The latter are discussed extensively in previous chapters, but for practical reasons a description of the sensory systems has been published separately (*Fundamentals of Sensory Physiology,*

R.F. Schmidt ed., 2nd Ed., Springer, 1981). In this chapter the relationships between the sensory and integrative functions of the nervous system will occasionally be mentioned; the reader is invited to refer to that companion volume where necessary.

9.1 Structure and General Physiology of the Cerebral Cortex; the Electroencephalogram

Functional Anatomy of the Cerebral Cortex. The two halves, or *hemispheres,* of the human cerebrum take up most of the volume of the cranial cavity; the cortex is highly convoluted, its surface consisting of convex folds, or *gyri,* separated by furrows called *sulci* (see Fig. 1-8). As Fig. 9-1 shows, the brainstem (for its subsections see Fig. 6-13) and diencephalon (with thalamus and hypothalamus as its chief subsections) are entirely surrounded by the cerebral cortex. Their relative positions can be understood by examining sections in the midline (median sections, Fig. 9-1B), horizontal sections (Fig. 9-1C), and frontal sections (Fig. 9-1D).

Each hemisphere can be roughly subdivided into four *lobes* (Fig. 9-1A) named, according to their position in the skull, the frontal, parietal, temporal, and occipital lobe. Each lobe is formed of several gyri, some of which (for example, the motor precentral gyrus of the frontal lobe; see Sec. 6.5 and Figs. 6-8 to 6-11) have been mentioned elsewhere in this volume. This subdivision is purely anatomic, and reveals nothing about the function of the brain tissue in the various lobes.

A simple but useful preliminary *functional subdivision of the cerebral cortex* can be derived from the efferent and afferent connections of its various areas. This subdivision corresponds to that obtainable from observation of the clinical deficits following brain damage. We have already learned in Sec. 6.5 that the precentral gyrus and its surroundings are *specific motor cortex;* interruption of its efferents in the internal capsule on one side produces spastic hemiplegia—that is, a severe impairment of motor ability on the opposite side.

The cortical regions in which the sensory pathways terminate (after relay in thalamic sensory nuclei) are termed the *specific sensory cortex.* These regions are shown in Fig. 9-2 for the hemispheres of four different mammals, including man. For example, the postcentral gyrus in the parietal lobe is the most important area of termination of the somatosensory pathways, in the occipital region the visual pathway terminates, and part of the temporal lobe contains the endings of the neurons of the auditory pathway; accordingly, these areas are re-

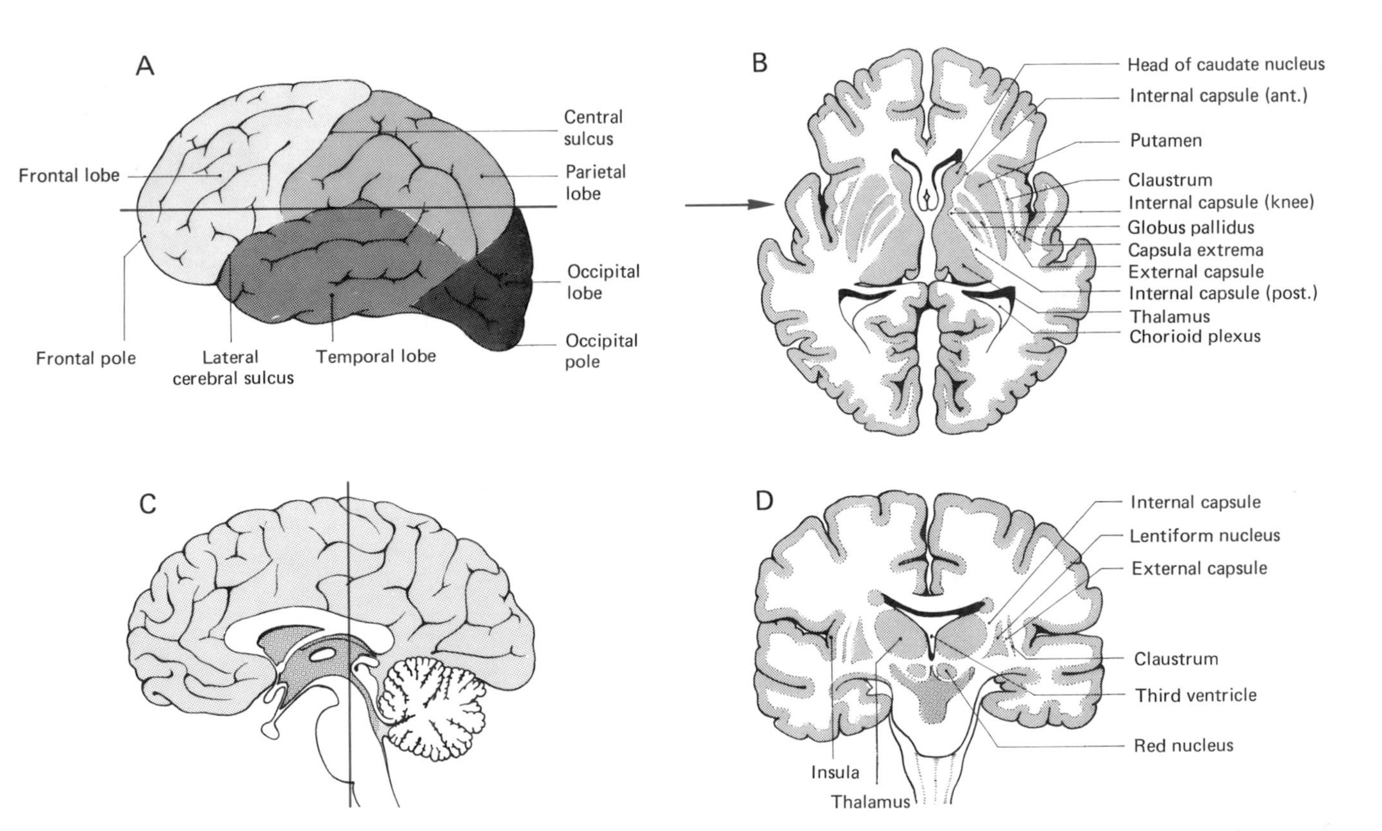

Fig. 9-1. Survey of the regions of the cerebral cortex and their spatial relationships with respect to subcortical structures. **A,** View from the side; **B,** section through the plane indicated by the red line in A; **C,** view of the medial surface after the brain has been sectioned along the midline (sagittal section); **D,** frontal section (in the plane indicated in *C*). Not all structures shown are discussed in the text.

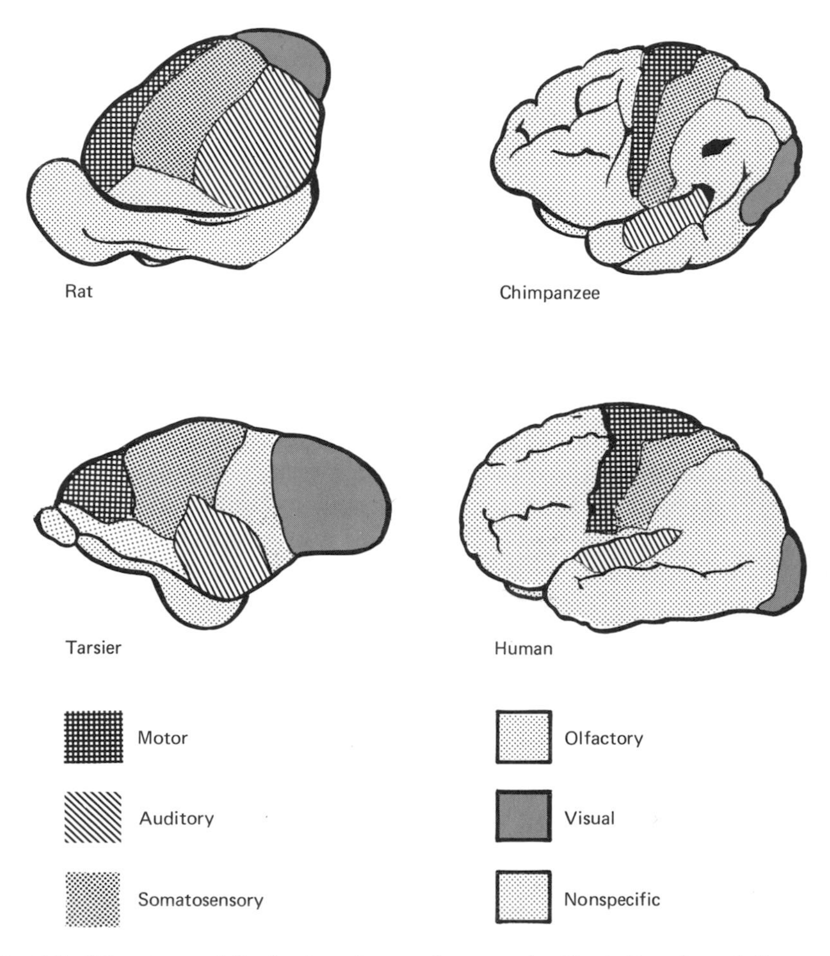

Fig. 9-2. Side views of the brains of several mammals, illustrating the relative areas occupied by motor, sensory, and association cortices. Note the massive increase in association cortex in the anthropoid ape and even more in the human brain. The considerable absolute difference in size of the different brains (not indicated) should also be kept in mind. From Stanley Coob.

ferred to as visual cortex, auditory cortex, and so on. Damage to these cortical areas results in deterioration of the associated sensations, which may, for example, be as severe as (central) blindness or deafness (so-called cortical blindness or deafness; for further information about the central sensory pathways, see *Fundamentals of Sensory Physiology*).

Not all sections of the cortex can be classified as specific sensory or motor areas, on the basis of termination of identified tracts. Since it had been suspected that these areas are mainly concerned with the interaction between motor and sensory cortical areas, they are usually called *association cortex* (sometimes *nonspecific cortex*).

The association cortex in humans, and even in the anthropoid apes, takes up appreciably more space than the sensory and motor cortices (Fig. 9-2). Moreover, in the course of evolution there has been an increase in absolute size of the brain. These two factors result in a pronounced absolute and relative increase in area of the association cortex. That is, the association cortex is apparently particularly significant with respect to the higher, integrative functions of the CNS, as defined at the beginning of this chapter. This point will be elaborated upon below.

Fiber Connections in the Cerebral Cortex (Fig. 9-3). The cerebral cortex receives afferents from subcortical structures almost entirely by way of *thalamocortical tracts*. The inputs to the cerebrum from all

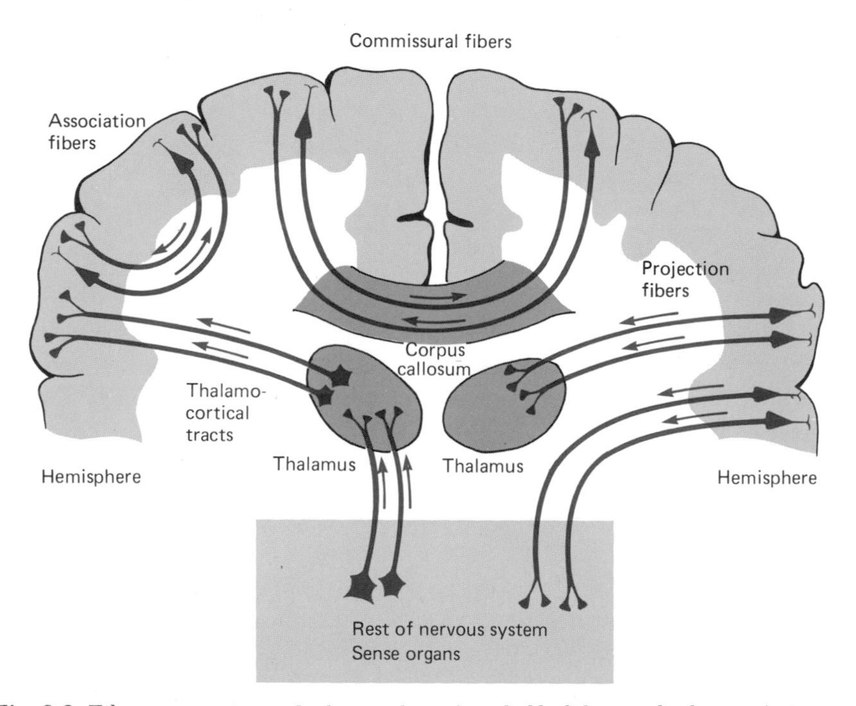

Fig. 9-3. Fiber connections of a hemisphere (one half of the cerebral cortex). Association fibers link different regions of a hemisphere with one another. Commissural fibers connect each hemisphere to the other (chiefly by way of the corpus callosum). The hemispheres are in communication with the rest of the nervous system by way of cortical efferents. Afferents to the hemispheres pass through thalamocortical tracts.

parts of the nervous system and the sense organs go through a final relay station in the thalamus before reaching the cortex (see Figs. 6-17 and 6-18 and *Fundamentals of Sensory Physiology*). The efferent connections of the cortex are called *projection fibers*. Examples of these were given in Figs. 6-10, 6-11, 6-17, and 6-18. The *association* and *commissural* fibers are of special interest—the former make connections within each hemisphere, and the latter connect the two hemispheres, chiefly by way of the corpus callosum. Association and commissural fibers are thus either efferent or afferent, depending upon the cortical region being considered.

Functional Histology of the Cerebral Cortex. The cerebral cortex is a thin layer of neuronal tissue with a surface area of about 2,200 cm^2 (the area of a square 47 cm × 47 cm) and a thickness varying between 1.3 and 4.5 mm in the various areas. Its volume is about 600 cm^3. It contains *10^9 to 10^{10} neurons* and a large but unknown number of glial cells. Within the cortex, layers containing primarily cell bodies alternate with layers predominantly occupied by axons, so that the freshly cut cortex has a striated appearance. Typically, *six layers* are distinguished on the basis of cell shape and arrangement (Fig. 9-4A); these will not be discussed further here.

Histologic and electrophysiologic observations have shown that *information processing in the cerebral cortex* basically occurs in circuits oriented perpendicular to the cortical surface. The arrangement is illustrated in Fig. 9-4B, in highly simplified, diagrammatic form. The superficial layers I–IV serve chiefly to receive and process the information flowing into the cortex. The neurons carrying efferent signals from the cortex (projection, association and commissural fibers; cf. Fig. 9-3) tend to have their cell bodies in the deeper layers, V and VI; these can therefore be regarded as the regions of origin of the cortical efferents.

Information processing perpendicular to the surface of the cerebral cortex was noted previously, in the discussion of the somatotopic subdivisions of the precentral gyrus (Fig. 6-9). This principle of organization appears again in the specific sensory areas of the cortex. The *postcentral gyrus*—that is, the somatosensory cortex (cf. Fig. 9-2)—is also *somatotopically arranged*, in close analogy with the precentral gyrus, and the other sensory cortices have comparable (though in some cases very complex) arrangements. Hence neurons that are closely associated in function are found in elongated groups aligned perpendicular to the surface; these are called *cortical columns*. Microstimulation and recording experiments in the motor cortex have shown that these functional cortical columns there have a diameter of

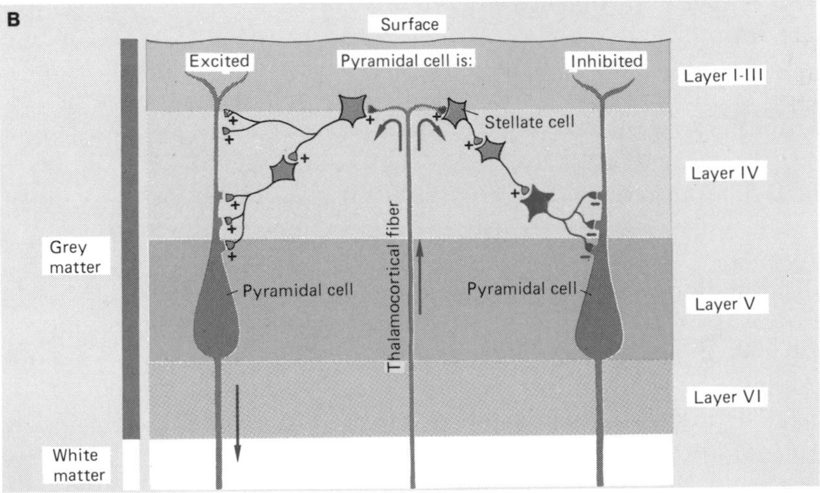

Fig. 9-4.A. Semidiagrammatic illustration of the layered structure of the cerebral cortex. The drawings are based on a Golgi-stained preparation, which emphasizes single neurons with their dendrites (*left*), a Nissl preparation, with only the somata stained (*middle*), and a myelin-sheath preparation showing the courses of the axons (*right*). Modified from Brodman and Vogt. **B.** Highly schematic illustration of the connections of the neurons and the information flow (*red arrows*) in the columns arranged perpendicular to the surface of the cerebral cortex. Only pyramidal and stellate cells are shown. Excitatory synapses are marked by plus signs, inhibitory synapses by minus signs.

about one millimeter. So far, however, we know very little about the detailed connectivity of the cortical neurons or the processing for which they are responsible.

The Electroencephalogram. If an electrode in the shape of a button is placed on the scalp and another ("indifferent" or reference) electrode is attached at a distant place (for example, an earlobe), then electrical potential differences can be recorded between the two in humans and other vertebrates. The continuous record of the smooth fluctuations in such a potential is called the *electroencephalogram* (*EEG*) (Fig. 9-5). The EEG exhibits fluctuations at frequencies in the range 1 to 50 Hz; the amplitudes of the fluctuations are of the order of 10 to 100 μV. The fact that such electrical activity of the human brain can be recorded was discovered by the German neurologist Hans Berger, who went on between 1929 and 1938 to lay the foundations for its clinical and experimental application.

Evaluation of the EEG has become a routine procedure in neurologic diagnosis. To facilitate comparison, the positions of the recording electrodes (Fig. 9-5, left) and the recording conditions (speed of the recording paper, frequency response of filters in the amplifier system) have been extensively standardized. Recording may be either *unipolar* (with a distant reference electrode as in Fig. 9-5) or *bipolar*, in which case the potential between two scalp electrodes is monitored. The records are evaluated primarily with respect to frequency, amplitude, and form of the fluctuations, as well as with respect to the distribution and commonness of occurrence of EEG waves of different types. This can be done either "by eye" or with analog and digital computers programmed for such analysis.

Frequency and amplitude of the EEG are determined by a number of factors. The effects of one of these, the recording site, are quite conspicuous in Fig. 9-5; the EEG waves recorded at the back of the head (when the eyes are closed) are considerably more pronounced then those over the frontal and parietal regions. The marked dependence of the form of the EEG upon the degree of wakefulness can also be seen in Fig. 9-5; when the eyes open, the large, slow waves instantly vanish and are replaced by high frequency waves of smaller amplitude. After the eyes are closed, the slow rhythm sets in again. This slow rhythm predominates in the healthy adult human in a waking but relaxed (eyes closed) state; it is most pronounced over the occipital brain and has frequency components of 8 to 13 Hz (10 Hz on the average). These waves are called *alpha waves*. Another aspect of these waves evident in Fig. 9-5 is that the alpha rhythm appears at many recording sites in about the same form (amplitude, frequency,

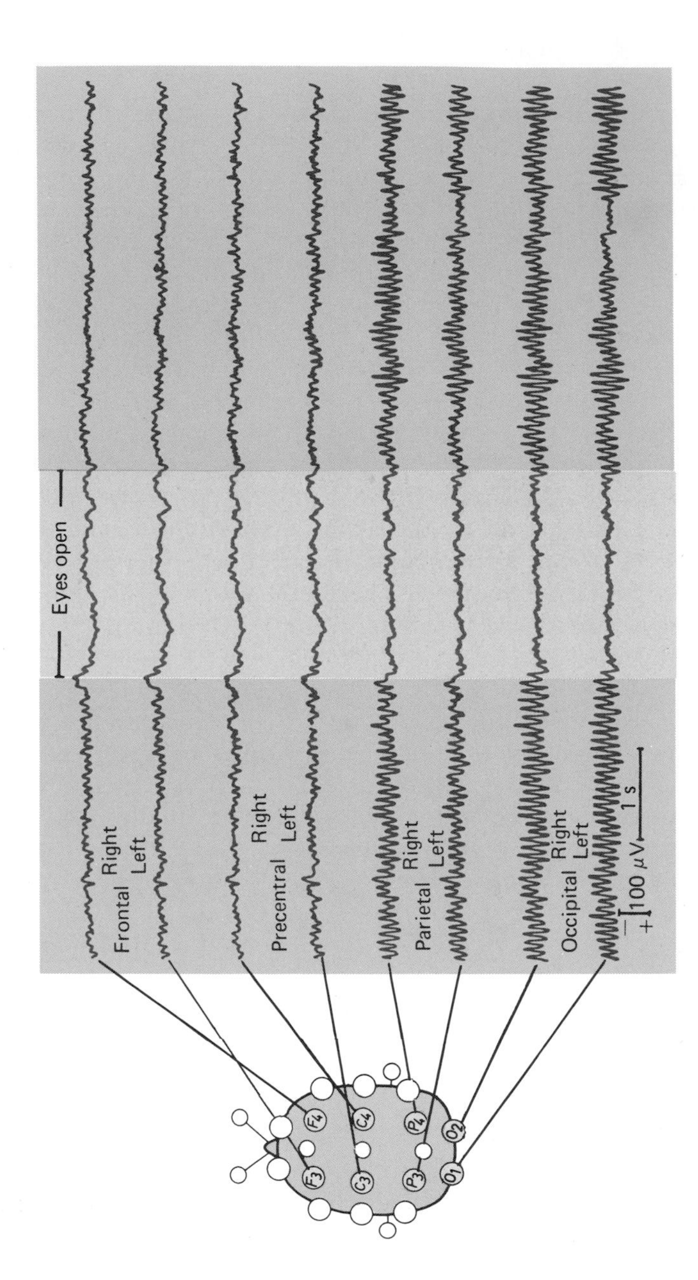

Fig. 9-5. Normal EEG of a resting, awake human. Simultaneous eight-channel unipolar recording from the sites on the skull indicated in the sketch at the left. An electrode was attached to each earlobe; the two together constituted the reference electrode. Opening the eyes *(center)* blocks the alpha rhythm. Modified from Richard Jung.

phase); for this reason, such a recording is also called a *synchronized EEG*.

When the eyes are opened (Fig. 9-5, middle part of the recording) or in the presence of other sensory stimuli or during mental activity, the alpha waves disappear. This disappearance is referred to as *alpha blocking*. In place of the alpha waves, there appear *beta waves* of higher frequency (14 to 30 Hz, average 20 Hz) and smaller amplitude. The EEG also becomes more irregular, and the recordings from different sites vary widely in amplitude, frequency, and phase; the *EEG is desynchronized*.

Two other important features of the EEG are slow, high amplitude waves—the *theta waves* (4 to 7 Hz, average 6 Hz) and the *delta waves* (0.3 to 3.5 Hz, average 3 Hz). Neither of these normally occurs in a waking adult. But they are observed, as will be described below (see Fig. 9-9), during sleep.

Apart from these two factors—recording site and degree of wakefulness—the frequency and amplitude of the EEG also depend very much upon the species of animal and its age. For example, in humans during childhood and adolescence, the EEG is distinctly slower and less regular than in adults; in young subjects, theta and delta waves can be recorded even in the waking state.

Which electrical processes in the cerebral cortex are responsible for the *generation of the EEG waves*? We know, for example, that there are conducted action potentials in the cortical neurons, and that there are also local, "slow-potential" fluctuations. In experiments on animals, the relationship of the EEG to these has been studied by recording both the EEG and the activity of single cortical neurons. The results show that the EEG apparently reflects chiefly the slow changes in the membrane potentials of cortical neurons, particularly *excitatory and inhibitory postsynaptic potentials* (*EPSPs and IPSPs*). Under normal conditions, little or no contribution to the EEG is made by the impulses conducted along neurons (or by cortical glial cells).

The electrodes used to record the EEG are relatively far from the sources of the EEG currents in the cortex (the movements of ions across the cell membrane during EPSPs and IPSPs; see Chapter 3). Accordingly, the amplitude of the EEG is smaller by a factor of 100 to 1,000 than that of the membrane potentials recorded intracellularly (compare the scales of the ordinates in Figs. 9-5 and 9-9 with those in Figs. 3-10 and 3-11). But if the EEG is recorded directly from the surface of the cortex (*electrocorticogram*) during animal experiments or brain surgery on humans, it is about ten times larger than when recorded from the scalp. In both cases, of course, the EEG reflects simultaneous potentials from a large population of nerve cells. It has been estimated that an electrode with a surface area of 1 mm^2 in direct contact with the cortical surface samples the activities of about 100,000 neurons, down to a depth of about 0.5 mm. When the recording is done with the skull intact, about ten times as many neurons contribute to it. For this reason alone it is understandable that the large amplitude waves that appear in the EEG require synchronous activity of a fairly large percentage of the neurons under the electrode.

This synchrony—the *rhythmic activity of the cortex,* and the alpha rhythm in particular—arises not in the cortex itself but in the thalamus. It is transmitted from thalamus to cortex, as the following experiments show. Interruption of the thalamocortical pathways, or removal of the thalamus, results in a disappearance of the alpha waves. On the other hand, the *rhythmic activity of the thalamus* survives transection of the thalamocortical connections or removal of the cortex (decortication). The rhythmic activity of the thalamus in turn is modified by inputs to the thalamus. Brainstem *reticular-formation structures* in particular have both synchronizing and desynchronizing effects—that is, they act to enhance or suppress, respectively, the rhythmicity of the thalamic potentials. This situation is discussed in more detail in Sec. 9.2, in the context of the waking—sleeping cycle.

The *clinical applications of the EEG* (chief among which is its great value in diagnosing seizures) will not be treated here. Generalized extinction of the EEG (a "flat" or isoelectric EEG) is currently becoming widely accepted as a criterion of death in cases of doubt. That is, when modern techniques of resuscitation succeed in bringing a patient out of circulatory and respiratory failure with no sign of a return of consciousness or of spontaneous respiration, one may well suspect that the cerebral cortex and brainstem have been irreversibly damaged by ischemia (interruption of circulation). In such cases of *brain death,* particularly those occurring in young people as a result of an accident, other organs that can remain functional longer without perfusion and hence without an oxygen supply (for example, the kidneys and heart) can in certain circumstances be used for *organ transplants.*

Evoked Potentials. If a peripheral afferent nerve (for example, a cutaneous nerve; Fig. 9-6A) or sense organ is stimulated at above-threshold intensities, potential fluctuations can be recorded from the associated sensory cortex after a short delay (about 10 ms). These are called evoked potentials (Figs. 9-6B and C). The first, positive potential change is called the *primary evoked potential* (Fig. 9-6B). This is found only in a narrowly circumscribed region of the cortex, the cortical projection area of the peripheral stimulus site (in the case of the cutaneous nerve, the corresponding somatotopic area of the postcen-

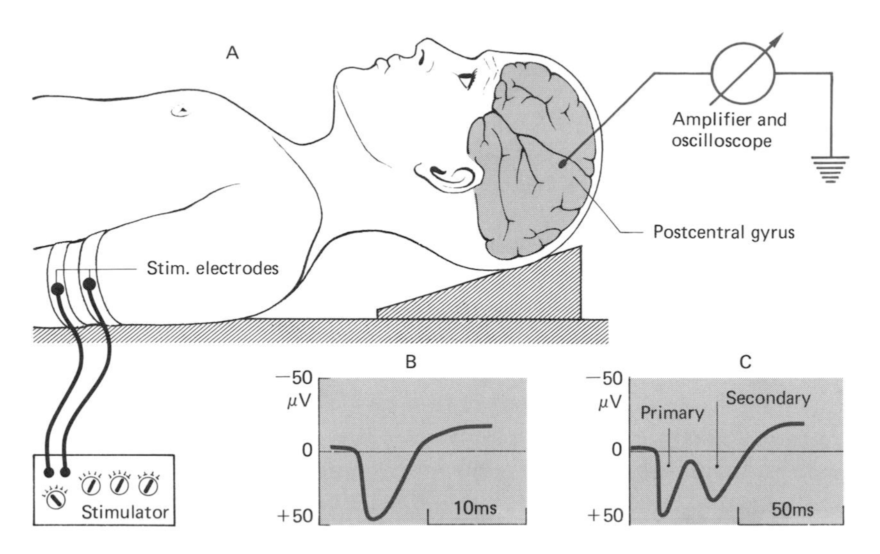

Fig. 9-6. Elicitation and recording of evoked potentials from the human brain. **A,** Experimental arrangement; the stimulus shown is electrical stimulation of the skin, but other stimuli (for example, mechanical or thermal) can be used. The potentials are recorded by an EEG electrode on the scalp. **B,** Primary evoked potential from the corresponding projection field in the postcentral gyrus. **C,** primary evoked and subsequent secondary evoked potential. Note the different time scales in **B** and **C**.

tral gyrus). The later response, which follows just after the primary potential and lasts appreciably longer (Fig. 9-6C), is called the *secondary evoked potential.* This potential is recorded from a more extensive region of the cortex.

With respect to the *mechanism* underlying evoked potentials, most neurophysiologists agree that these (like the EEG waves) reflect primarily *slow synaptic activity* rather than the impulse activity of neurons. Here, again, the electrode records *summed potentials* due to the extracellular currents of many neurons.

Evoked potentials can be recorded not only after stimulation of peripheral nerves or receptors, but also in response to stimulation of central tracts, nuclei, or cortical areas. Measurement of evoked potentials is thus an *important electrophysiologic method* for studying the connections between peripheral and central structures and among different central structures. By averaging the responses to repeated stimuli (a process facilitated by modern computing devices), it is often possible to detect even very small evoked potentials superimposed on the EEG waves. A particularly elegant example of the application of this method is represented by the evoked readiness potentials associated with voluntary movement, illustrated in Fig. 6-19. Such *averaged evoked potentials* have also found clinical application. For example, recording of the evoked potentials over the auditory cortex in response to sound stimuli is used to diagnose certain hearing difficulties in children; the evoked potentials provide an objective measure of the type of deficiency and its progression as the child grows.

Brain Activity, Metabolism, and Blood Flow. A person at rest consumes about 250 ml oxygen/min. The brain accounts for a disproportionately large fraction of this for an organ of its weight, ca. 20%. That is, ca. 50 ml oxygen/min is used for the metabolism of the neurons and glia of the brain. The cerebral cortex has the highest requirement, ca. 8 ml oxygen/100 g tissue per minute; the white matter below it has been found to consume only about 1 ml O_2/100 g/min. The permanently high oxygen requirement of the cerebral cortex is also reflected in the fact that an interruption of oxygen transport (i.e., of blood circulation) causes unconsciousness with only 8–12 s (cf. p. 7).

Superimposed on this life-long high baseline oxygen consumption are additional, function-related demands. Whenever a *particular region of cortex is active,* within seconds the local oxygen consumption increases, and as a result the *local blood flow increases* simultaneously. The increase in blood flow can be measured by the method diagrammed in Fig. 9-7A, developed by D.H. Ingvar and N.A. Lassen. A small amount of radioactive xenon gas (^{133}Xe), which is completely harmless, is injected into the carotid artery. Its arrival in various regions of the brain is detected by Geiger counters (as many as 254 of them) mounted on the side of the head. The radiation intensity detected is directly dependent on the local blood flow, which can be calculated by computer from the total oxygen consumption of the brain and the radiation distribution.

Measurements of this type have been done on healthy volunteer subjects; the results are summarized in the diagram of Fig. 9-7B. In a person at rest, with a typical alpha-wave EEG, the rate of blood flow through the frontal regions of the brain is distinctly higher than elsewhere. Nonpainful stimuli to the skin of the opposite (in this case the right) hand cause little change in the pattern of blood flow (diagram labeled "touch"). With slightly painful stimuli ("pain") the total blood flow (percentage written above each brain outline) clearly rises, with no fundamental change in the distribution of the maxima and minima. Voluntary rhythmic opening and closing of the opposite hand ("hand movement") also raises total blood flow, with a simultaneous rise in flow through the somatosensory postcentral gyrus and the adjacent parts of the parietal cortex. Speaking and reading bring about a

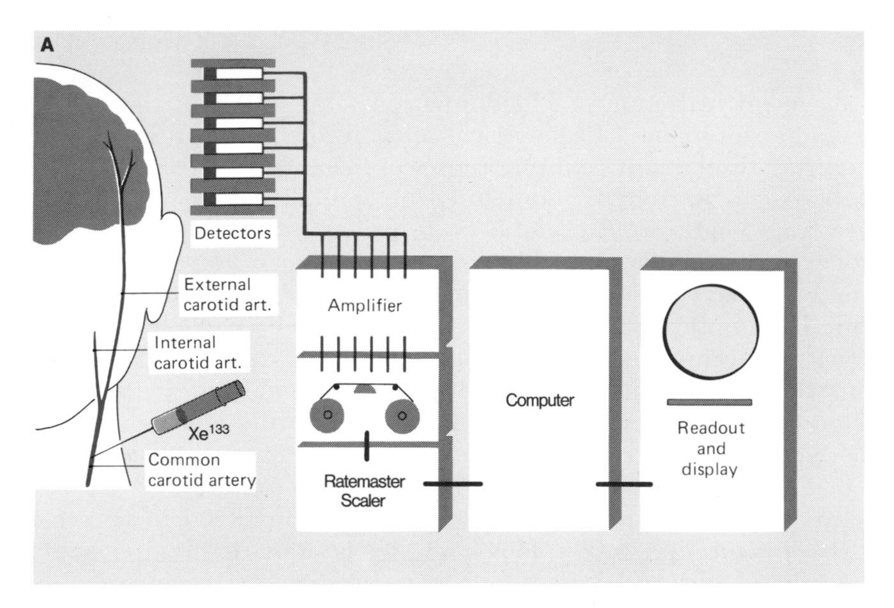

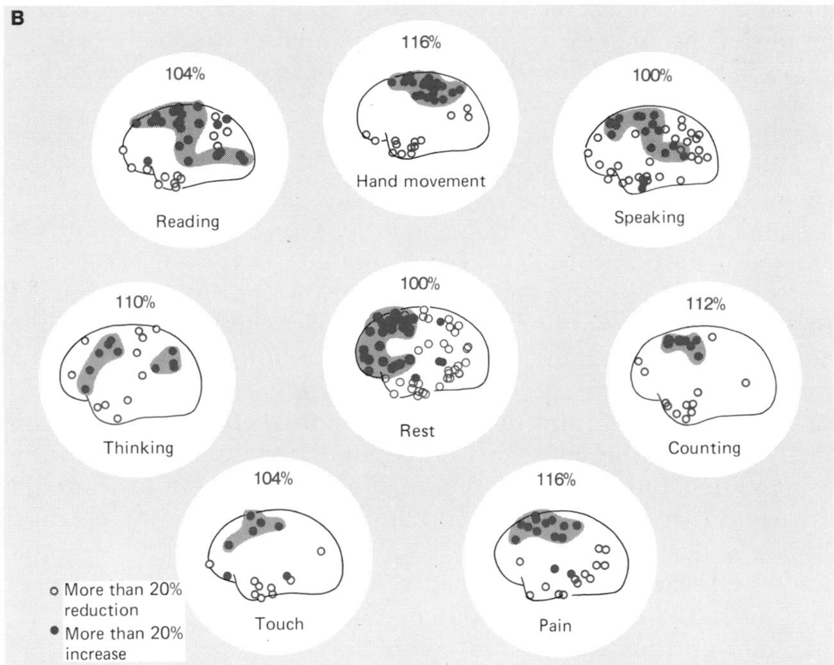

Fig. 9-7. Measurement of regional blood flow through the brain by intraarterial injection of radioactive xenon (^{133}Xe). **A.** Diagram of the method. **B.** Maxima and minima of the regional blood flow on the speech-dominant (left) side, at rest and during seven different types of brain activity. The total blood flow through the resting brain was designated 100%. Only regions differing from the resting flow by over 20%, whether more (*filled red circles*) or less (*open red circles*), are indicated. Measurements of D.H. Ingvar and coworkers.

Z-shaped distribution of blood-flow maxima, which in the case of reading extends back to the visual areas of the occipital lobe. Tests requiring thinking or counting cause an increase in total blood flow, with maxima in front of and behind the central sulcus.

In other words, *any special activity of the brain,* whether sensory, motor, or a particular form of thinking, produces a change in either the total blood flow through the brain, the distribution of blood flow, or both. The implication is that the altered, regionally elevated neuronal activity is accompanied by *increased metabolic activity of the neurons;* the acid metabolic products thereby released cause local vasodilation and hence increased blood flow. The opposite effect also seems to occur. Without a continual supply of energy, immediately increased to meet the demands of higher activity, neurons cannot function. This is true of all neurons, even those intimately involved in the *most abstract mental processes.* This inference is supported by Ingvar and co-workers' observations of unconscious, comatose, severely demented or schizophrenic patients; they found a remarkable correlation of deficits in sensory, motor, and mental performance with decreases in total blood flow through the brain as well as in flow through the associated brain region.

The conclusion is obvious: all conscious and unconscious mental achievements of our brain are possible only if the neuronal networks responsible for these achievements are in operating condition. This statement in itself says nothing about the form in which "mind" and neuronal activity are linked to one another. At present, there is no decisive experimental means of discovering this link. But from the experiments just described, it is clear that measurable—that is, determinable from the actions or statements of a human subject—*mental performance* is always accompanied by certain highly specific forms of *neuronal activity* and *does not occur in their absence.* Patients sometimes describe conscious perceptions remembered from times during which, although their neuronal activity was not measured, an absence of neuronal activity could be inferred—for example, when a "clinically dead" person is subsequently resuscitated. Such reports should be regarded very skeptically. Many factors can cause perceptions not based on reality (hallucinations) in a damaged brain as it sinks into and awakes from a completely nonfunctional state. The striking similarity among the various descriptions of a "life after death" reported by such patients probably results from the fact that all brains are similar in structure and, when their energy supply is cut off and then restored soon enough, resume functioning in a similar way, just as a television set lights up and goes out with typical patterns when the electricity is turned on and off. Evidence in favor of this interpretation is also provided by the hallucinatory disturbances of vision (the so-called aura) accompanying certain forms of migraine and epilepsy; these, too, are quite uniform on the whole. It is well established that in cases of ophthalmic migraine the patient often hallucinates zig-zag lines reminiscent of the outline of a fortress. Some mystics, for example, Saint Hildegard of Bingen (1098–1179), evidently took these "fortification illusions" to be visions of the heavenly city.

Q 9.1 The precentral gyrus in humans is located
a. in the frontal lobe.
b. in the parietal lobe.
c. in the temporal lobe.
d. in the occipital lobe.
e. in the cerebellum.

Q 9.2 Those nerve fibers that link the two hemispheres are called
a. projection fibers.
b. association fibers.
c. commissural fibers.
d. mossy fibers.
e. climbing fibers.

Q 9.3 In a healthy adult human, awake but relaxed (with eyes closed), the following EEG waves are recorded over the occipital lobe:
a. delta waves (0.3 to 3.5 Hz).
b. theta waves (4 to 7 Hz).
c. alpha waves (8 to 13 Hz).
d. beta waves (14 to 30 Hz).

Q 9.4 The electroencephalogram usually exhibits rhythmic potential fluctuations. The underlying rhythmic neuronal activity of the cortex has its pacemaker chiefly in
a. the cortex itself.
b. the thalamus.
c. the reticular formation.
d. the cerebellum.
e. the basal ganglia.

Q 9.5 The amplitudes of the EEG waves are of the order of
a. 100 to 1,000 mV.
b. 10 to 100 mV.
c. 1 to 10 mV.
d. 10 to 100 μV.
e. 1 to 10 μV.

9.2 Waking, Sleeping, Dreaming

Circadian Periodicity in Humans; the Basis of the Waking–Sleeping Cycle. Almost all organisms, from protozoans to humans, display various rhythmic changes of state of their organs and their behavior.

These are usually coupled to the 24-hr period of the earth's rotation; as a result, it was once a widespread belief that diurnal rhythms in animals and man were passive responses of the organism to one or more periodically changing variables in the environment. But countless experiments have now shown conclusively that such cyclic behavior continues even when all known environmental factors are held constant. The basis of the periodicity is thus not the environment, but certain processes within the organism (endogenous processes). The actual pacemaker processes—the exact nature of which is not yet known—are called *biological clocks.*

The free-running *period of the biological clock,* measured under constant experimental conditions, is usually somewhat shorter or longer than 24 hr. That is, it corresponds only approximately (*circa*) to the natural duration of a day (*dies*); for this reason such processes are called *circadian rhythms.* In natural conditions, circadian periodicity is synchronized with the 24-hr rhythm of the planet by external entraining signals (also designated by the German word *Zeitgeber*). Among humans, *social factors are the most important entraining signals* for synchronization of the circadian rhythm. By comparison, other factors, such as the light–dark alternation between day and night, play only a modest role. When the rhythm of one's work differs from that of the general social environment, as happens in the case of shift workers, the periodicity of the social environment is usually retained. One consequence is that performance falls off unavoidably at certain times (especially in the early hours of the morning), and at these times the danger of accidents is particularly great.

Diurnal rhythms in humans have been shown to affect more than 100 measured variables associated with particular organs and functions. A well-known example is the daily fluctuation in body temperature, with a minimum in the early morning and a maximum, about 1 to 1.5°C higher, in the evening. But the most striking diurnal fluctuation is the waking–sleeping cycle. It is not surprising that the many adjustments the organism makes when entering the state of sleep—the fall in body temperature, heartbeat rate, and respiratory rate, for example (Fig. 9-10)—were once thought to be causally related to sleeping. But the daily periodicity in these and many other physical and psychologic variables is maintained even during sleep deprivation. Humans (like other highly organized metazoans) therefore appear to have a *number of biologic clocks* (oscillators), synchronized partly by one another and partly by external entraining signals.

When a human is *isolated from the environment* (for example, in underground bunkers and caves) his circadian rhythm becomes *freerunning;* its duration changes, in most experimental subjects

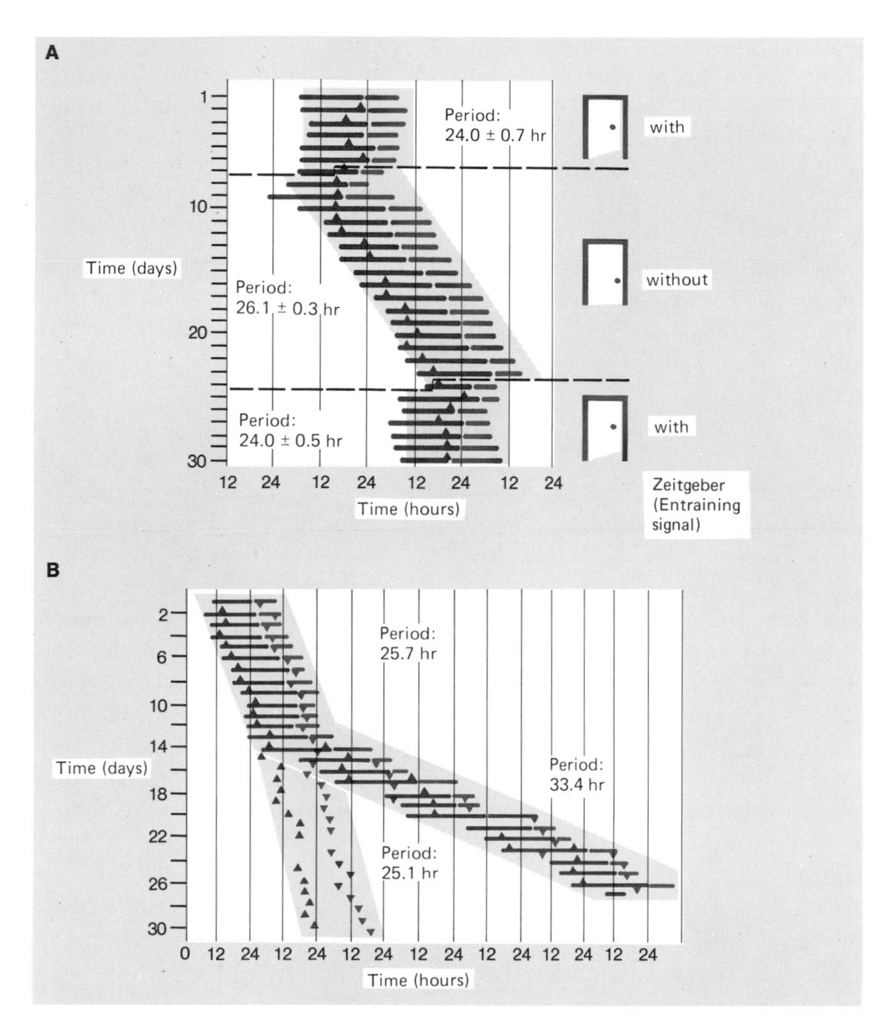

Fig. 9-8. Human circadian periodicity. **A.** Rhythm of waking (*red parts of bars*) and sleeping (*black parts of bars*) of a subject in the isolation room with the door open (i.e., with a social Zeitgeber or entraining signal) and closed (no entraining signal). The triangles mark the time of highest body temperature. With the door open the period was always exactly 24 hr (mean daily deviation ±0.7 and ±0.5 hr), but in isolation it was 26.1 ± 0.3 hr. **B.** Activity rhythm of a subject isolated in a bunker, whose temperature rhythm (maxima = *red upward-pointing triangles*, minima = *gray downward-pointing triangles*) became uncoupled from the waking-sleeping rhythm on the 15th day, and thereafter continued with a period of 25.1 hr. On this day the waking-sleeping rhythm (activity rhythm) jumped, for unknown reasons, to a period of 33.4 hr. Measurements of Prof. J. Aschoff, Seewiesen, and co-workers.

becoming somewhat longer than 24 hr (Fig. 9-8A). Under such conditions, measurements of period can even suggest a measure of independence of individual oscillators. For example, the body-temperature maxima (red triangles in Fig. 9-8A) occurred at the usual time, near the end of the waking part of the cycle, when the subject was synchronized to 24 hr, but in the free-running condition they shifted to its beginning. This finding in itself indicates that the normal daily fluctuation in temperature is not a consequence of the rhythm of activity and rest. On the other hand, it is consistent with the view expressed in the preceding paragraph, that the two functions are controlled by two different clocks coupled together. This hypothesis is further supported by the observation that the two rhythms can become completely uncoupled. In the experiment illustrated in Fig. 9-8B, on the 15th day the subject, for unknown reasons, suddenly extended the activity rhythm to 33.4 hr. The "temperature clock," evidently less flexible than the "activity clock," could not follow this extremely long period; it separated from the sleeping-waking rhythm, retaining a period of about 15 hr. Such cases of entirely internal desynchronization have also been described by French researchers. They found that subjects living in caves, completely isolated from the outside world for weeks and months, sometimes exhibited sleeping-waking rhythms with 48-hr periodicity. These *bicircadian rhythms,* consisting of 14 hr of sleep and 34 hr of waking, were experienced by the subjects as entirely normal, as though they were 24-hr days. Under these conditions the autonomic functions (body temperature, heart rate, respiratory frequency, etc.) were fully uncoupled and continued with the original period of 25 hr.

If there is a *single disruption of the rhythm of the external entraining signal*—for example, if the day is shortened by a flight toward the east or lengthened by a westward flight—the circadian systems often go through several cycles before reassuming their normal phase with respect to the local entraining signal. The various functions differ in the time required for resynchronization. This disturbance surely contributes to "jet-lag"—the temporary deterioration of performance people experience after long distance flights.

The *biologic significance of circadian rhythms* has generally been underestimated. In the diagnosis and treatment of diseases, more attention should be paid to the diurnal fluctuations of nearly all organ functions. Circadian periodicity is inherited and should be regarded as a phylogenetic adaptation to the temporal structure of our environment. For practical purposes, it amounts to an "internal copy" of the schedule of events in the outside world. With the aid of this internal schedule, the organism can adjust itself at the right time—that is, in advance—to predictable changes in the environment. For example, the body temperature and the blood-plasma level of many hormones begin to rise long before a person wakes up. Similarly, for animals in general circadian periodicity offers considerable advantage. Minimally, it facilitates the performance of certain actions at particular times of day; at the other extreme, the "internal clock" can be used for genuine time measurement—an ability required, for example, by animals that use the sun as a compass for in-flight orientation. Viewed in this way, the *waking-sleeping rhythm* is not the cause but rather one of the *outward manifestations of the endogenous circadian periodicity.*

Human Sleep. A waking human is in active contact with the environment, responding to stimuli with appropriate actions. During sleep, contact with the environment is to a great extent interrupted. But just as the degree to which attention is directed outward can vary greatly in a waking person, different *stages of sleep* can be distinguished. The simplest and oldest *measure of the depth of sleep* is the *intensity of the stimulus required to awaken* the subject. The deeper the sleep, the higher the threshold to such stimuli.

Today, the *EEG* is usually used to monitor the depth of sleep. On the basis of such recordings, four or five stages (Fig. 9-9) are distinguished by criteria that have been standardized by general agreement. On the whole, the EEG becomes progressively slower (more "synchronized") as sleep deepens, and there appear special groups of waves such as sleep spindles and K complexes (Fig. 9-9A, Stages C and D). The deepest state of sleep (Stage E) is characterized by slow, large amplitude delta waves.

During the *course of a night,* a person passes *several times through the various stages of sleep*—three to five times on the average (Fig. 9-9B, 9-10). In general, the maximal depth of sleep reached in each cycle is reduced toward morning, so that in the later cycles Stage E is no longer reached. Some of the numerous *autonomic functions with circadian periodicity* (for example, body temperature) are not affected by these rhythmic fluctuations in depth of sleep; in others, phasic fluctuations are superimposed on the diurnal rhythms (for example, heartbeat rate and respiration in Fig. 9-10). These phasic components appear particularly in association with Stage B during the night. Other responses (for example, penis erection; Fig. 9-10) can be observed only during these repeated B stages.

Another remarkable correlate of the repeated B stages is measurable in the behavior of the motor system. During Stage B, as in deep sleep, the tonus of the peripheral musculature nearly disappears (see

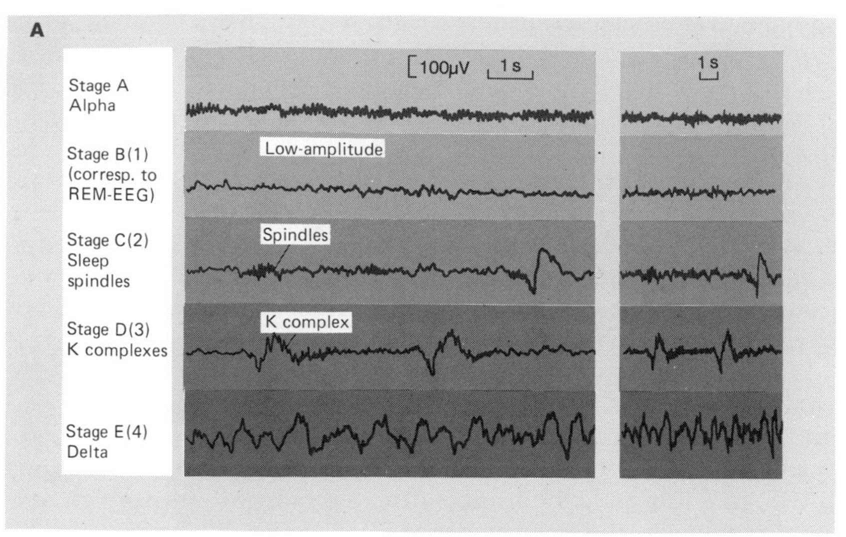
A
100µV
1 s
1 s
Stage A
Alpha
Stage B(1)
(corresp. to
REM-EEG)
Low-amplitude
Stage C(2)
Sleep
spindles
Spindles
Stage D(3)
K complexes
K complex
Stage E(4)
Delta

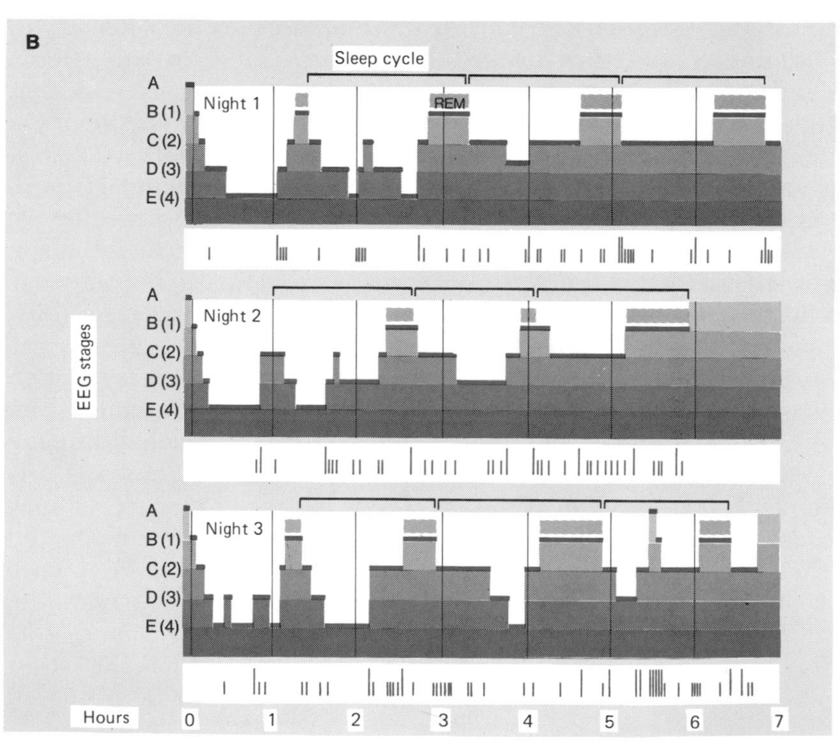
B
Sleep cycle
EEG stages
A
B(1)
C(2)
D(3)
E(4)
Night 1
REM
Night 2
Night 3
Hours
0
1
2
3
4
5
6
7

electromyogram in Fig. 9-10). By contrast, however, there are *salvos of rapid eye movement* (see electrooculogram in Fig. 9-9). These are so characteristic of this stage that is is called the *REM* (rapid eye movement) *stage* of sleep. The threshold for an awakening stimulus is about as high in REM sleep as in deep sleep, whereas the EEG is like that recorded from a person just falling asleep. Because of this conjunction of symptoms, REM sleep has also been called *paradoxic or desynchronized sleep*. All the other stages of sleep are sometimes grouped under the heading *NREM sleep* (non-REM sleep). As Fig. 9-10 shows, REM stages appear about every 1.5 hr during normal sleep. They last on the average 20 min, and their duration increases during the course of the night.

Over the human lifespan, the time spent sleeping per night decreases. Moreover, the *proportion of REM sleep* during the total period becomes considerably *smaller*. These trends are illustrated in Fig. 9-11. The sequence and length of the individual stages of sleep during a night (not shown in Fig. 9-11) are also distinctly different in infants and small children than in later life. The high proportion of REM sleep in infants and small children has suggested that these periods of increased neuronal activity (with desynchronized EEG resembling that during attentiveness; see alpha blocking in Fig. 9-5) are important in maturation of the brain of the young human; the external stimuli affecting these individuals are relatively few. Further evidence for this interpretation is the fact that dreams (that is, a special form of conscious activity of the brain) apparently occur predominantly during REM sleep.

Fig. 9-9. Classification of the stages of sleep in humans on the basis of the EEG (**A**), and the occurrence of these stages during a night (**B**). Two types of nomenclature in current use (*A–E* and *W*, *1–4*) are given. The depth of sleep increases from top to bottom. Stage A or W: relaxed sleepiness just before going to sleep. Alpha rhythm dominates. Stage B or 1: falling asleep; alpha rhythm disappears and is replaced by shallow theta waves. Stage C or 2: light sleep. Further decrease in frequency, to delta waves, between which 12- to 15-Hz sleep spindles appear. Stage D or 3: moderately deep sleep, delta waves and K complexes. Stage E or 4: deep sleep, EEG dominated by large, slow delta waves. With regard to the EEG, the REM stage corresponds approximately to stage B. There are no sharp transitions between the various stages. The EEG in **A** was recorded at two paper speeds (see time scales). **B** shows the cyclic fluctuations in depth of sleep of a single subject on three consecutive nights, as indicated by the EEG. The REM stages are indicated by the *light-red bars*. Each of the square brackets at the top encloses a complete sleep cycle. The vertical lines below the graphs represent the number and strength of body movements. Part **A** from Snyder and Scott 1972; **B** modified from Dement and Kleitman 1957.

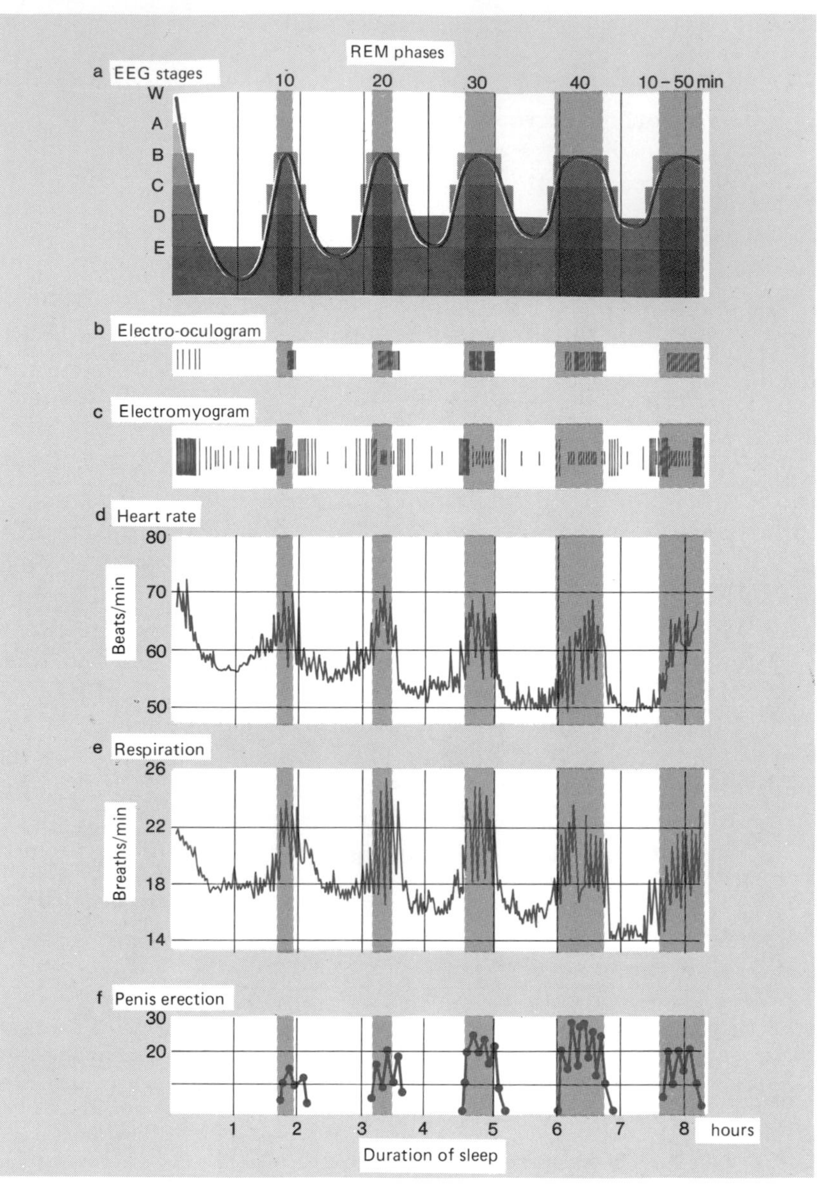

Fig. 9-10. Passage through the stages of sleep during a night, correlated with variations in certain autonomic functions. Shown are averages, highly schematic. **a.** The *red curve* shows the changes in depth of sleep according to the EEG, with the REM phases (here and in the graphs **b–f** below) marked by the *light red* bars. As the data from the electro-oculogram (**b**) show, rapid eye movements (symbolized by *vertical lines*) occur only during the REM phases. Only when the subject is falling asleep, on the left in **b**, are there a few extra eye movements. The data from the electromyogram of the neck muscles are shown in **c.** The heart rate is given in **d,** in pulse-beats/min. It increases transiently during the REM phases, as does the respiratory frequency, shown in **e.** In men, penis erection regularly occurs during the REM phases (**f**). Modified from Jovanovic 1971.

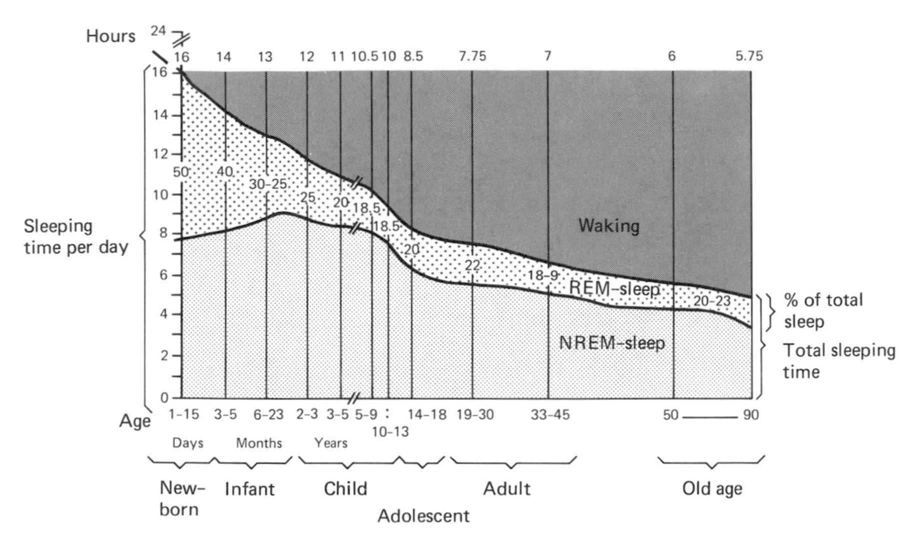

Fig. 9-11. Average times spent in waking and sleeping, showing the proportions of REM and NREM sleep, during a human lifetime. Apart from the reduction in total sleeping time, the most notable change as a child grows is the marked reduction in the duration of REM sleep. (Modified from Roffwarg, Muzio, and Dement 1966).

Physiologic Correlates of Dreams. If children or adults are awakened during or immediately after observation of a REM stage, they are considerably more likely to report having just dreamed than when they are awakened from NREM sleep. All who have made such surveys have found a high percentage (60 to 90%) of *reports of dreams upon awakening from REM sleep,* whereas the percentages of dream reports by people awakened from NREM sleep in the various studies were much lower on the whole and had a larger scatter (1 to 74%). Taken together, the results to date indicate that, while it is not justified to equate REM sleep simply with "dreaming sleep," dreams do occur very frequently or most commonly when a REM stage is in progress. With this reservation, the other changes in organ function accompanying REM sleep (see Fig. 9-10) can also be regarded as physiologic correlates of dreams.

The inference of a correlation between REM sleep and dreams is methodologically questionable in that it relies on subjective reports of dreaming, with all the potential sources of error in such data. Even the way the experimenter questions the subject can have a considerable effect on the statements that are made. Furthermore, other observations do not fit the notion of a strict correlation between dreaming and REM sleep. For example, newborn babies also have phases of sleep with rapid eye movements (Fig. 9-11); here the early state of development of the cerebral cortex would suggest that (visual) dream experiences are unlikely. The same is true of other newborn mammals. Second, salvos of rapid eye movements have been observed in humans blind from birth, even though these people have had no experience of vision. Finally, even chronically decerebrate cats go through phases of sleep in which all the peripheral signs of REM phases are exhibited. In view of these findings, it appears that REM sleep, with all its symptoms, is not a secondary phenomenon, the consequence of dreaming; rather, it may be more appropriate to say that once the brain has matured, *dreams occur primarily during the REM phases.*

If humans are awakened whenever a REM phase begins—that is, are put under conditions of *REM-sleep deprivation*—the proportion of REM phases in the subsequent nights of undisturbed sleep rises by about 50 to 80%. When REM-sleep deprivation is greatly prolonged (for up to 16 nights), a procedure always involving a considerable shortening of the total sleeping time, subjects become irritable; they display brief REM phases even with open eyes and during apparently wakeful behavior, and occasionally they suffer from hallucinations and anxiety states. It remains unclear which parts of this syndrome result from sleep deprivation and which from the lack of dreams *per se.* It should be noted, however, that no severe, long-lasting mental or physical damage has been found to result in subjects deprived of sleep, whether it is REM sleep, NREM sleep, or a balanced combination of the two.

Causes of the Waking-Sleeping Cycle. A routine part of the daily life of both man and animals is the periodic recurrence of an *irresistible need to sleep.* This observation has suggested the plausible idea that tiredness and sleep result primarily from the periodic accumulation, depletion, or specific production of metabolic substances that circulate in the blood and must be eliminated, decomposed, or replaced during sleep. This *chemical theory of waking and sleeping* is certainly not correct in this simple formulation. Evidence against the theory, apart from the failure to demonstrate the existence of such substances, is provided above all by Siamese twins with shared circulation but separate nervous systems. There is no reciprocal influence between the sleeping–waking cycles of such twins! Moreover, in experiments on animals in which the cerebrum is separated from the rest of the CNS, or in which the two hemispheres have been separated by a sagittal section, symptoms of sleep and wakefulness appear independently in the disconnected parts of the brain.

Other attempts to understand the waking–sleeping cycle have been based on the *differences in neuronal activity* between a waking and a sleeping brain; efforts have been made to determine the processes that bring about the transition between the sleeping and waking states. It should be kept in mind that sleep has by no means been shown to be a sort of "rest" of the brain—that is, the simple absence of the activity patterns associated with waking. On the contrary, the neurophysiologic data clearly demonstrate that the complexity of neuronal activity during the different stages of sleep resembles that in the waking state. For example, EEG recordings (see Figs. 9-5 and 9-9) are no less varied during the various sleep stages than when the subject is awake. The existence of dreams also demonstrates that the brain does not simply "turn off" during sleep. Sleep reflects the presence of *alternative modes of operation of the brain,* not the absence of coordinated neuronal activity; the latter does, however, occur in deep narcosis or coma.

The brainstem is a central element in the control of sleeping and waking. High frequency electrical stimulation of the reticular formation of the brainstem produces arousal of a sleeping animal, a response accompanied by the desynchronization of the EEG. Conversely, destruction of this area or of the pathways ascending from it to the cerebrum (which, to distinguish them from the sensory, specific pathways, are often called nonspecific projections) leaves the experimental animal in a coma. These findings led to the formulation of the *reticular theory of waking and sleeping.* This theory assigns to the reticular formation the function of producing an excitatory level necessary for the waking state by means of ascending activating impulses. The term *ascending reticular activating system,* abbreviated *ARAS,* has been employed. Relatively large fluctuations in the intensity of the ascending reticular activation are thought to be responsible for the transition from sleeping to waking and back. Subtle behavioral changes during the waking state (for example, alteration in the degree of attentiveness) are ascribed to small fluctuations in activity of the ARAS. This simplified view of the reticular formation as the critical center for wakefulness is opposed by a number of recent findings, so that the theory—at least in its original form—can hardly be retained. For example, even a chronically isolated brain lacking a reticular formation exhibits a sleeping–waking rhythm; the reticular formation is, therefore, not essential for sleep and waking.

The central role of the brainstem in the sleep–waking phenomenon is also emphasized in neuropharmacologic research; liberation of the monoaminergic transmitters serotonin (5-HT) and noradrenaline by certain brainstem nuclei occurs at regular times during the sleeping–waking cycle. As a result, a *biochemical theory of waking and sleeping* is beginning to take shape. The most important observations in animal experiments are as follows. (a) Neurons in the *raphe nuclei*, a group of cells in the brainstem, contain large amounts of *serotonin*. When this serotonin is depleted (for example, by poisoning the reactions by which it is synthesized), there is severe insomnia; both REM and NREM sleep are affected. The insomnia can be alleviated by administration of 5-hydroxytryptophan, the precursor of serotonin (the latter cannot cross the blood–brain barrier). (b) Neurons of the *locus coeruleus*, another cell group in the brainstem, contain large amounts of *noradrenaline*. Bilateral destruction of the locus coeruleus eliminates REM sleep entirely, but has no effect on NREM sleep. (c) If the stores of both serotonin and noradrenalin are exhausted, by administration of reserpine, both sorts of sleep are eliminated [as would be expected from result (a)]; the animal suffers extreme insomnia. The subsequent administration of 5-hydroxytryptophan restores NREM sleep but not REM sleep, which is consistent with observation (b). These findings indicate that the two substances are intimately involved in sleep, serotonin with NREM sleep and noradrenaline with REM sleep, and that normally REM sleep is possible only when preceded by NREM sleep (Fig. 9-11).

Studies of humans, however, suggest reversed roles of serotonin and noradrenaline. The amount of REM sleep is said to be larger the higher the serotonin and the lower the noradrenaline level. The reason for this discrepancy is not yet known.

Q 9.6 Synchronization of the circadian rhythms of human organ function with the 24-hr diurnal rhythm is brought about primarily by
a. the light–dark alternation of day and night.
b. the fluctuations of body temperature in the course of a day.
c. the time and duration of periods of work.
d. the dream phases associated with REM sleep.
e. the daily rhythm of life in the social environment.

Q 9.7 Which two of the following are the best measures of the depth of sleep?
a. Measurement of the intensity of a stimulus capable of awakening a subject.
b. Determination of the fluctuations in body temperature during a night.
c. Measurement of the intensity and the frequency of penis erections during sleep.
d. Recording of the electroencephalogram.
e. Measurement of pulse and respiratory rate during sleep.

Q 9.8 The EEG can be used to distinguish four or five stages of sleep in the healthy adult. In the course of a night, the individual normally passes through each of these stages
a. once.

b. twice.
c. three to five times.
d. more than ten times.
e. None of the above is correct.

Q 9.9 Which of the following lists best describes normal human REM sleep?
a. Synchronized EEG, average duration 60 min, rapid eye movements, penis erection.
b. Desynchronized EEG, average duration 20 min, rapid eye movements, large fraction of total sleeping time in infants and small children.
c. Deep-sleep EEG, average duration 1 to 2 min (max. 5 min), rapid eye movements, often accompanied by dreams.
d. Deep-sleep EEG, duration very variable (1 to 60 min), slow eye movements, increased heartbeat and respiratory rates, very low threshold to arousing stimuli.
e. Desynchronized EEG, average duration 20 min, slow eye movements, very high arousal threshold, accompanied by dreams, does not occur in infants.

Q 9.10 In experiments on animals, depletion of the serotonin and noradrenaline stores of the brainstem (by administration of reserpine) produces
a. absence of REM sleep only, with normal NREM sleep.
b. absence of NREM sleep, with much-prolonged REM sleep.
c. perpetual sleep, with regular alternation between REM and NREM phases.
d. shortening of the total duration of sleep to 70 to 90% of normal, primarily by decrease in REM sleep
e. extreme insomnia, with almost complete absence of both NREM and REM sleep.

9.3 Consciousness and Speech: Structural and Functional Prerequisites

Consciousness in Man and Animal. The most striking change of state that we experience daily is the return of consciousness when we awake from sleep. Consciousness is a state which, in all its subtle shadings, can only be appreciated introspectively but is surely the essential fact of human existence. Many attempts to interpret it have been made in both physiologic and psychologic terms; some of these interpretations are in blatant contradiction to others, and the whole topic is still very much in a state of flux. Physiology can contribute to the discussion by specifying the boundary conditions, from the viewpoint of the natural sciences, within which consciousness appears to be possible. Some of the observable aspects of human and animal behavior that can be taken as *indications of the presence of consciousness* are the following:

1. Attentiveness and the ability to direct attention in a purposeful way.

2. The creation and use of abstract ideas, and the ability to express them by words or other symbols.
3. The ability to estimate the significance of an act far in advance, and thus to have expectations and plans.
4. Self-recognition and the recognition of other individuals.
5. The presence of esthetic and ethical values.

These characteristics, of course, differ greatly in their importance as criteria, and some of them are primarily or only to be found in man (for example, ordinary speech). But if they were accepted, at least provisorily, as behavioral characteristics of consciousness, they would imply the interesting proposition that *consciousness occurs in both humans and other animals.*

Not all animals have consciousness as just defined. There can hardly be any doubt that higher vertebrates (birds, mammals) with greatly differentiated nervous systems exhibit at least a few of the characteristics of conscious behavior listed above, but in animals with simpler nervous systems such behavioral traits are either nonexistent or rare and rudimentary. *Hence consciousness is bound to complex neuronal structures and cannot exist apart from these structures.* There is admittedly no sharp dividing line between animals with and without consciousness, as the above considerations may already have made clear. It seems that consciousness develops approximately in parallel with the phylogenetic development of the nervous system. In other words, *in the animal kingdom there are many gradations and widely differing forms of consciousness,* of which the human consciousness undoubtedly represents the most highly differentiated form.

This view—that consciousness requires an appropriately differentiated nervous system—suggests the idea that in the course of evolution (phylogenesis) consciousness in one form or another always appeared when simpler forms of neuronal activity (e.g., reflexes) were no longer adequate to control and direct the organism. If this is so, then the development of consciousness is a necessary evolutionary step, an indispensable means of optimally adapting higher organisms to their environment.

Neuronal Basis of Human Consciousness. As far as human consciousness is concerned, one can as yet make only very simple and quite inadequate statements about its functional prerequisites—that is, about the neuronal activity underlying it. For example, purely from the point of view of neuronal activity, we know that consciousness is associated with an intermediate level like that indicated by a desynchronized, waking EEG. Too little neuronal activity, such as prevails

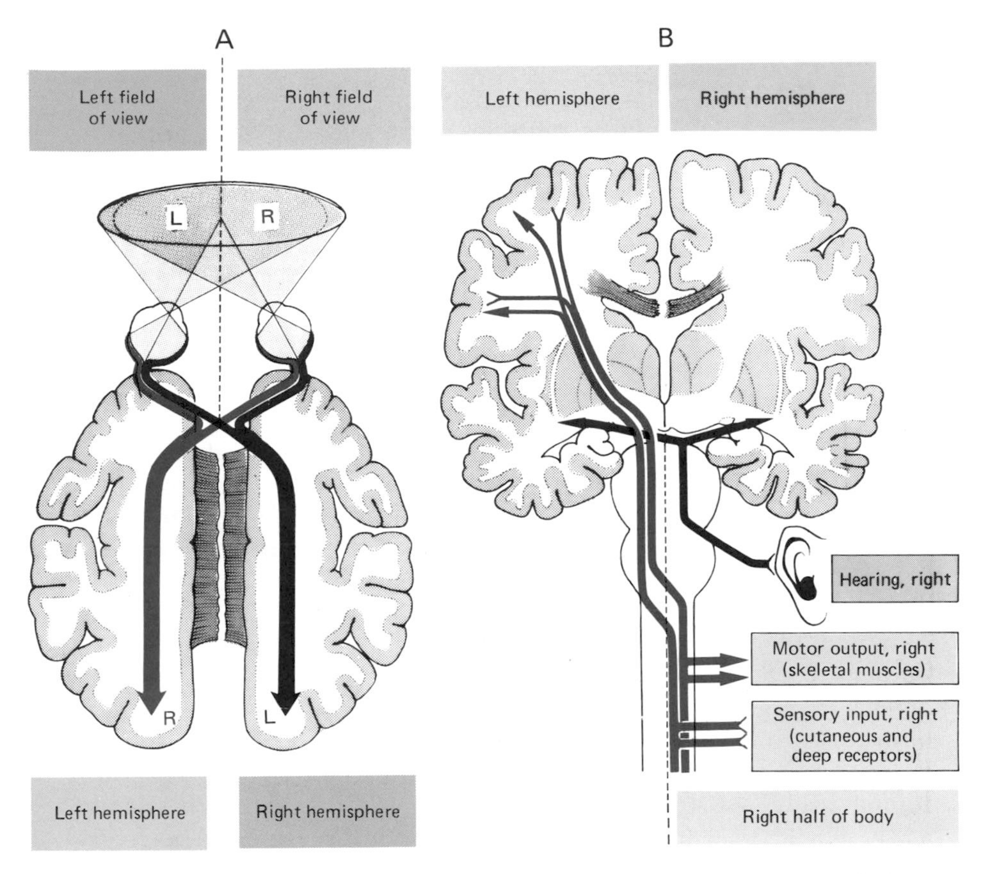

Fig. 9-12. Somatosensory, motor, visual, and auditory connections in split-brain patients. **A,** Horizontal section; **B,** frontal section. The *left* hemisphere makes somatosensory (afferent) and motor (efferent) connections only with the *right* half of the body, and vice versa. The *right* half of the visual field of each eye projects to the visual cortex of the *left* hemisphere and conversely. By contrast, each ear is connected to the auditory cortex of both hemispheres, even in split-brain patients.

under anesthesia or in coma, as well as excessive activity like that measured during an epileptic attack or electroshock, are inconsistent with consciousness. It also seems clear that consciousness is possible only when there is *interaction between cortical and subcortical structures*. Each of these levels alone is incapable of sustaining consciousness.

Recently, as a serendipitous result of a therapeutically required brain operation (transection of the corpus callosum, through which commissural fibers connect the two cerebral hemispheres), *split-brain patients* have given valuable insights into the neuronal bases of human consciousness. In these patients, because of the numerous ascending and descending pathways that cross in the brain (Fig. 9-12), the left half of the cerebrum provides motor and somatic innervation primarily to the right side of the body, whereas the right half of the cerebrum is "responsible" for the left side of the body. Similarly, the special way in which the optic nerves cross in the chiasm results in projection of the right half of the visual field to the left hemisphere and conversely. On the other hand, the central auditory pathways are partly crossed and partly uncrossed, so that each hemisphere receives inputs from both the ipsilateral and the contralateral ear (for a more detailed description of the central sensory pathways, see *Fundamentals of Sensory Physiology*).

The postoperative situation in split-brain patients reflects the above divisions in very subtle and informative ways. In going about their *daily routine,* they are inconspicuous, and there appears to be no intellectual change. At most, one can observe reduced spontaneous activity in the left side of the body (of right-handed people) and a reduction or absence of response to stimuli on the left (for example, bumping into the edge of a table). But special tests designed for such patients, particularly those using experimental arrangements like that shown in Fig. 9-13, have revealed considerable differences in the abilities or repertoires of the two halves of the brain. Such an arrangement enables the experimenter to present visual signals (flashes of light, writing) separately to the right or the left half of the visual field, so that the signals are received only by the left or right (respectively) hemisphere, and the other hemisphere is ignorant of their existence. Moreover, the right or left hand can be used, without visual control, to identify objects by feel, or to write. Here too, in accordance with Fig. 9-12, the right (or left) hand has motor and sensory connections only with the left (or right) hemisphere. The most important results of these experiments are as follows.

If objects are shown only to the right half of the visual field, the split-brain patient can name them (for example, key, pencil) or pick them out from among other objects with the right hand. If words are projected into this half of the visual field, he can read them aloud, write them down, and pick out the appropriate object (again with the right hand). If objects are laid in the right hand, the results are consistent with the above: the patient can name the objects and write their names down. In other words, the patient behaves like a normal subject in this situation.

If, however, objects are shown in the left half of the visual field, the split-brain patient cannot name them. He is able to select them from

Fig. 9-13. The experimental setup used by Sperry and his co-workers to examine split-brain patients. The patient is seated in front of a translucent screen onto which pictures or writing can be projected so that they are visible in the left, the right, or both fields of view. The patient is told to fixate on a dot in the middle of the screen. When the visual stimuli are presented briefly (0.1 s) the direction of gaze does not change during presentation, preventing stimulation of the other field of view. By reaching under the screen, the patient can perform manual tasks without visual control; these are recorded on film. Auditory signals are presented through the earphones and are also heard by the experimenter.

other objects with his left hand, when he is asked to do so. Even then, after his search has been successful, he cannot name the object. Nor can he do so when the object is laid in his left hand. If words are projected into the left visual field he cannot read them aloud, but if the words refer to everyday objects he can select the appropriate object with his left hand (Fig. 9-14). Again, having found the object, he cannot name it. In this experimental situation, then, the patient can carry out certain higher level tasks but cannot express himself verbally or in writing, even when requested to do so. In other words, he behaves as though the acts performed with the help of his right hemisphere had never occurred!

The conclusions that may be drawn from these results are as follows. The *performance of the left hemisphere* cannot be distinguished from that of the intact brain, either from the subjective viewpoint of the patient or on the basis of his everyday behavior and his responses under the test conditions just described. This is true in

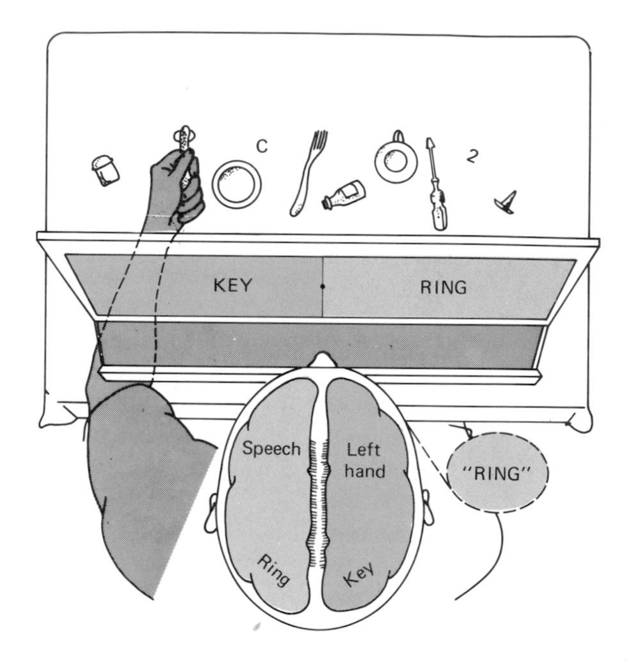

Fig. 9-14. Behavior of a split-brain patient during a test by Sperry and his colleagues. The patient reports (by way of his left, speaking hemisphere) that he has read the word RING in the right field of view. He denies having seen the word KEY in the left field, nor can he name objects laid in his left hand. But he can use his left hand to select the correct object, though he says he has no knowledge of the object. If asked to name the object he has selected, the speaking hemisphere calls it "RING."

particular of the introspectively experienced consciousness of the patient and the verbally expressed statements he makes about it. The left hemisphere, together with the associated subcortical structures, even in the normal brain, must, therefore, be regarded as the crucial neuronal substrate for specific human consciousness, and the associated linguistic ability.

The *right hemisphere isolated* from the left cannot express itself verbally or in writing. The patient is evidently not conscious of the sensory, integrative, and motor processes it carries out. When separated from the left hemisphere, the right hemisphere leads an *independent life*, which the patient knows about only indirectly, by way of the sensory channels of the left hemisphere. At the same time, the *performance of the right hemisphere is remarkable;* it has a memory, can recognize visual and tactile shapes, is capable of abstraction, and can understand speech to a certain extent (verbal commands are carried out and simple words are read; see Fig. 9-14). In several respects, the right hemisphere actually outperforms the left—for example, with regard to musical understanding and the ability to deal with spatial concepts. Altogether, the achievements of the right hemisphere are certainly better than those of the brains of other animals, even those of anthropoid apes (for example, chimpanzees). Returning to our earlier question of consciousness in higher animals, the *consciousness of the isolated right hemisphere is highly developed* as measured by the listed behavioral characteristics. But since it lacks the ability to express itself in language, it is no more capable than the lower animals of communicating that state of consciousness directly.

A further hint about the *nature of the consciousness of the right hemisphere* lies in the following observation. Split-brain patients uniformly report that they do not dream following their operation. Correspondingly, the sleeping EEG of the left hemisphere no longer exhibits REM phases. On the other hand, the sleeping EEG of the right hemisphere includes distinct REM phases. It therefore seems likely that the isolated right hemisphere dreams during sleep, even though it cannot make this experience known to us. The implication is that normally dream events originate in the right half of the brain and spread into the left half by way of the corpus callosum.

Neurophysiologic Aspects of Language. The study of the split-brain patients just described has also revealed that the brain regions necessary and responsible for linguistic ability are found, as a rule, only in the left hemisphere. This correlation had already been inferred from older clinical observations of brain-damaged subjects. For this reason the left hemisphere is also called the dominant hemisphere. But since evidence is accumulating, particularly as a result of split-brain studies, that in certain special respects the right hemisphere outperforms the left, it is more accurate to speak of a mutually complementary *specialization of the two hemispheres,* in which the left is ordinarily *language-dominant.*

Within the cortex of the left hemisphere, certain regions that are particularly significant regarding language can be delimited and, therefore, are called *speech areas.* In Fig. 9-15A these are labeled the frontal, temporal, and tertiary speech areas. A common property of all three areas is that electrical stimulation (performed during therapeutically required brain operations on cooperative, locally anesthetized patients) produces a failure of speech (aphasia) lasting for the duration of stimulation, which cannot be overcome by the patient. On the other hand, words or sentences are never elicited by stimulation of these regions.

The frontal speech area is also called *Broca's area;* over a hundred

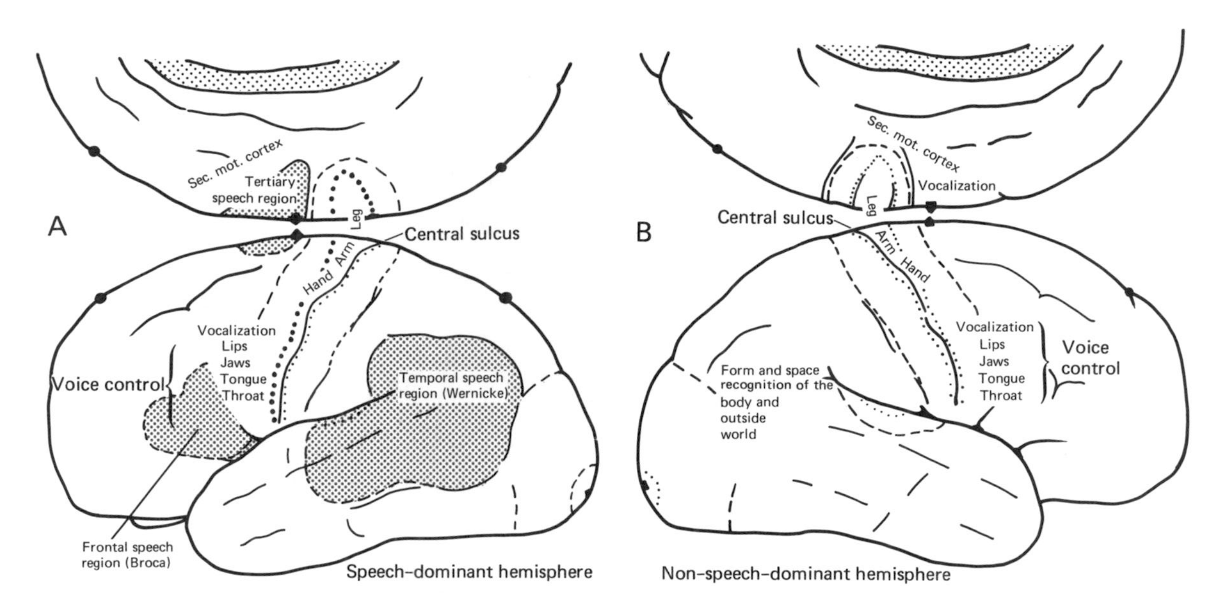

Fig. 9-15. Speech regions (*red*) in the speech-dominant (*left*) hemisphere (**A**), and the corresponding areas of the non-speech-dominant (*right*) hemisphere (**B**), as established by electrical stimulation of the exposed cortex of adult patients by Penfield and his co-workers. Control of the vocal muscles is localized on both sides of the brain, in the precentral gyri. In contrast to the cortical motor control for the rest of the body, both the right and the left parts of the musculature used in speech are controlled bilaterally. Modified from Penfield and Roberts.

years ago Broca pointed out that damage to this region causes a failure of language ability in which the expressive aspect of speech is particularly impaired. This kind of disability, in which the understanding of language is retained but the affected person says almost nothing spontaneously, and on demand can only with difficulty produce short sentences, is called *motor aphasia*. Under these conditions the muscles necessary for speaking are not paralyzed; they are readily used for other tasks (eating, drinking, swallowing).

The temporal speech area is also called *Wernicke's area*, because Wernicke (at about the same time as Broca) observed that when it is damaged there is severe disturbance of the understanding, or receptive aspect, of language. This disability is called *sensory aphasia*. Damage in this region generally results in more profound and prolonged (often permanent) language impairment than does damage in the frontal or tertiary regions (the latter was discovered by Penfield and his co-workers during the stimulation experiments described above, which produced short-term aphasia).

The functions originally ascribed by Broca and Wernicke to the areas they discovered have proved to be great oversimplifications, which hold only in a first approximation. Disturbances of expressive (motor) and receptive (sensory) linguistic performance and the associated abilities (for example, writing, reading, and calculating) almost never occur in sterotyped forms, but rather in many different combinations. Thus, attempts to localize the various forms of aphasia to particular brain areas are as a rule impossible, in view of recent clinical and psychologic findings. We shall not give a detailed description here of the symptoms of these aphasias or the recent work in the direction of characterizing and classifying them.

In right-handed people the speech regions are practically always found in the left hemisphere. Surprisingly, left-handers usually also have them on the left, though in a few they are on the right and occasionally even on both sides. There is now a test, as simple as it is revealing, to determine which of the two hemispheres contains the speech regions. A very short-acting anesthetic, the barbiturate amytal, is injected into the common carotid artery on the side of the neck. The hemisphere on that side is almost immediately anesthetized, in isolation. The other hemisphere remains awake. If the patient can speak with the physician, the waking hemisphere contains the speech regions; if not, they are on the other side. This procedure detects even the rare cases of bilateral speech regions. The main use of this test is as a preliminary to difficult brain operations, to be absolutely certain that the patient's ability to speak will not be destroyed.

Individual Development and Plasticity of the Speech Areas. Once a child has learned to speak, accidental destruction of the speech area in the left hemisphere produces total failure of speech. But after about a year the child begins to speak again. Now, however, the ability is represented in the corresponding regions of the right hemisphere (see Fig. 9-15). This *transfer of speech dominance* from the left hemisphere to the right is one of the most astonishing capacities of the brain; it illustrates the brain's *plasticity*, its ability to adjust after structural damage.

The transfer of speech dominance from the left to the right hemisphere is possible only for a limited time, until about the tenth year of life at the latest. At this age the option of establishing speech mechanisms in either hemisphere is lost, probably for two reasons. First, the formation of the basic neuronal patterns necessary for speech (which are also used when a second language is learned) is no longer possible. Second, the corresponding regions of the hemisphere that is not speech-dominant by this time have taken on other tasks, in particular that of spatial orientation, including the spatial positioning of the body itself in its environment.

Use of the plasticity of the brain illustrated in the above example is accompanied by disadvantages. Patients in which, because of damage to the left hemisphere during childhood, the right hemisphere has taken over the control of speech are not only less skilled in language than average but show less general intelligence as well.

Damage to the *right parietal and temporal cortical areas*—those corresponding to the speech areas in the left hemisphere (compare A and B in Fig. 9-15)—produces impairments of spatial orientation known as *spatial agnosia*. The symptoms are varied. For example, such patients get lost even in familiar surroundings, or they are quite incapable of making three-dimensional drawings of simple objects like cubes.

Q 9.11 Which of the following behavioral characteristics of consciousness is found only in humans (choose the most appropriate answer)?

a. Attentiveness and the ability to direct attention purposefully.
b. The expression of abstract ideas in words.
c. The ability to have expectations and plans.
d. Self-recognition.
e. The presence of esthetic and ethical values.

Q 9.12 Which of the following statements about split-brain patients is/are *false*?

a. Split-brain patients report having particularly lively and prolonged dreams.
b. Split-brain patients behave inconspicuously in everyday life.
c. Split-brain patients must learn to read and write again after the operation.

d. Split-brain patients can no longer see objects in the left half of the visual field.
e. Split-brain patients can recognize objects held in the right hand but not name them.

Q 9.13 Assign to each of the speech areas in *List 1* the most appropriate terms in *List 2*.

List 1	*List 2*
1. Broca's area	a. precentral gyrus
2. Wernicke's area	b. temporal speech area
	c. tertiary speech area
	d. occipital lobe
	e. frontal speech area

Q 9.14 When the temporal speech area ceases to function the result is (choose the most appropriate answer):
a. primarily a disturbance of receptive (sensory) language ability.
b. an isolated failure of writing, reading, and calculating ability.
c. a disturbance of expressive (motor) language ability in particular.
d. a transient, usually very brief, aphasia.
e. spatial agnosia.

9.4 Learning and Memory

The receipt, storage, and retrieval of information are general properties of neuronal networks that serve to adapt individual behavior to the environment. Without the ability to learn and to retain and retrieve items by way of memory it would be possible neither to repeat successes according to plan nor to avoid failure intentionally. Accordingly, these processes have received much attention in recent decades from neurobiologists and researchers in other disciplines, particularly psychology. Nevertheless, no satisfactory theories have as yet been formulated. The mechanism of retrieval from storage (recollection or remembering) is still utterly obscure, whereas somewhat more is known about the acquisition of information (learning) and its storage in memory.

Memory in Humans. Among the few established facts about human memory is the almost trivial observation that we store only a very small fraction of the events of which we are conscious. It has been estimated that of all the information flow available to the consciousness (which itself is only a small section of all the sensory inputs) only about 1% is selected for long term storage. Moreover, we forget a large part of the information that has been stored. The two mechanisms, selection and forgetting, protect us from being overwhelmed with data—a situation that would be just as detrimental as the absence of all learning and memory.

A second important point is that is is easier to remember a short list (of nonsense syllables, for example) than a long one. As banal as this assertion may seem, it serves to show that our memory does not operate like a magnetic tape, which takes up information until the available capacity is exhausted or the process of adding to the store is stopped. Also relevant in this connection is the general rule that we store generalizations, rather than details. When you have read this sentence the message it contains, that *concepts are stored,* will remain in your memory. But the actual verbatim formulation of this thought will be quite forgotten. When retrieval is desired, the reverse mechanism operates; we recollect the concept, and the language mechanisms provide us with the necessary details of syntax. In this respect, too, human memory processes differ distinctly from those of electronic data storage.

The ability of humans to *verbalize both abstract and concrete ideas* and store them in this form distinguishes human memory from that of animals, even the anthropoid apes. It must at least be assumed that the storage of verbally coded material, which is possible for humans, is a process supplementary to the one available to both man and other animals, that of nonverbal information storage. Because of this complication it is difficult to apply the results of experiments on animals to the interpretation of human memory processes.

Finally, there are good reasons to suspect that *storage in memory occurs in several steps* that can be demarcated experimentally even though the underlying mechanisms are still largely unknown. According to these findings, our memory comprises at least two stages, *a short and a long term memory.* Information in the short term memory (for example, a telephone number one has just looked up) is soon forgotten unless it is transferred to the long term memory by *practice.* Once in the latter, it is available for repeated reference even after quite a long time; the memory trace it has formed, the so-called *engram,* is reinforced every time it is used. This fixing of the engram, so that an item of memory becomes steadily less likely to be lost, is called *consolidation.*

The long familiar concept of short and long term memory is inadequate in the light of present knowledge, at least with respect to *verbal human memory.* As Fig. 9-16 illustrates, the material to be stored is initially taken into the *sensory memory* for fractions of a second. From there it is transferred, after verbalization, into a *primary memory,* which represents the short term memory for verbally coded information. The average duration of stay in the primary memory is brief, and amounts to only a few seconds. But this duration can be extended by *practice*—that is, by attentive repetition. This practice also facilitates

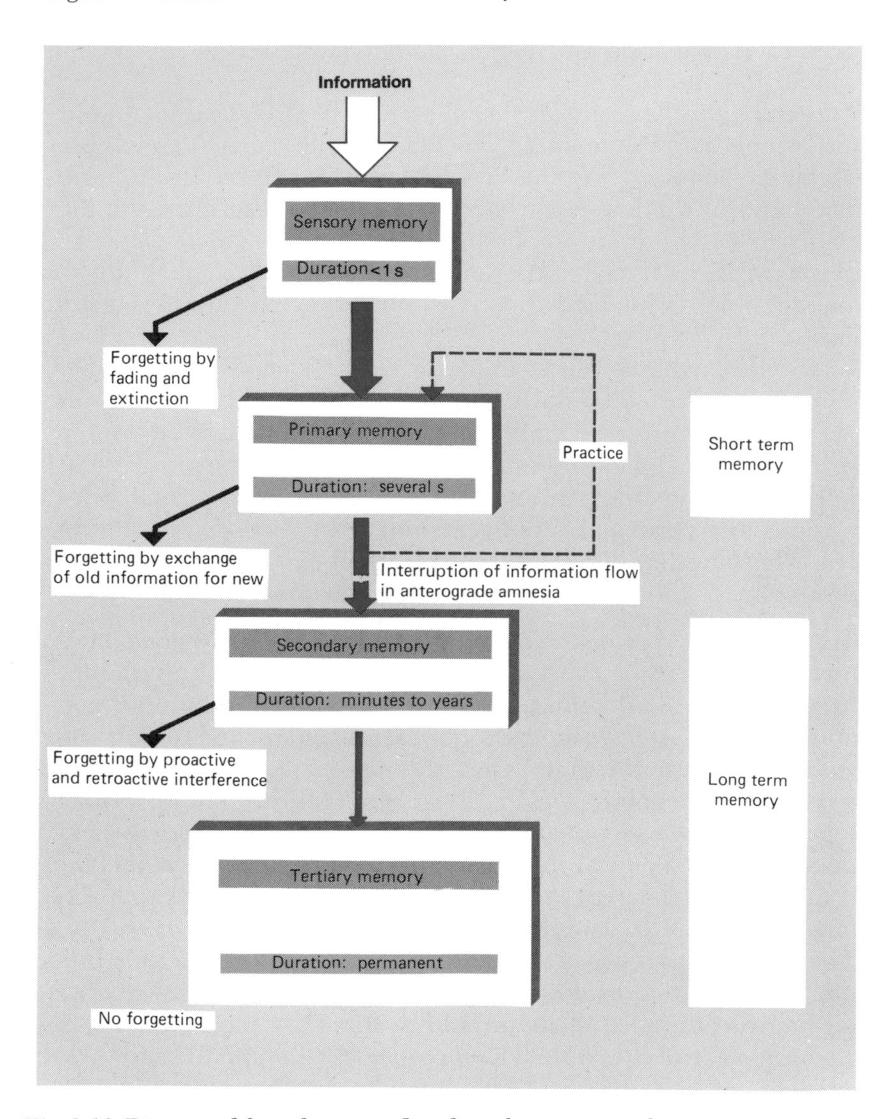

Fig. 9-16. Diagram of the information flow from the sensory to the tertiary memory. The duration of storage in each memory is indicated, as is the mechanism of forgetting. Only part of the material stored at each level reaches the next (more stable) level. Repetition (practice) facilitates transfer from the primary into the secondary memory; but it is not an indispensable requirement, nor does it guarantee transfer. Derived from Waugh and Norman.

transfer of information into the permanent, large storage system of the *secondary memory.*

Forgetting in the secondary memory appears to be based largely on interference with the learning process by other things learned previously or subsequently. In the first case, the term *proactive inhibition* is used and in the latter, *retroactive inhibition.* Proactive inhibition appears to be the more important factor, since a greater amount of previously learned material is available. Viewed in this way, the blame for much of our forgetting must be put on what we have learned before!

There are engrams—for example, one's own name or the ability to read and write or other skills employed daily—which as a result of years of practice are essentially never again forgotten, even if through disease or injury the entire remaining content of the memory is erased (see below; retrograde amnesia). These engrams appear in a sense to be stored in a particular form of memory, the *tertiary memory* (Fig. 9-16). The long term memory referred to above corresponds, in this formulation, to the secondary and tertiary memories together.

Disturbances of Memory. The inability to learn freshly presented information—that is, to store it permanently and keep it available for reference—is called *anterograde amnesia.* The patients (often chronic alcoholics) have a nearly normal secondary and tertiary memory with respect to the time before the illness, and the primary memory is also functional. However, they cannot transfer information from the primary to the secondary memory (Fig. 9-16). It appears that the hippocampus and other limbic structures play a key role in this transfer, since bilateral damage to or surgical removal of these structures in humans gives rise to complete and irreversible anterograde amnesia.

The inability to retrieve items stored in memory in the time before normal brain function was impaired is called *retrograde amnesia.* Well-known examples of its possible causes are mechanical shocks (concussion), stroke (apoplexy), electroshock (therapeutic and by accident), and narcosis. All are associated with a fairly generalized disruption of function of the brain, and it is not yet known which particular structural and functional disturbances give rise to retrograde amnesia. But certain findings suggest that it is basically a matter of *interference with access to the secondary memory* and less of loss of the content of memory (for example, during the recovery phase there is a narrowing of the forgotten time span). The tertiary memory, as a rule, is unaffected even in severe cases of retrograde amnesia.

Neuronal Mechanisms of Memory. A simple and visualizable inference about the neuronal bases of learning is that an item of information is first stored as a *dynamic engram* in the form of reverberating excitation (see Fig. 4-5) in a spatiotemporally arranged pattern. This circulating excitation could then bring about structural changes at the synapses involved and thus effect consolidation into a *structural engram.* Conceivably, the content of memory could then be retrieved by a corresponding activation of such synapses.

The *concept of reverberating excitation* is consistent with the subjective experience that we must *practice* material to be learned—that is, let it pass repeatedly through our consciousness—in order eventually to retain it. Morphologic and electrophysiologic findings are available that indicate at least the possibility of such circulation of excitation. But it is still an entirely open question whether such events are related to learning processes.

The *changes in synaptic efficacy* during and after tetanic stimulation have been discussed extensively (Sec. 4.1 and Fig. 4-6). Particularly in cases of posttetanic potentiation, which may persist for many hours or even longer at certain excitatory synapses (for example, in the hippocampus), it has long been thought that the altered synaptic properties may reflect changes in the nervous system associated with formation of the structural engram. This view is consistent with the fact that in the spinal cord, where only relatively brief posttetanic potentiation occurs, no persistent learning is observed. Another factor favoring this concept is that in the visual cortex of the mouse there are histologic and functional signs of degeneration at synapses when the animals are prevented from using their eyes from birth (either by surgical means or by raising the animals in the dark). In this case, the efficacy of synapses is diminished as a consequence of insufficient use.

The study of *changes in the EEG* and other brain potentials during learning has so far produced an abundance of data, but on the whole little insight into the neuronal mechanisms of learning and memory. Some interpretations of the findings are quite controversial. We shall not have room to discuss them here.

Biochemical (Molecular) Mechanisms of the Engram. The success of the search for the way the genetic memory is coded in the desoxyribonucleic acids (DNA), and comparable results in the study of the immunologic memory, have provided an incentive to look for molecular changes underlying the *neuronal memory*, which might be considered the *basis of the engram.* For example, these might involve special proteins incorporated in the cell or the cell membrane.

A number of experiments have been concerned with the question whether learning can bring about *changes in the ribonucleic acids*

(RNA) of neurons and glial cells (RNA plays an important role in the synthesis of protein, which continues in cells throughout life; a change in the RNA would, therefore, result in a change in the composition of the cell proteins). Microtechniques that permit measurement of both the amount of RNA and the proportions of the four nucleotide bases of which RNA molecules are built have in fact shown that there are changes in the proportions of these bases during certain learning processes. But it cannot be ruled out—indeed, it is likely—that these changes are entirely nonspecific. To overcome this obvious objection, further attempts have been made to demonstrate the possibility of *transfer of learned behavior,* by extracting RNA from the brains of trained populations of animals and injecting the extract into control animals. These experiments have so far provided no convincing evidence, whether for simple organisms like flatworms (planarians) or for fish and mammals.

Two further attempts to reveal the biochemical bases of neuronal memory deserve mention. First, in an approach converse to those just described, an attempt has been made to interfere with the formation of a structural engram in the cell or in the cell membrane by *inhibiting RNA or protein synthesis* (for example, by application of actinomycin or puromycin). To the extent that these experiments have succeeded, they too face the objection that a general inhibition of protein synthesis leads not only to a disturbance of engram formation but also to a general disruption of all cellular function.

Second, from the brains of rats trained to avoid dark places (contrary to their natural preference) by punishment with electric shocks, a polypeptide has been isolated which, when injected into normal rats (or into mice or goldfish), causes them to spend more time in the light. This polypeptide, called *scotophobin* (from the Greek *scotos,* darkness, and *phobos,* fear), comprises 15 amino acids; it has since been synthesized. The implications of this finding are not yet clear. In any case, these transfer experiments have not yet been confirmed, although the results have been known for more than 12 years. Moreover, no other macromolecule has been isolated as an "information carrier" associated with the learning of another kind of behavior. Skepticism with regard to scotophobin was further increased when it was discovered that part of its amino-acid chain resembles one of the hypophyseal hormones, adrenocorticotropic hormone (ACTH). ACTH is known to enhance the general level of wakefulness and attentiveness of the organism, which in itself would offer an explanation of improved performance.

Despite all the claims that have been made for so-called memory-promoting substances—for example, glutamic acid (glutamate), cholinesterase as well as cholinergic and anticholinergic substances, strychnine, picrotoxin, tetrazole, caffeine, and ribonucleic acid—it has so far been impossible to achieve a direct and specific improvement in intelligence and the performance of memory-requiring tasks by pharmacologic means. But there are those who try to exploit our wishes for faster and better learning, retention, and recollection for their own gain; one is well advised to beware of their pseudoscientific praises of some useless nostrum.

Q 9.15 Estimates of the percentage of the available information selected by the human brain for long term storage indicate that, of the total information flow through the consciousness, about the following fraction is permanently stored:
a. nearly 100%.
b. 10%.
c. 1%.
d. 0.1%.
e. 0.01%.

Q 9.16 The short term memory for verbally coded information is also called (choose the most appropriate answer):
a. primary memory.
b. recent memory.
c. secondary memory.
d. ancient memory.
e. tertiary memory.

Q. 9.17 Forgetting for the reason that previously learned things interfere with the learning of new material is called
a. retrograde amnesia.
b. proactive inhibition.
c. anterograde amnesia.
d. engram formation.
e. consolidation.

Q 9.18 Which of the following mechanisms is regarded as a possible neuronal basis of a dynamic engram?
a. Posttetanic potentiation.
b. Retroactive inhibition.
c. Occlusion.
d. Reverberation of excitation.
e. Synaptic degeneration.

Q 9.19 Which of the mechanisms listed in Question 9.18 could best be taken as reflecting the formation of a structural engram?

Q 9.20 In cases of anterograde amnesia, such as often occur among chronic alcoholics, the disturbance affects primarily
a. transfer from the secondary to the tertiary memory.
b. access to the secondary memory.
c. transfer from the sensory into the primary memory.
d. the formation (synthesis) of scotophobin.
e. transfer from the primary to the secondary memory.

9.5 The Frontal Lobes

As already mentioned in Sec. 9.3, the parietal and temporal regions of the association cortex (see Fig. 9-2) participate partly in language processes, and partly in form and spatial recognition with respect to the body and the outside world. There are interesting differences in the degree to which left and right hemispheres are involved in each of these functions (see Fig. 9-15). It is at present impossible to make such detailed statements about the functions of the frontal lobes (see Fig. 9-1). Their extensive reciprocal connections with the limbic system (which participates especially in the control of species-specific behavior; see Sec. 8.6) suggest that one of the roles of the frontal lobe is related to the *learned control of innate behavior patterns.* This idea is supported by the fact that many patients with frontal-lobe damage are said to be unusually impulsive, uninhibited, irritable, euphoric, or psychologically labile in other ways.

Frontal-Lobe Injury in Humans. Certain inferences about the functions of the frontal lobes may be drawn from clinical observations of people with frontal-lobe injury. These patients have normal scores on most of the standard intelligence tests. But often they exhibit personality changes such as *lack of drive* and the *absence of firm intentions and plans based on foresight.* Moreover, they are often unreliable, crude or tactless, frivolous or irascible; as a result, despite their normal "intelligence," they can become embroiled in social conflicts (for example, while at work).

In tests with tasks involving movement, these patients are inclined to persist in an act they have begun, even when the rules of play have long since demanded that they do something else. For example, in the task illustrated in Fig. 9-17 the patients are told, after each drawing, which geometric figure they should draw next. Even though they understand this instruction (and can repeat it if asked), they frequently proceed to draw again a figure already drawn one or more times. Such *persistence in what has been begun* is called *perseveration.* This tendency to perseveration is also reflected in learning experiments, in which patients have difficulty distinguishing a stimulus in a series from those that preceded it. Their behavior gives the impression that the preceding memory trace cannot make room for the next, rapidly enough—that their problem is an enhanced proactive inhibition.

Thus, frontal-lobe patients find it difficult to *change their behavior when the circumstances require it.* The effectiveness of external constraints appears to be weakened, and when several external and inter-

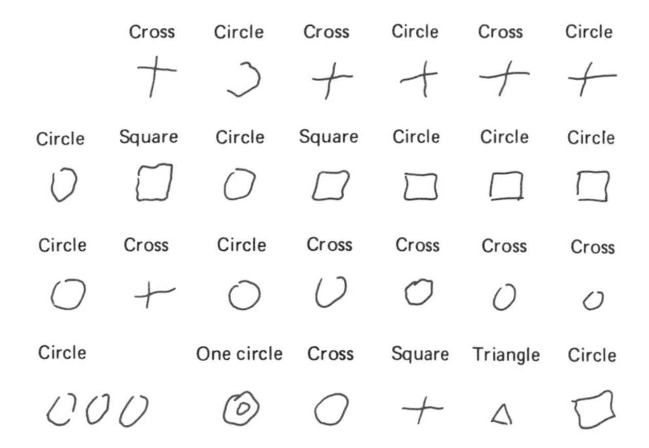

Fig. 9-17. Perseveration in the performance of motor tasks by four patients with frontal-lobe damage. Each line shows the patient's drawings in red, and above them the instructions given by the examiner. The first, second, and fourth patients had tumors of the left frontal lobe, and the third had an abscess in the right frontal lobe. Modified from Luria.

nal motivations are in competition, it is hard for the patients to change from one to another rapidly and appropriately. This conclusion from observations of behavior is consistent with that drawn above, in the context of the anatomic connections, that the frontal lobes are involved in the learned control of inborn behavior patterns and in harmonizing external and internal motivations. Experiments on chimpanzees and other mammals have given similar results. Altogether, the findings in man and animals have suggested the hypothesis that the frontal lobe plays a leading role in the *development of behavioral strategies.*

Psychosurgery. Before effective psychoactive drugs were introduced, transection of the connections between frontal lobe and thalamus (leukotomy) was used to treat certain forms of neuropsychiatric illness. This procedure was always controversial and is no longer justifiable today. But when it was first used, around 1940, it marked the beginning of psychosurgery—that is, the *deliberate attempt to affect human behavior permanently by the destruction or removal of brain tissue.* In a broader sense, electroshock treatments, long term therapy with psychoactive drugs, and the introduction of electrodes into the brain should be regarded as psychosurgery, since these too may produce permanent alterations of brain tissue.

In view of the extent of our ignorance about the way the brain works and the functions of its individual components, the grounds for using psychosurgical procedures today are more empiric than theoretical. The operation most often employed currently appears to be cingulotomy, the isolation of the cingulate gyrus (which lies above the corpus callosum and belongs to the limbic system) in order to relieve intractable pain and a number of psychologic disturbances such as depression and severe anxiety states. The destruction of another nucleus in the limbic system, the amygdaloid nucleus (amygdalotomy), is used as a last resort to control aggressive behavior, though grave misgivings are expressed about such profound and irreversible interference with personality. Here—as with every scientific and technical advance—it is up to society, developing and applying the appropriate norms, to ensure that newly acquired knowledge is used only for the benefit of mankind.

Q 9.21 Which of the following are regarded as functions of the frontal lobes? (Choose the two you think most appropriate).
a. Regulation of vegetative organ function.
b. Learned control of inborn behavior patterns.
c. Coordination of supportive and directed movement.
d. Development of behavior strategies.
e. Long term storage of information.

Q 9.22 Which of the following symptoms is/are particularly characteristic of frontal-lobe injury?
a. Anterograde amnesia.
b. Asynergy.
c. Loss of drive.
d. Psychic instability.
e. Perseveration.

Q 9.23 Which of the following operations counts as the first widely attempted method of influencing human behavior by destruction of brain tissue?
a. Amygdalotomy.
b. Cingulotomy.
c. Decerebration.
d. Decortication.
e. Leukotomy.

10

SUGGESTED READINGS

The references given below are to assist the reader in further studies. In keeping with the introductory character of this book, emphasis has been placed on textbooks, monographs, and review articles. The references were chosen by the authors of the individual chapters.

General Literature

BRODAL A (1981) Neurological anatomy in relation to clinical medicine, 3rd edn. Oxford University Press, New York London Toronto, pp 1–1053

ECCLES JC (1957) The physiology of nerve cells. Hopkins, Baltimore, pp 1–270

ECCLES JC (1973) The understanding of the brain. McGraw-Hill, New York St. Louis San Francisco Düsseldorf, pp 1–238

ECCLES JC (1979) The human mystery. Springer, Berlin Heidelberg New York, pp 1–255

ECCLES JC (1980) The human psyche. Springer, Berlin Heidelberg New York, pp 1–279

Handbook of Physiology (1977/1981) Section 1: The nervous system, vol I: Cellular biology of neurons (2 books), (1977). Vol II: Motor control (2 books) (1981) American Physiological Society, Bethesda

KANDEL ER, SCHWARTZ JH (eds) (1981) Principles of neural science. Elsevier, Amsterdam, pp 1–731

KUFFLER SW, NICHOLLS JG (1976) From neuron to brain. Sinauer, Sunderland, Mass, pp 1–486.

MCGEER PL, ECCLES JC, MCGEER EG (1978) Molecular neurobiology of the mammalian brain. Plenum, New York, pp 1–644

MOUNTCASTLE VB (ed) (1980) Medical physiology, vol I, 14th edn. Mosby, Saint Louis, pp 1–948

RUCH TC, PATTON HD (eds) (1979) Physiology and biophysics. I. The brain and neural function, 20th edn. Saunders, Philadelphia, pp 1–743

SCHMIDT RF (ed) (1981) Fundamentals of sensory physiology, 2nd edn. Springer, Berlin Heidelberg New York pp 1–286
SCHMIDT RF, THEWS G (eds) (1983) Human physiology. Springer, Berlin Heidelberg New York, pp 1–725
SCHMITT FO, WORDEN G (eds) (1979) The neurosciences, Fourth Study Program, MIT, Cambridge Massachusetts London, pp 1–1185
SHERRINGTON CS (1961) The integrative action of the nervous system. Yale University Press, New Haven (1st edn 1906) pp 1–413
WORDEN FG, SWAZEY JP, ADELMAN G (eds) (1975) The neurosciences: Paths of discovery. MIT, Cambridge, Mass., pp 1–622

Chapter 1

BRADBURY MW (1979) The concept of a blood-brain barrier. Wiley, Chichester, pp 1–465
DAVSON H (1976) The blood-brain barrier. J Physiol (Lond) 255: 1–28
FAWCETT DW (1966) An atlas of fine structure: the cell, its organelles and inclusions. WB Saunders, Philadelphia London
GRAFSTEIN B, FORMAN DS (1980) Intracellular transport in neurons. Physiol Rev 60: 1167–1283
Handbook of Physiology (1977) Section 1: The nervous system, vol I: Cellular biology of neurons. William & Wilkins, Baltimore, pp 1–1238 (in two books)
KUFFLER SW (1967) Neuroglial cells: physiological properties and a potassium mediated effect of neuronal activity on the glial membrane potential. Proc R Soc 168: 1
MARAN TH (1980) The cerebrospinal fluid. In: Mountcastle V (ed) Medical physiology, vol II. 14th ed. Mosby, St. Louis, p 1218
PETERS A, PALAY SL, WEBSTER H DEF (1976) The fine structure of the nervous system. WB Saunders, Philadelphia London Toronto
SCHARPER E (1944) The blood vessels of the nervous tissue. Q Rev Biol 19: 308
WATSON WE (1974) Physiology of neuroglia. Physiol Rev 54: 245
WAXMAN SG (ed) (1978) Physiology and pathobiology of axons. Raven, New York

Chapter 2

ARMSTRONG CM (1981) Sodium channels and gating currents. Physiol Rev 61: 644
HODGKIN AL, HUXLEY AF (1952) Quantitative description of membrane current and its application to conduction and excitation in nerve. J Physiol (Lond) 117: 500
HOPPE W, LOHMANN W, MARKL H, ZIEGLER H (1983) Biophysics. Springer, Berlin Heidelberg New York Tokyo
KATZ B (1966) Nerve, muscle and synapse. McGraw Hill, New York
NOBLE D (1966) Applications of Hodgkin-Huxley equations to excitable tissues. Physiol Rev 46: 1
RUCH TC, PATTON HD (1974) Physiology and biophysics. Saunders, Philadelphia
SCHMIDT RF, THEWS G (eds) (1983) Human physiology. Springer, Berlin Heidelberg New York, pp 1–725
ULBRICHT W (1977) Ionic channels and gating currents in excitable membranes. Ann Rev Biophys Bioeng 6:7

Chapter 3

Burgen A, Kosterlitz HW, Iversen LL (eds) (1980) Neuroactive peptides. The Royal Society, London, pp 1–195

Ceccarelli B, Hurlbut WP (1980) Vesicle hypothesis of the release of quanta of acetylcholine. Physiol Rev 60: 396–441

Cooper JR, Bloom, FE, Roth RH (1978) The biochemical basis of neuropharmacology, 3rd edn. Oxford University Press, New York, pp 1–327

Cottrell GA, Usherwood PNR (eds) (1977) Synapses. Blackie, Glasgow, pp 1–384

DeFeudis FV, Mandel P (eds) (1981) Amino acid neurotransmitter. Raven, New York, pp 1–572

Eccles JC (1964) The physiology of synapses. Springer, Berlin Göttinger Heidelberg New York

Eccles JC (1982) The synapse: from electrical to chemical transmission. Ann Rev. Neurosci 5:325–339

Katz B (1966) Nerve, muscle and synapse. McGraw-Hill, New York

Kravitz EA, Treherne JE (eds) (1980) Neurotransmission, neurotransmitters, and neuromodulators. J Exp Biol 89: 1–286

Loewenstein WR (1981) Junctional intercellular communication: The cell-to-cell membrane channel. Physiol Rev 61: 829–913

Schmidt RF (1971) Presynaptic inhibition in the vertebrate central nervous system. Ergeb Physiol Biol Chem Exp Pharmacol 63: 20–101

Stjärne L, Hedqvist P, Lagercrantz H, Wennmalm Å (eds) (1981) Chemical neurotransmission. Academic Press, New York, pp 1–562

Taxi J (ed) Ontogenesis and functional mechanisms of peripheral synapses. Elsevier, Amsterdam, pp 1–196

The Synapse (1976) Cold Spring Harbor Symp. Quant Biol 40

Tsukahara N (1981) Synaptic plasticity in the mammalian central nervous system. Ann Rev Neurosci 4:351–379

Vincent A (1980) Immunology of acetylcholine receptors in relation to myasthenia gravis. Physiol Rev 60: 756–824

Zaimis E (ed) (1976) Neuromuscular junction. Springer, Berlin Heidelberg New York, pp 1–746

Chapter 4

Eccles JC (1969) The inhibitory pathways of the central nervous system. The Sherrington Lectures IX. Thomas, Springfield/Ill., pp 1–135.

Fearing F (1930) Reflex action. A study in the history of physiological psychology. Williams & Wilkins, Baltimore

Feinstein B, Lindegaard B, Nyman, E, Wohlfahrt G (1955) Morphologic studies of motor units in normal human muscles. Acta Anat (Basel) 23: 127

Fulton JF (1943) Physiology of the nervous system. Oxford University Press, London New York Toronto

Handbook of Physiology (1981) Section 1: The nervous system, vol II: Motor control,

Sherrington CS (1961) The integrative action of nervous system. Yale University Press, New Haven (1st edn 1906), pp 1–413

Schmitt FO, Worden FG (eds) (1979) The neurosciences. Fourth Study Program. MIT, Cambridge, Mass., pp 1–1185

Taylor A, Prochazka A (eds) (1981) Muscle receptors and movement. Macmillan, London, pp 1–446

Chapter 5

Bourne GH (ed) (1972) The structure and function of muscle, 2nd edn, vol I–III. Academic Press, London New York
Carlson FD, Wilkie DR (1974) Muscle physiology. Prentice-Hall, Englewood Cliffs, New Jersey
Hoppe W, Lohmann W, Markl H, Ziegler H (1983) Biophysics. Springer, Berlin Heidelberg New York Tokyo
Huxley AF (1974) Muscular contraction. J Physiol 243: 1
Schmidt RF, Thews G (eds) (1983) Human physiology. Springer, Berlin Heidelberg New York, pp 1–725

Chapter 6

Boyd JA, Davey MR (1968) Composition of peripheral nerves. Livingstone, Edinburgh London
Brodal A (1981) Neurological anatomy in relation to clinical medicine, 3rd edn. Oxford University Press, New York London Toronto, pp 1–1053
Creutzfeldt O, Schmidt RF, Willis WD (eds) (1984) Sensory-motor integration in the nervous system. Exp Brain Res Suppl 9, Springer, Berlin
Desmedt JE (ed) (1983) Motor control mechanisms in health and disease. Raven Press, New York
Eccles JC, Ito M, Szentágothai J (1967) The cerebellum as a neuronal machine. Springer, Berlin Heidelberg New York
Granit R (1970) The basis of motor control. Academic Press, London New York
Handbook of Physiology (1981) Section 1: The nervous system, vol II: Motor control, Part 1, pp 1–733, Part 2: pp 735–1480. American Physiological Society, Bethesda (Part 1, pp 1–733; part 2, pp 735–1480)
Ito M (1984) The cerebellum and neural control. Raven Press, New York
Kemp JM, Powell TPS (1971) The connexions of the striatum and globus pallidus: synthesis and speculation, Phil Trans 262: 441
Larsell O, Jansen J (1972) The comparative anatomy and histology of the cerebellum. The human cerebellum, cerebellar connections, and cerebellar cortex. University of Minnesota Press, Minneapolis
Matthews PBC (1972) Mammalian muscle receptors and their central actions. Arnold, London
Penfield W, Rasmussen T (1950) The cerebral cortex of man. Macmillan, New York
Schmidt RF (1973) Control of the access of afferent activity to somatosensory pathways. In: Iggo A (ed) Somatosensory system. Springer, Berlin Heidelberg New York (Handbook of sensory physiology, vol II, p 151)
Schmidt RF, Thews G (eds) (1983) Human physiology. Springer, Berlin Heidelberg New York, pp 1–725
Taylor A, Prochazka A (eds) (1981) Muscle receptors and movement. Macmillan, London, pp 1–446

Chapter 7

Granit R (1970) The basis of motor control. Academic Press, London New York
Grodins FS (1963) Control theory and biological systems. Columbia University Press, New York
Homma S (1976) Understanding the stretch reflex. Progr Brain Res 44
Houk J (1980) Principles of system theory as applied to physiology. In: Mountcastle VB (ed) Medical physiology, vol 1. 14th Ed. Mosby, St. Louis, pp 225
Matthews PBC (1972) Mammalian muscle receptors and their central actions. Arnold, London
Smith JM (1968) Mathematical ideas in biology. Cambridge University Press, Cambridge New York
Wiener N (1948) Cybernetics. Freymann, Paris New York

Chapter 8

Brodal A (1981) Neurological anatomy in relation to clinical medicine, 3rd edn. Oxford University Press, New York London Toronto
Cannon WB (1929) Bodily changes in pain, hunger, fear and rage, 2nd edn. Appleton & Co., New York
Cannon WB (1939) The wisdom of the body, 2nd edn. Norton, New York
Davson H, Segal MB: Introduction to physiology vol 2: Basic mechanisms (Part 2) (1975), vol 3 (1976), vol 5: Control of reproduction (1980) Academic Press, London Toronto Sydney; Grune & Stratton, New York San Francisco
Folkow B, Neil E (1971) Circulation. Oxford University Press, New York London Toronto
Gabella G (1976) Structure of the autonomic nervous system. Chapman & Hall, London
Johnson RH, Spalding JMK (1974) Disorders of the autonomic nervous system. Blackwell, Oxford London Edinburgh Melbourne
Masters WH, Johnson VE (1966) Human sexual response. Little, Brown & Co., Boston
Monnier M (1968) Functions of the nervous system. General physiology: autonomic functions (neurohumoral regulations) Vol. I. Elsevier, Amsterdam

Chapter 9

Andersen P, Andersson SA (1968) Physiological basis of the alpha rhythm. Appleton-Century-Crofts, New York, pp 1–235
Bindman L, Lippold O (1981) The neurophysiology of the cerebral cortex. Arnold, London, pp 1–495
Brodal A (1981) Neurological anatomy in relation to clinical medicine, 3rd edn. Oxford University Press, New York London Toronto, pp 1–1053
Damasio AR, Geschwind N (1984) The neural basis of language. Ann Rev Neurosci 7:127–147
Eccles JC (ed) (1966) Brain and conscious experience. Springer, Berlin Heidelberg New York
Eccles JC (1979) The human mystery. Springer, Berlin Heidelberg New York, pp 1–255
Eccles JC (1980) The human psyche. Springer, Berlin Heidelberg New York, pp 1–279
Gazzaniga MS (ed) (1979) Neuropsychology. Handbook of behavioral neurobiology, vol 2. Plenum, New York, pp 1–566

GAZZANIGA MS, LEDOUX JE (1978) The integrated mind. Plenum, New York, pp 1–168
JOVANOVIĆ UJ (1971) Normal sleep in man. Hippokrates, Stuttgart
MCGEER PL, ECCLES JC, MCGEER EG (1978) Molecular neurobiology of the mammalian brain. Plenum, New York London, pp 1–644
MILLS JN (1966) Human circadian rhythms. Physiol Rev 46: 128–159
MILNER B (1970) Memory and the medial temporal regions of the brain. In: PRIBRAM KH, BROADBENT DE (eds) Biology of memory. Academic Press, New York London, p 29
MORUZZI G (1972) The sleep-waking cycle (Neurophysiology and neurochemistry of sleep and wakefulness). Ergeb Physiol Biol Chem Exp Pharmakol 64: 1–165
PENFIELD W, ROBERTS L (1959) Speech and brain mechanisms. Princeton University Press, Princeton/N.J.
ROSE SPR (1981) What should a biochemistry of learning and memory be about? Neuroscience 6: 811–821
SCHMITT FO, WORDAN FG, ADELMAN A, DENNIS SG (eds) (1981) The organization of the cerebral cortex. MIT, Cambridge, Mass, pp 1–592
SPERRY RW (1969) A modified concept of consciousness. Psychol Rev 76: 532–536
THOMPSON RF, BERGER TW, MADDEN IV J (1983) Cellular processes of learning and memory in the mammalian CNS. Ann Rev Neurosci 6:447–491
SQUIRE LR (1982) The neuropsychology of human memory. Ann Rev Neurosci 5:241–273
WEITZMAN ED (1981) Sleep and its disorders. Ann Rev Neurosc 4: 381–417
WEVER RA (1979) The circadian system of man: Results of experiments under temporal isolation. Springer, New York Berlin Heidelberg, pp 1–276
WOLMAN BB (ed) (1979) Handbook of dreams. Research, theories and applications. Van Nostrand Reinhold, New York, pp 1–447
ZIPPEL HP (ed) (1973) Memory and transfer of information. Plenum, New York London, pp 1–582

11
ANSWER KEY

Chapter 1

Q 1.1: b, c, e
Q 1.2: c
Q 1.3: as in Fig. 1-1
Q 1.4: as in Fig. 1-3
Q 1.5: b
Q 1.6: a
Q 1.7: d
Q 1.8: e
Q 1.9: a, b, c
Q 1.10: c
Q 1.11: b
Q 1.12: Nerve cells as in Fig. 1-1; connections between them as in Fig. 1-3.
Q 1.13: d
Q 1.14: Somatic afferents, motor efferents, autonomic efferents
Q 1.15: a, e
Q 1.16: a, b, d

Chapter 2

Q 2.1: Fig. 2-5
Q 2.2: Na^+ = 1/5–15
Cl^- = 1/20–100
reciprocally asymmetrical
Q 2.3: b, c
Q 2.4: a, c

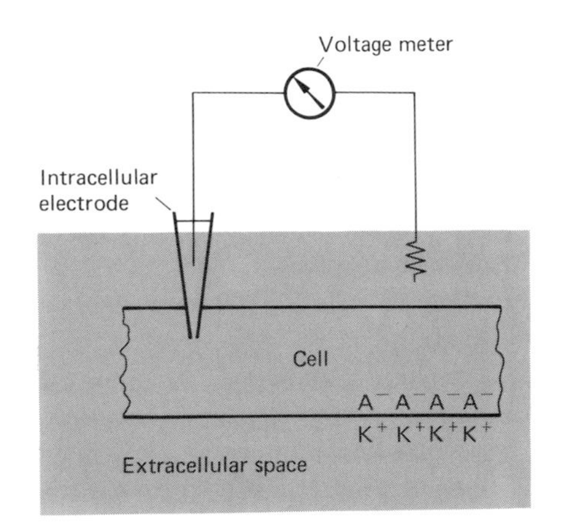

Fig. 2-5. Membrane charge and the measurement of potential; diagram in answer to Q 2.1.

Q 2.5: $I_{Cl}/(E_{Cl} - E)$
Q 2.6: $E_{Na} = +65$ mV
Q 2.7: b, c, d
Q 2.8: b, c
Q 2.9: a, d, e
Q 2.10: c
Q 2.11: Fig. 2-11
Q 2.12: b, c
Q 2.13: b
Q 2.14: b, d
Q 2.15: b
Q 2.16: Fig. 2-16
Q 2.17: a, c
Q 2.18: Fig. 2-19*B*
Q 2.19: b, c, d

Q 2.20: e
Q 2.21: d
Q 2.22: Fig. 2-22, bottom and top curves
Q 2.23: c
Q 2.24: c, e
Q 2.25: a, b, c

Chapter 3

Q 3.1: a, b, d
Q 3.2: d
Q 3.3: c
Q 3.4: a, c
Q 3.5: c, d
Q 3.6: c, e
Q 3.7: d
Q 3.8: b
Q 3.9: c
Q 3.10: c
Q 3.11: a
Q 3.12: e
Q 3.13: b
Q 3.14: a, d
Q 3.15: b
Q 3.16: a, c, e
Q 3.17: d
Q 3.18: c
Q 3.19: b

Chapter 4

Q 4.1: b
Q 4.2: Neither; there is summation or addition of the two excited populations
Q 4.3: b, e
Q 4.4: c
Q 4.5: e
Q 4.6: b, c
Q 4.7: 12 + 15 + 13 = 40 ms
Q 4.8: a
Q 4.9: Nutritional reflexes b, d
Protective a, c, e, f
Q 4.10: c
Q 4.11: c
Q 4.12: b, e

Chapter 5

Q 5.1: Fig. 5-4
Q 5.2: b, d, f
Q 5.3: Fig. 5-2
Q 5.4: (1) myosin, (2) actin, (3) Ca^{++}, (4) adenosine triphosphate (sequence arbitrary)
Q 5.5: b, c
Q 5.6: c
Q 5.7: b, d
Q 5.8: c, d
Q 5.9: a, b, c
Q 5.10: b, c, d
Q 5.11: a, c
Q 5.12: b
Q 5.13: a, c
Q 5.14: d, e
Q 5.15: b, d

Chapter 6

Q 6.1: b
Q 6.2: a, b
Q 6.3: d
Q 6.4: d
Q 6.5: (a) the muscle spindles, (b) the tendon organs
Q 6.6: b, e, g
Q 6.7: b
Q 6.8: b, d, e
Q 6.9: c
Q 6.10: a
Q 6.11: d
Q 6.12: b, d, e
Q 6.13: d
Q 6.14: a, c
Q 6.15: a, d

Q 6.16: c
Q 6.17: e
Q 6.18: b, e
Q 6.19: b, e
Q 6.20: Neck afferents: a, b, c; labyrinth: d
Q 6.21: c
Q 6.22: e
Q 6.23: f
Q 6.24: f; there are no inhibitory synapses between basket cells and the *dendrites* of the Purkinje cells
Q 6.25: b, c, d

Chapter 7

Q 7.1: d, e
Q 7.2: b, d
Q 7.3: a
Q 7.4: b, c, d
Q 7.5: a, c, d
Q 7.6: e
Q 7.7: b, c
Q 7.8: b, d, e

Chapter 8

Q 8.1: b, c
Q 8.2: c, e
Q 8.3: b, c, d
Q 8.4: c, d
Q 8.5: a, d
Q 8.6: b, d
Q 8.7: a, d
Q 8.8: a, d, e
Q 8.9: (a) seconds; (b) 50; (c) more slowly than; (d) the same
Q 8.10: c, d
Q 8.11: a, d
Q 8.12: a, d
Q 8.13: as in Fig. 8-11
Q 8.14: b, e
Q 8.15: c, e
Q 8.16: c, e
Q 8.17: b, d
Q 8.18: (a) increases; (b) decreases; (c) increases; (d) decreases
Q 8.19: b, e
Q 8.20: a, e
Q 8.21: a, d
Q 8.22: a, d
Q 8.23: a, d
Q 8.24: d, f
Q 8.25: a, c, d
Q 8.26: b, c, d
Q 8.27: as in Fig. 8-23
Q 8.28: d, e

Chapter 9

Q 9.1: a
Q 9.2: c
Q 9.3: c
Q 9.4: b
Q 9.5: d
Q 9.6: e
Q 9.7: a, d
Q 9.8: c
Q 9.9: b
Q 9.10: e
Q 9.11: b
Q 9.12: a, c, d, e
Q 9.13: le, 2b
Q 9.14: a
Q 9.15: c
Q 9.16: a
Q 9.17: b
Q 9.18: d
Q 9.19: a
Q 9.20: e
Q 9.21: b, d
Q 9.22: c, d, e
Q 9.23: e

INDEX